Integrated Devices for Artificial Intelligence and VLSI

Scrivener Publishing
100 Cummings Center, Suite 541J
Beverly, MA 01915-6106

Publishers at Scrivener
Martin Scrivener (martin@scrivenerpublishing.com)
Phillip Carmical (pcarmical@scrivenerpublishing.com)

Integrated Devices for Artificial Intelligence and VLSI

Edited by

Balwinder Raj

Suman Lata Tripathi

Tarun Chaudhary

K. Srinivasa Rao

and

Mandeep Singh

Scrivener Publishing

WILEY

This edition first published 2024 by John Wiley & Sons, Inc., 111 River Street, Hoboken, NJ 07030, USA and Scrivener Publishing LLC, 100 Cummings Center, Suite 541J, Beverly, MA 01915, USA

For more information about Scrivener publications please visit www.scrivenerpublishing.com.

Wiley Global Headquarters
111 River Street, Hoboken, NJ 07030, USA

For details of our global editorial offices, customer services, and more information about Wiley products visit us at www.wiley.com.

Library of Congress Cataloging-in-Publication Data

ISBN 978-1-394-20435-9

Front cover images supplied by Wikimedia Commons
Cover design by Russell Richardson

Set in size of 11pt and Minion Pro by Manila Typesetting Company, Makati, Philippines

Printed in the USA

10 9 8 7 6 5 4 3 2 1

Contents

Preface

The current research trend in Integrated Semiconductor Devices for Artificial Intelligence and Very Large Scale Integration (VLSI) is exploring the potential of device and artificial intelligence for emerging engineering applications. In the current Integrated Circuits, switching time is on the order of nanoseconds, minimum feature size is on the order of nanometers, transistor count is on the order of billions and cost is on the order of a few dollars. Scaling of conventional MOSFETs towards smaller sizes is near to end because of limited mobility, large Short Channel Effects (SCEs), high leakage currents and power dissipation. To mitigate these challenges of conventional MOSFET devices and to scale the semiconductor device further into nanoscale domain, alternative materials and device structures are required. There are numerous challenges in the development of models for distributed intelligence systems for environmental monitoring, but these can be mitigated with artificial intelligence, machine learning, and deep learning. This book will help identify the areas where we encounter numerous challenges and apply diverse solutions for the development of advanced Integrated Devices for Artificial Intelligence and VLSI.

Chapter 1: The chapter 1 offers a thorough comparison of the two well-known transistor designs, MOSFET and FinFET, which have revolutionized integrated circuit architecture. The chapter offers a thorough investigation of the core ideas, design factors, and operational traits between the two transistor techniques. It describes the development of MOSFET technologies and how it has significantly advanced current electronics. The drawbacks of conventional MOSFET scaling are brought to light, necessitating the use of FinFET and other innovative transistor architectures. FinFET, revealing its ground-breaking design with a channel shaped like a fin and many gates. It examines the special operational benefits of FinFET, such as greater device scalability, improved electrostatic control, and lower leakage current. The article also looks at difficulties in designing and producing FinFETs, such as increasing process complexity and unpredictability. The outcomes of MOSFET and FinFET are compared and contrasted over a number of important factors. These factors involve switching speed, noise immunity, power efficiency, power dissipation, leakage current, on-state current, subthreshold slope, threshold voltage control, and device scaling.

Chapter 2: Improvements in the VLSI industry have always been striving to justify the Moore's law by implanting, twice count transistors from the existing one. This law has made a significant improvement in the trends of FET family starting with the straightforward MOSFET with one gate controlling the channel, then next followed by FET with Dual gates, later Tri-gate/Fin FET and now persisting Gate-All-Around Field Effect Transistor. Each era of technology contributed its own advantages and disadvantages. Presently the Gate-All-Around FETs are ruling the FET industry because of its major advantage of improved gate electrostatic integrity over the channel from all the directions, with reduced overall size of the FET. In this research work, the vertically stacked GAA Nanosheet FET is simulated at the device-level using Visual TCAD – 3D Cogenda tool. Research is conducted to substantiate the influence of geometrical variations with respect to thickness and width on the performance of the FET. Parameters analyzed to in the research are ION, IOFF, Switching ratio (ION/IOFF), Subthreshold swing (SS), DIBL, and Threshold voltage (Vth). Device is optimal if it offers better ON current, minimum OFF current. To evaluate this optimal performance the thickness of the nanosheet (NT) is varied from 5nm to 9nm, and the width is varied from 10nm to 50nm. Visual TCAD - 3D Cogenda tool is used in analyzing the outputs generated and conclude the best geometric dimension for optimum characteristics.

Chapter 3: The standard MOSFET used in modern IC technology is expected to be replaced by new tunnel fieldeffect transistor (TFET), which is seen to be the utmost viable contender. Researchers have been very interested in it as a result of it's capacity to attain a steep sub-threshold slope, a stronger resilience to shortchannel effect's, and a reduced standby power dissipation. The current transport technique, band to band tunnelling (BTBT), causes TFET's two main bottlenecks, lower Ion current and ambipolar conductivity, despite the fact that it promises several benefits over other MOSFET competitors. This article provides a thorough overview of the many methods researchers have proposed to increase the ON-State current and subthreshold swing in Tunnel FETs.

Chapter 4: The unique characteristics of nanowire field effect transistors (NW FETs), such as high electron mobility, low power consumption, and scalability, have made them a promising technology for upcoming electronic devices. An overview of the underlying concepts, methods of fabrication, and performance traits of NW FETs are given in this chapter. The chapter discusses the benefits and drawbacks of NW FETs, including how contact resistance, defect density, and surface effects affect device performance. The chapter also highlights the potential uses of NW FETs in several industries, including sensing, computing, and energy harvesting. Overall, this chapter offers a thorough reference for researchers and engineers interested in this innovative technology since NW FETs are anticipated to play a significant role in the creation of next-generation electronic devices.

Chapter 5: The Moore law for scaling of technological nodes has led to the major advancement in chip design and functionality. With the down-scaling of technological nodes, the efficiency of gate becomes insignificant compared to the interconnect performance. The reliability issue play a major role at nano-scaled technological nodes since future interconnects requires higher current density with compact cross-sectional dimensions. Formerly, copper was employed as interconnect due to high current density and less resistivity but as technology downscales below 45nm its resistance increases due to the reduction of MFP. The scaling of interconnect dimensions effects the delay time as well as power dissipation. The electronics circuit operational speed is affected by the signal delay at the output of wire. The power dissipation is also a key factor in VLSI ICs. In today's world, everyone demands higher power standby and speed in electronic circuits. The motivation for minimizing delay in signal and power dissipation at global lengths for various technological nodes varies according to applications. The overall performance of electronic circuit depends on the product of power dissipation and delay. Both are independent parameters but to estimate actual performance of an interconnect their product must be taken into account for elite integrated circuits. This chapter has proposed MLGNR as suitable and alternative material in on-chip interconnects for future technological nodes compared to copper interconnects.

Chapter 6: We examine the recent development of ferroelectric memory, along with ferroelectric random access memory (FRAM). Sophisticated non-volatile memories including resistive random access memory (ReRAM), phase-change random access memory (PRAM), and magneto-resistive random access memory (MRAM) have all been expanded, but FRAM is the first of these to be commercially available. Currently, a few Mb of extremely dependable FRAM is accessible. FRAM has been used in sophisticated smartcards, RFID tags, and other electronic devices because of its superior electric features, which include fast read/write speeds around 50 ns, low power consumption, and high switching durability around 10^{13}. We also discuss recently created materials, manufacturing techniques, and circuit technologies that are anticipated to solve the scalability issue with FRAM.

Chapter 7: Very large-scale integration (VLSI) is the process of combining a large number of transistors on a single chip to form an integrated circuit. VLSI development is rapidly moving toward the efficient design of a single chip. However, VLSI design faces numerous challenges, including power optimization, design productivity, manufacturing variability and testability. Conventional methodologies are primarily manual, timeconsuming and resource-intensive. Artificial Intelligence (AI) and Machine Learning (ML) techniques, on the other hand, provide automated learning approaches for dealing with large amounts of data in complex chip design. Furthermore, incorporating machine learning algorithms can improve system-on-chip performance. AI/ML is becoming more important in VLSI design due to its numerous applications in both logical and physical design. As a result, AI/ML has become an essential component in the VLSI industry. This chapter discusses popular ML algorithms, their mathematical foundations, performance metrics, and how AI and ML applications can transform VLSI design by allowing for high-speed, intelligent and efficient implementations.

Chapter 8: A computing device is deemed rudimentary if its operational scope is confined to elementary high-school-level mathematical tasks. Conversely, a mechanism that incorporates components of biotic origin, like as DNA units, as opposed to conventional electrical elements, closely mirrors the intricacies of the human brain. The instantiation of a computation system that operates neurons and synapses in parallel, exhibiting high capacity and low power consumption, constitutes a neuromorphic computing system (NCS). Neurons communicate through the transmission of both chemical and electrical signals. A neural network is established through minuscule junctions, termed "synapses," where each neuron is intricately linked to others, facilitating signal propagation. Neuromorphic systems seamlessly integrate computing systems and memories, thereby mitigating the separation between them and overcoming the challenges associated with the "memory wall. This chapter provides an exhaustive perspective on neuromorphic computing, commencing with an examination of recent challenges in achieving effective computing systems capable of supporting diverse novel applications. Subsequently, device-level perspectives are scrutinized to enhance the efficacy of neuromorphic solutions. The comprehensive applications of neuromorphic computing are delineated to maximize exposure for research groups and the scientific community engaged in this field.

Chapter 9: Neuromorphic computing is a rapidly developing field that seeks to emulate the neural structure and function of the human brain using hardware and software technologies. In recent years, the development of neuromorphic computing has been fueled by advancements in semiconductor technology and the need for more efficient and intelligent computing systems. This review chapter provides an overview of the state-of-the-art in neuromorphic computing, including the principles and concepts that underlie the technology, its key applications, and the challenges and opportunities that lie ahead. In this chapter discussion about the potential of neuromorphic computing to enable a wide range of applications, including sensory processing, robotics, machine learning, and cognitive computing. In this chapter discussions are made on some of the key challenges associated with the development of neuromorphic computing systems, including scalability, power consumption, and programming models. Overall, this review chapter provides a comprehensive overview of the current state of the art in neuromorphic computing and its potential impact on the future of computing and artificial intelligence.

Chapter 10: This chapter presents the production and characterization of a sol-gel composite nanoparticle composed of titanium dioxide (TiO_2) and zinc aluminate ($ZnAl_2O_4$)as a prototype microstrip patch antenna. Initially, the molecular composition was used with the chemical formula $xTiO_2(1\text{-}x)$ $ZnAl_2O_4$ and defined as $0.08TiO_2$ for x = 8%. X-ray diffraction (XRD) was used to investigate the generated composite nanoparticles, which indicated the dominating peaks of $ZnAl_2O_4$ and TiO_2. The crystallite size of the nanoparticles was calculated to be 16.9 nm. Fourier transform infrared spectroscopy (FTIR), field-emission scanning electron microscopy (FESEM), and energy dispersive spectroscopy (EDS) were also used to evaluate the functional groups, shape, and elemental content of the composite nanoparticles.These nanoparticles were also used to fabricate a prototype patch antenna in order to test its performance. Finally, a fabricated patch antenna was suitable for the X-band applications with the operating frequency 8.49 GHz, return loss (RL) -19.97 dB and VSWR<2.

Chapter 11: Advances in the semiconductor era have increased the overall performance and capabilities of chipsets while reducing size and cost. Additionally, advances in AI frameworks and libraries are enabling more AI with less assistance on IoT devices at the edge. Sensors are essential in today's environment, providing seamless data for the development of new smart applications. An example is a smart home situation with many connected devices. In such a place, there is easy and fast access to web services and personal statistics, including calendar, notes, email, reminders, bank and more., users connect 0.33 Birthday Celebration talents or skills from the Amazon store to their smart speaker. In addition, smart home devices such as smart security cameras, video doorbells, smart stores, smart carbon monoxide video indicators and smart door locks have come very close to smart homes. With smart speakers that add special requirements. Considering that the smart machine interacts with the products and devices used by the user account, every day people who own a physical device can use it. They have the most advanced technology available for smart speakers via voice commands. In this case, poverty, home security and other new aspects of the consumer suffer. Recently, Tensor Cam brought AI cameras, and Toshiba's Symbio and Facebook's Portal are smart speakers that support digital cameras and AI capabilities. Although they support cameras, they currently do not have an authentication process. This article discusses how smart speakers face cyber security risks due to the lack of authentication mechanisms, and discusses improvements to the modern, digital camera-enabled, microphone array-based Alexa smart speaker prototype.

Chapter 12: Recently, nanostructure-based drug delivery systems have become the center of attraction for researchers. In the last decade, various studies have revealed that fullerenes can be effectively utilized as carriers to transport various drugs to the affected cells. Thus, the potential application of borospherene (B_{40}) as a carrier for delivery of anti-cancer drug, 5-Fluorouracil, to prostate cancer cells is investigated using the density functional theory. To achieve targeted drug delivery, anisamide molecule is attached as ligand to the nano-carrier. It is deduced that the drug prefers to interact with the boron fullerene via its fluorine and oxygen atoms. The interaction of the drug and ligand at both the 6- and 7-membered rings of the borospherene is considered. The complex with ligand attached at the heptagonal ring and drug at the hexagonal ring is found to be the most stable with adsorption energy of -0.76eV. In order to get more insight, electronic properties viz. Density of states and Eigen states are also calculated for the most stable complex. This complex is found to be stable even in the aqueous medium. A pH stimulated drug release phenomenon is also proposed based on the knowledge that the cancer cells have lesser pH in comparison to the healthy cells. Thus, this study reveals that borospherene can be efficiently utilized as a potential nano-carrier for delivery of 5-Fluorouracil drug to prostate cancer cells.

Chapter 13: When a rectangular waveguide is in free space, the waveguide's characteristics and field distribution for different modes are computed using HFSS software. The analysis is performed for several parameters. The electric and magnetic field intensities, as well as the fundamental modal distributions, are investigated within a rectangular waveguide. Simulated analysis for the X-band is performed on a single size that has dimensions of 22.76, 10.15, 0.2, and 60 mm for breadth, width, wall thickness, and waveguide lengths. The performance of the rectangular waveguide is examined together with characteristic profiles. The waveguide is simulated, and the outcomes are confirmed. An HFSS (High Frequency Structural Simulator) simulation platform is shown in the figure for the analysis of aperture radiation from a rectangular waveguide made of dielectric material or air.

1

Comparative Analysis of MOSFET and FinFET

Mandeep Singh[1], Tarun Chaudhary[1], Balwinder Raj[1*], Girish Wadhwa[1,2] and Suman Lata Tripathi[3]

[1]Dr. B. R. Department of Electronics and Communication Engineering, Ambedkar National Institute of Technology Jalandhar Punjab, India
[2]Chitkara University Institute of Engineering and Technology, Chitkara University, Punjab, India
[3]Lovely Professional University Phagwara, Punjab, India

Abstract

The current chapter offers a thorough comparison of the two well-known transistor designs, MOSFET and FinFET, which have revolutionized integrated circuit architecture. The chapter offers a thorough investigation of the core ideas, design factors, and operational traits between the two transistor techniques. It describes the development of MOSFET technologies and how it has significantly advanced current electronics. The drawbacks of conventional MOSFET scaling are brought to light, necessitating the use of FinFET and other innovative transistor architectures. FinFET, revealing its ground-breaking design with a channel shaped like a fin and many gates. It examines the special operational benefits of FinFET, such as greater device scalability, improved electrostatic control, and lower leakage current. The article also looks at difficulties in designing and producing FinFETs, such as increasing process complexity and unpredictability. The outcomes of MOSFET and FinFET are compared and contrasted over a number of important factors. These factors involve switching speed, noise immunity, power efficiency, power dissipation, leakage current, on-state current, subthreshold slope, threshold voltage control, and device scaling.

Keywords: FET, BJT, MOSFET, FinFET

**Corresponding author*: rajb@nitj.ac.in

Balwinder Raj, Suman Lata Tripathi, Tarun Chaudhary, K. Srinivasa Rao and Mandeep Singh (eds.) Integrated Devices for Artificial Intelligence and VLSI, (1–24) © 2024 Scrivener Publishing LLC

1.1 Introduction

Bipolar junction transistors (BJTs) have a significant problem with static power dissipation, which means that power is lost even when the circuit isn't switching. The vacuum tubes were replaced with BJT. A silicon component having three regions—emitter, base, and collector—the BJT was a current-controlled device. Before this, the current-controlled MOSFET was developed. A FET (field effect transistor) is so-called since the gate switches are switched on or off through the transistor with an electric field crossing the gate oxide, as opposed to a basic MOS transistor, which utilized metal as the gate material, SiO_2 as insulator, and semiconductor as the substrate [1–5]. The drain, gate, source, and body of a MOS transistor are its four terminals. NMOS and PMOS are two complementary architectures that may be used to describe MOS transistors. In overall, poly-silicon is utilized as the gate material and contains heavy n or p type doped materials. Silicon serves as the insulator, and donor dEffects are implanted on each side to provide the source and drain. If two regions are biased with distinct possibilities, the less potentially area will function as the source and the more significant potential area as the drain. The "insulated gate" MOS transistors that were previously employed in IGFETs or MOSFETs (metal oxide semiconductor field effect transistors) are no longer around. IGFETs and MOSFETs are voltage-controlled electronics with metal oxide layers separating the gates from the bodies. The input resistance is significantly increased by this layer. There is not any current passing through the gate because it is electronically insulated from the primary current-carrying channel joining the drain and the source [6–10].

1.1.1 Scaling Issue

The Moorey's law, which was first proposed by Gordon E. Moore Sr. in 1965, states that the total number of transistors in a chip shall double on an annual basis. Considering its low power dissipation, speed, packing density, efficiency cost proportion, advancement in computing innovations, and most importantly, scalability, CMOS has greater characteristics than other semiconductors [11–16]. Technological scaling refers to the shrinking of the transistor chip's horizontal and vertical widths as well as the supply voltage V_{dd} that minimizes power loss and prevents oxide breakdown. In this circumstance, the threshold voltage (Vt) is correspondingly decreased in order to equalize the transistor's outputs. Therefore, it is crucial for nanoscale design to minimize either dynamic or leakage power [17–22].

The difficulty was in coming up with methods to decrease both of these elements while extending the life of the cellular devices' batteries. Continuing scaling has caused the number of dopants to drop from thousands to just a few several dozen, which also lowers the doping's expensive cost in CMOS devices and gives us better operational densities. The most intriguing novel technologies that have been launched, such as double gate MOSFET, FinFET, and Carbon Nanotube Field Effect Transistor, can replace traditional CMOS semiconductors. Planar technologies had a scaling issue that was present in planar technologies beneath 32 nm, which led to the introduction of non-planar devices [23–28].

1.1.2 Problems in MOSFET

The MOSFET's channel length has a significant impact on how well it works. In contrast to short channel devices, long channel devices have a very long channel that connects the source and drain. The transistor densities on a single chip are increased by the low operational voltage requirements and fast processing rate of short channel MOS. Even with all these benefits, short channel devices have significant drawbacks [29–34]. A few of which are briefly discussed below:

i. **Leakage Current:** That is the unwanted flowing of current while the transistor is meant to run in its off state, is a problem that MOSFETs may encounter. Leakage current can result in wasted energy and decreased efficiency, especially in low-power applications because cutting down on energy use is crucial [35–37].
ii. **Subthreshold Leakage:** Whenever a MOSFET is functioning in the sub threshold zone, wherein the gate voltage is lower than the threshold voltage, sub threshold leakage relates to the tiny amount of current that passes through the device. In low-power applications wherein the transistor must be fully off to save power, sub threshold leakage might be troublesome [38–40].
iii. **Threshold Voltage Fluctuation:** Throughout production, processing differences cause MOSFETs to display threshold voltage fluctuation. The voltage that causes a transistor switches on or off is known as the threshold voltage. The efficiency and dependability of circuits can be negatively impacted by large fluctuations in threshold voltage, which can result in instability and probable failure [41–43].

iv. **Self-Heating:** As a MOSFET conducts current, heat is produced as a result of power loss within the channel. The MOSFET's temperatures may increase if its heat is not adequately drained, which might result in efficiency deterioration or possibly thermal runaway.

v. **Hot Carrier Effects**: Whenever MOSFETs run at high voltages, the carriers—electrons or holes—in the channel may acquire significant kinetic energy, which can result in collision against the silicon lattice. The performance of the gadget may eventually deteriorate as a result of these accidents damaging the channel [44–50].

1.2 Double Gate

A variation of the multigate transistor known as the double gate MOSFET (DG MOSFET) was created from the incredibly popular MOSFET technologies. In contrast to the typical MOSFET design, which only has one gate regulating the channel, DG MOSFETs have two gates. For nanoscale size transistors in very dense IC design, the DG MOSFET shows significant potential [6]. The DG MOSFET's fundamental design requires placing two gate contacts towards the top and bottom of the channel within an ultra-thin body to enable efficient channel control via the two gates. Enhanced gate control improves drain to source current during the "ON" mode while preventing leakage current from flowing across the source and drain connections in the "OFF" mode. In this manner of lowering SCE, it also has a larger ON-OFF ratio than a typical MOSFET [7, 8]. Due to the employment of two gates to regulate the channel from each side, the DG MOSFET has better electrostatic properties than a Single Gate MOSFET (SG MOSFET). Larger switching signals arise from the two gates controlling nearly three times much current as a single gate. Each gate has the ability to regulate one device half, and each gate's function is totally distinct from each other. Since DG MOSFETs have double the current carrying capacity of planar CMOS, they can function at significantly lower inputs as well as threshold voltages. The DG MOSFET has a roughly two-times scaling benefit. The DG MOSFET has two gates that enable two different operating modes. Two configurations of DG MOSFET are possible: symmetrically driven double gate (SDDG) and independently driven double gate (IDDG). The bias of the gates is the primary distinction among both of them. Front and rear gates are coupled together when using SDDG method, but they get independent biasing in IDDG method [51–55]. The DG MOSFET idea has been studied extensively various academics during the past ten years, particularly as a remedy for the very short channel length technology. The DG MOSFET has established itself as a transistor's forerunner. The development of the DG MOSFET has opened up a vast array of potential study areas. It additionally acts as the primary driver of our academic initiatives as we transition from the simple design of devices to their practical uses in the digital sector [56–62]. Therefore, adopting a DG MOSFET circuit to create a logic gate structure that is more effective and has better gate voltage control would be a step forward. The general block diagram of DG MOSFET is shown in Figure 1.1.

A double gate MOSFET is a type of MOSFET that has two gates instead of one. The schematic diagram of a double gate MOSFET is similar to that of a traditional MOSFET, but with an additional gate. The double gate MOSFET has a source and a drain, which are connected by a channel. The channel is made of a semiconductor material, such as silicon, and is covered by a thin insulating layer, typically made of oxide. This oxide layer is known as the gate oxide. In a double gate MOSFET, there are two gates: the front gate (also known as the control gate or the top gate) and the back gate (also known as the body gate or the bottom gate). Both gates are separated from the channel by a thin oxide layer. The front gate is the primary gate and controls the flow of current through the channel. It is connected to the input signal and is used to turn the MOSFET on and off. The back gate is used to control the threshold voltage of the MOSFET. The threshold voltage is the voltage at which the MOSFET starts to conduct current. By applying a voltage to the back gate, the threshold voltage can be adjusted, which can be used to optimize the MOSFET for specific applications [9–11]. The DG MOSFET is a more advanced version of the traditional MOSFET and has several advantages, including improved performance, lower power consumption, and better scalability. It is commonly used in high-performance

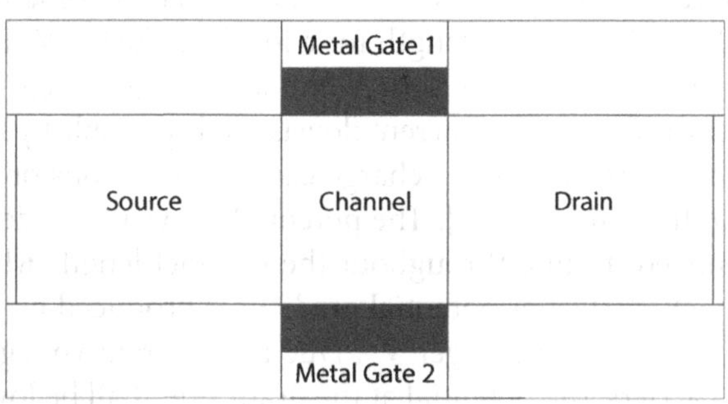

Figure 1.1 Double gate MOSFET.

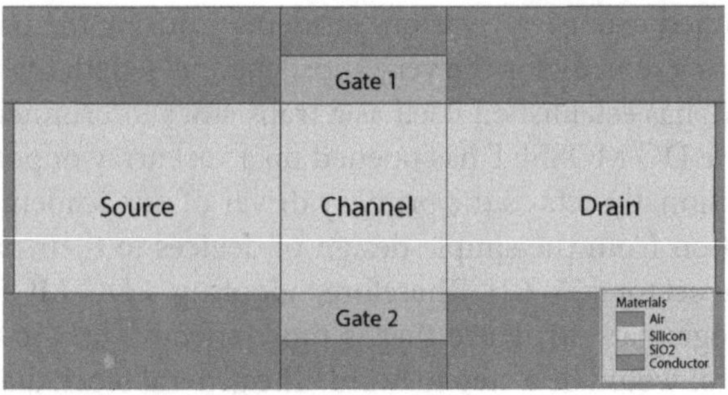

Figure 1.2 Schematic of double gate MOSFET.

electronics, such as microprocessors and memory chips, as well as in other applications that require low power consumption and high speed [63–68]. The Schematic of DG MOSFET is shown in Figure 1.2.

Because it establishes the fundamental doping characteristic of the semiconductor material, the acceptor concentration in a double gate MOSFET is typically constant across the substrate or channel area. The amount of doping in a material affects its conductivity as well as other electrical properties. In a MOSFET, neither the biasing conditions nor the applied voltage (V_{ds}) have a direct impact on the acceptor concentration [12]. It is a material characteristic that is established throughout the production processes and doesn't change until the doping profile is purposefully altered. It is important to keep in mind that the acceptor concentration might indirectly affect the efficiency and behavior of the double gate MOSFET, considering aspects such threshold voltage as well as sub threshold properties [69–72]. The acceptor concentration curve is shown in Figure 1.3 at different V_{gs}.

The electric potential of a double gate MOSFET is commonly shown across the source to the drain through the channel length. The supplied V_{ds}, gate voltages, and biasing circumstances all have an impact on the potential curve. The potential curve could display a somewhat straight drop throughout the channel length for a low V_{ds}, like 0.5 V. The potential would be stronger close to the source end and progressively decline as it moved nearer the drain side. Current flow is made possible by the potential gradient, which aids in drawing charge carriers (electrons or holes) from the source to the drain [13–15]. The potential curve might demonstrate a more pronounced decline throughout the channel length when V_{ds} rises to 1 V and 1.5 V. A steeper potential gradient is produced by the stronger electric field caused by the larger V_{ds}. Due to the drain voltage's ability to draw charge carriers, the potential at the drain side shall be lower than the source side. It is significant to consider that other elements, including as

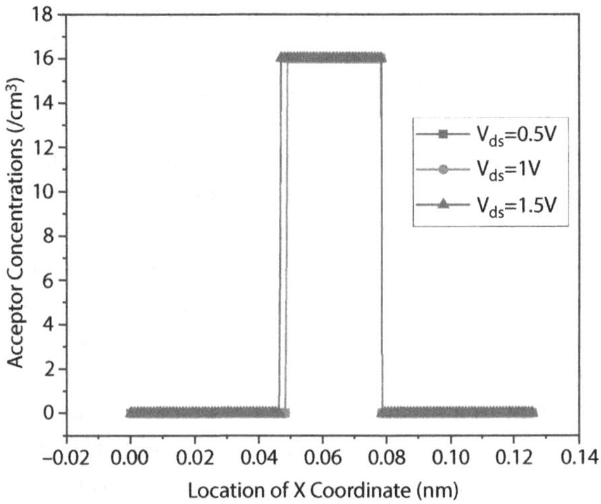

Figure 1.3 Acceptor concentration curve of DG MOSFET.

gate voltages, doping profiles, and device dimensions, may have an impact on the appearance and properties of the electric potential chart [73–77]. Additionally, the potential distributions in modern MOSFET architectures may be impacted by device scaling effects and technological improvements. Electric potential graph of DG MOSFET is shown in Figure 1.4.

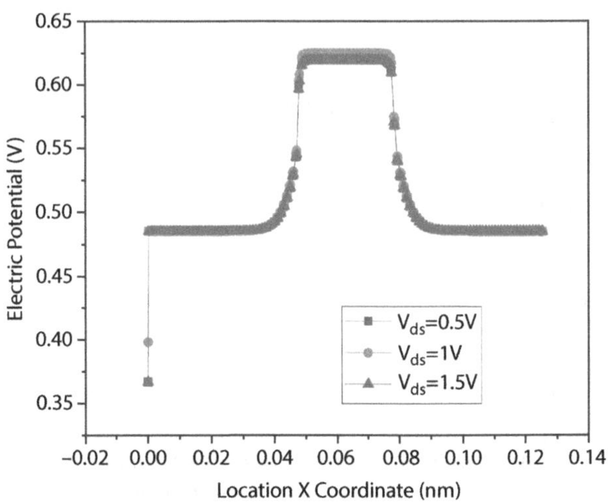

Figure 1.4 Electric potential graph of DG MOSFET at different V_{ds} = 0.5 V, 1 V, 1.5 V.

A double gate MOSFET's electric field curve will vary depending on the amount of V_{ds}, the device shape, the biasing setup, and the operational parameters. The electric field of a double gate MOSFET is commonly depicted across the source to the drain across the channel length. The gate voltages and the channel's potential profiles have an impact on the electric field distribution. The electric field variation might display a somewhat uniform distribution over the channel length at a low V_{ds}, such 0.5 V. The electric field would be comparatively small and consistent, both gates would be biased to regulate the channel area, and the transistor behavior would be good [16–20]. The electric field distributions may show more dramatic fluctuations when V_{ds} approaches 1 V and 1.5 V. The higher drain voltage might result in a greater electric field close to the drain side of the channel [78–83]. Larger electric field intensity might arise from this, might lead to hot carrier effects and deterioration of the transistor's efficiency. Scientists strive to reduce the peak electric field of a well-constructed double gate MOSFET to prevent reliability problems. To lessen the impacts of high electric fields, methods may be used including modifying the doping profile, enhancing channel length, and managing gate voltages. It's crucial to remember that a double gate MOSFET's exact electric field distribution might change depending on the device's size, the material's characteristics, and the biasing setup. Device designers frequently do thorough simulations and analyses to comprehend and optimize the behavior of the electric field in particular device architecture [21, 22]. Electric field graph of DG MOSFET is shown in Figure 1.5.

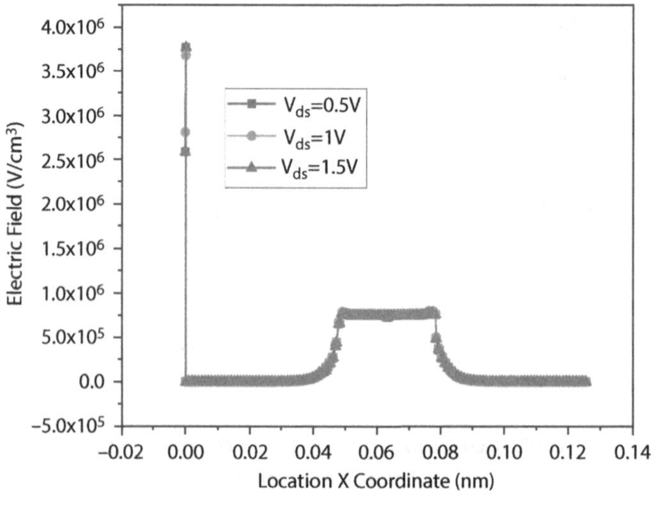

Figure 1.5 Electric field graph of DG MOSFET at different V_{ds} = 0.5 V, 1 V, 1.5 V.

The conduction current density in a MOSFET is generally shown from the source to the drain along the channel length [84–90]. It indicates the movement of charge carriers across the channel area, such as electrons in an n-channel MOSFET or holes in a p-channel MOSFET. The conduction current density curve could demonstrate a somewhat linear growth throughout the channel length at a low V_{ds}, like 0.5 V. When the bias voltage provided to the gate rises, the current density rises steadily, enabling additional charge carriers to pass across the channel. The conduction current density curve could display a more prominent non-linear behavior when V_{ds} rises to 1 V and 1.5 V. This is brought about by a greater electric field and a larger potential drop throughout the channel length as a result of the increased drain voltage [23]. As a consequence, the current density may reach saturation or show signs of velocity saturation, wherein subsequent rises in V_{ds}, are unlikely to result in a rise in current density that is proportionate [91–94]. Conduction current density chart of DG MOSFET is shown in Figure 1.6. Conduction current density chart features and form in a double gate MOSFET can change depending on the device's size, the material's properties, and the biasing circumstances.

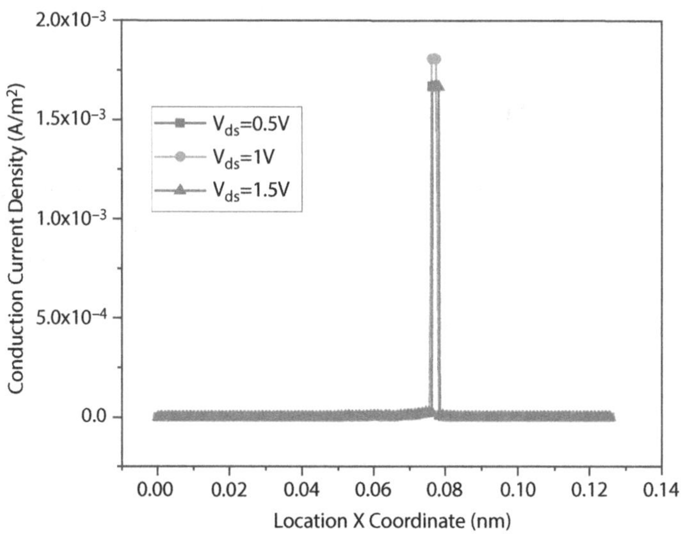

Figure 1.6 Conduction current density of DG MOSFET at different V_{ds} = 0.5 V, 1 V, 1.5 V.

1.3 Advantages and Disadvantage of MOSFET

MOSFETs are frequently employed in electronics because they offer a number of benefits, yet they also have certain drawbacks. Let's examine the benefits and drawbacks of MOSFETs:

High switching speed: MOSFETs have a fast on/off transition rate, making them suited for high-speed switching applications including power supply, motor drives, as well as digital electronics. They are able to manage and operate electrical equipment effectively thanks to their rapid switching.
Low power consumption: MOSFETs operate efficiently because they dissipate less power during switching. For rechargeable gadgets and purposes wherein power efficiency is essential, this is extremely advantageous [95–97].
High input impedance: MOSFETs have a low input current need due to their high input impedance. They are suitable for low-power control signals produced by computer chips and controllers owing to this feature.
Compact size: MOSFETs may be produced in tiny dimensions, enabling high chip integration densities. This makes it possible to construct sophisticated electrical devices with a more compact form factor and more capabilities [98–101].
Noise immunity: MOSFETs have excellent noise immunity because they are significantly less prone to interfering and electrical noise. Due to this characteristic, they are excellent for systems like speakers and communication systems wherein noise rejection is crucial capabilities.

1.4 MOSFET Drawbacks

Restricted voltage dealing ability: In comparison to other power electronic devices like IGBTs (Insulated Gate Bipolar Transistors), MOSFETs often have a restricted voltage handling capacity. Their use in high-voltage applications is thus constrained.
Electrostatic discharge (ESD): ESD is a threat that can harm or even destroy MOSFETs because of their sensitivity to it. ESD must be avoided when handling and assembling components.

1.5 FinFET

The term "FinFET" refers to a sophisticated transistor design that has grown significantly in significance in the area of semiconductor technologies.

Contrasting to conventional planar transistors, it is a three-dimensional in nature transistor shape that offers higher performance as well as energy efficiency. Contemporary integrated circuits (ICs) have been built on the foundation of FinFET technology, allowing for the continual scaling and miniaturization of electronic components. In order to overcome the constraints of conventional transistor as semiconductor manufacturing techniques advanced to nanoscale levels, FinFETs were developed as a key goal [24]. Scaling down planar transistors exacerbates problems such as leakage current and power dissipation, which reduces efficiency and increases the creation of heat. FinFETs were created to address these issues and provide better leakage current regulation. The insertion of a fin-shaped channel that rises vertically upward from the silicon substrate, as opposed to the flat channel in planar transistors, is what makes a FinFET unique. The conducting path among the transistor's source and drain regions is supplied by this fin. To improve electrostatic control over the channel, the gate electrode spans across all three faces of each fin. FinFETs significantly minimize leakage current by enhancing the transistor's electrostatic control through the use of a vertical fin structure [102–106]. The "gate-all-around" construction, which surrounds the fin from both ends with the gate electrode, enables more control over the transistor's on and off phases. This is accomplished, as a consequence, in comparison with planar transistors, FinFETs may achieve reduced power consumption and better performance. Moore's Law, which asserts that the number of transistors in an integrated circuit doubles about once every two years, was successfully advanced by the semiconductor sector in large part because to the switch to FinFET technologies [25]. In order to provide enhanced reliability, cost-effectiveness, and all-around usefulness, FinFETs have been incorporated into a variety of electronic products, including smart phones, laptops, computer systems, and various other electronic devices for consumers.

1.6 SOI-FinFET

The SOI-FinFET is virtually identical to the FinFET as shown in Figure 1.7. The SOI-FinFET substrate has been entirely filled by layers of oxide or buried oxide layer (BOX), which is the only little change. However, when using a FinFET, the source and drain are linked directly to the substrate thus there isn't an oxide layer across them. Minimal leakages will take place in contrast to MOSFET and FinFET because of the BOX layer. Compared with other technologies likes MOSFET and FinFET, SOI-FinFET switches more quickly. Comparing to the MOSFET and FINFET, the gate end has

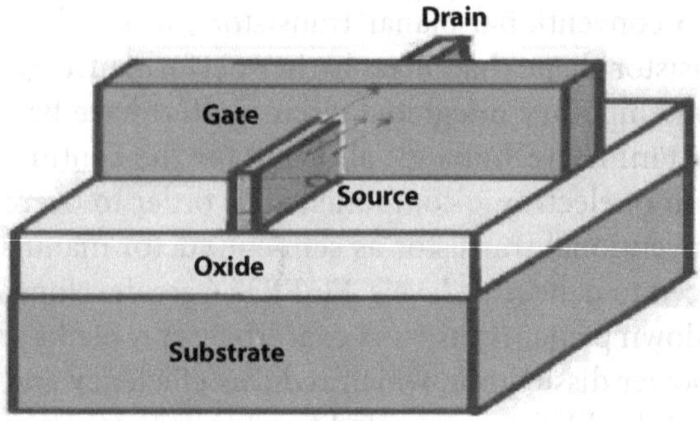

Figure 1.7 SOI-FinFET.

more control over the channel. In comparison to MOSFET and FinFET, drain current is additionally pulled higher at the output, while doping impurities are much less [26]. In comparison to the MOSFET and FinFET, only few stages are needed during the manufacturing process [107–109].

1.7 Issues with FinFET-Based Technology

Multi-gate short-circuited gate type and an independent gate type 3D transistor are features of a sophisticated FinFET technology [10]. When contrasted with CMOS's large, bulky planar the transistors, FinFETs operate better and consume less power. Additionally, it has improved electrical characteristics and far less both static and dynamic power [27–30]. The 10 nm FinFET process poses several significant design issues for improved FinFET technologies. These come from the views of analogue, digital, parasitic, and signoff [110–113].

- For accurate releasing using under 20 nm technologies during the fabrication of the FinFET device, extremely masking is required.
- Architecture developing at beneath 28 nm gets more challenging due to the necessity of extra terminals.
- 50 times or higher possibilities are needed to account for the metallic layers' resistance.
- Electro-migration will significantly boost efficiency all of the low circuit nodes operations.
- More complex fabricating-based design criteria;
- Longer time to commercialization due to designing a billion gates for manufacturing.

1.8 Advantage of FinFET

i. **Better Performance:** When contrasted with planar transistors, FinFETs operate better. The transistor's switching rate is accelerated by the gate-all-around design and vertical fin framework, which improve electrostatic control by lowering leakage current. The efficiency and quickness of electrical gadgets as a consequence are improved.
ii. **Lessened electricity Consumption:** FinFETs are renowned for using less electricity. Leakage current can now be better controlled, leading to more effective power utilization that decreases power loss and increases the lifespan of batteries in devices that are portable. It's crucial in the energy-conscious society of nowadays.
iii. **Scalability:** FinFET methodology has demonstrated good scaling, enabling ongoing scaling of transistor and the miniaturization of integrated circuits. By cramming additional transistors onto fewer dimensions, FinFETs offer a practical way to keep up with Moore's Law as semiconductor manufacturing methods develop.
iv. **Superior regulate of Short Channel Issues:** FinFETs have a better ability to regulate short channel effects like leakage current and poor transistor performance brought on by shorter channel lengths. These effects are lessened by the vertical fin construction, which leads to enhanced transistor behavior even at reduced aspect sizes.

1.9 Drawbacks of FinFET

i. **Sophisticated production method**: When contrasted with conventional planar transistors, FinFET technology necessitates more complex production procedures. The addition of the fin design and the exact position of the gate electrodes complicate manufacturing, raising manufacturing expenses and raising the possibility of yield problems.
ii. **Enhanced Design Complexity:** Because of the transistor's three-dimensional structure, designing circuits utilizing FinFETs might be more difficult. FinFET-specific design guidelines and limitations must be taken into account

by circuit designers, therefore may raise the difficulty of design as well as creation time.

iii. **Higher Cost**: FinFETs' complicated process of manufacture and more sophisticated design might result in greater manufacturing expenses. The availability and price of electrical products using FinFET technology may be impacted by this pricing issue.

1.10 Applications of FinFET Technology

FinFET technology has been used in a wide range of electrical devices and gadgets. Following are some significant applications for FinFETs:

High-performance processors: The layout of CPUs and GPUs in today's computers has been completely transformed by FinFET technology. They are perfect for high-performance computational tasks due to their exceptional performance, minimal power consumption, and significant scalability. More computing power can be delivered while preserving energy efficiency through FinFET-based CPUs.
Portable devices: FinFETs are ideal for mobile electronics like tablets and smart phones because to their high power efficiencies and low leakage current. The majority of the time, these electronic gadgets runs on batteries, and FinFET technology helps increase the lifespan of batteries while offering powerful computational and multimedia features.
Telecommunications and Networking: FinFETs are often utilized in both networking components and telecommunications systems. In switches, routers, broadband connections, and other networking elements, they provide high-speed information processing, effective signal enhancement, and enhanced power consumption.
Automotive electronics: FinFET innovation is useful in a number of automotive-related areas. FinFETs are used in engine control units (ECUs), multimedia frameworks, advanced driver assistance systems (ADAS), and power control modules. They are essential for the rigorous needs of automobile electronics because of their great efficiency and reliability.
Internet of Things (IoT): FinFETs are essential for allowing connection, processing power, and power conservation in IoT devices as the IoT continues to grow. FinFET technology helps to optimize power usage while maintaining superior efficiency and integration capability for everything from smart home appliances to industrial instruments.

Cloud computing and data centers: Comparative analysis of MOSFET and FinFET is given in Table 1.1. FinFET technology has significantly improved the effectiveness along with efficiency of the architectures for cloud computing and information centers. Because FinFET-based computers have more computing capability and consume less power, these kinds of devices can process and store data more effectively.

Table 1.1 Comparative analysis of MOSFET and FinFET.

Feature	MOSFET	FinFET
Gate Structure	Single gate	Multiple gates (fin-shaped structure)
Channel Control	Voltage-controlled	Voltage-controlled
Substrate Type	Planar substrate	Non-planar substrate
Gate Length	Usually in the range of 10–100 nm	Usually below 10 nm
Gate Leakage Current	Can be higher due to short channel effects	Significantly reduced due to improved electrostatic control
Scaling Limitations	Limited scalability due to short channel effects	Enhanced scalability and improved control over short channel effects
Power Efficiency	Lower power efficiency due to leakage current	Higher power efficiency due to reduced leakage current
Performance	Moderate performance in terms of speed and power	Improved performance with higher speed and lower power consumption
Manufacturing Complexity	Relatively simpler manufacturing process	More complex manufacturing process
Applications	Widely used in various applications, including CPUs, memory, and analog circuits	Preferred for advanced technologies, such as high-performance microprocessors

Aerospace and defense systems: Systems for the aerospace and defense industries, especially satellites, radar, avionics, other circuits with military-grade specifications, depend on FinFET technology. In difficult situations, FinFETs increase performance, dependability, and power efficiency.

1.11 Conclusion

The powerful transistor known as the MOSFET has been essential to the semiconductor sector. It has gained widespread use due to its adaptability, fabrication simplicity, and advanced architecture. Yet, due to the scaling constraints of MOSFETs, FinFETs have become an attractive option. Increased electrostatic control, lower leakage current, and increased device scaling are provided by the multi-gate architecture and special architecture of FinFETs. The FinFETs are used in many circuits to address difficulties like excessive latency, increased power consumption, and short channel effects brought on by MOSFETs that are getting smaller. Its validity has been established via analysis. The FinFETs can be scaled down to the nanoscale and yet operate superior to MOSFETs.

References

1. Takur Vaishali, Pahadiya Pallavi, Upadhyay Akhilesh, Tiwari Nidhi "A Review on MOSFET and its limitations: New era of transistor "FinFET Technology" *Pramana Res. J.* Volume 10, Issue 5, 2020.
2. Ruchika, Tripti Sharma, Sharma Krishna Gopal "Double Gate MOSFET Circuit Design" *IEEE International Conference on Recent Advances and Innovations in Engineering (ICRAIE-2014)*, May 09-11, 2014, Jaipur, India.
3. G. Baccarani, S. Reggiani, "A Compact Double-Gate MOSFET Model Comprising Quantum-Mechanical and Nonstatic Effects", *IEEE Trans. Electron Devices*, vol. 46, no. 8, pp. 1656-1666, 1999.
4. P. Francis, A. Terao, D. Flandre and F. Van de Wiele. "Moderate Inversion Model of Ultra-Thin Double-Gate nMOS/SOI Transistors", *Solid-State Electron.*, vol. 38, no. 1, pp. 171-176, 1995.
5. O. Moldovan, A. Cerdeira, D. Jiménez, J-P. Raskin, V. Kilchystka, D. Flandre. N. Collaert, B. Iñiguez, "Compact model for highly-doped double-gate SOI MOSFETs targeting baseband analog applications", *Solid-State Electron.*, vol. 51, pp. 655-661, 2007.
6. A. Ortiz-Conde, F. J. Garcia-Sanchez, J. Muci, S. Malobabic, and J. J. Liou, "A Review of Core Compact Models for Undoped Double-Gate SOI MOSFETs," *IEEE Trans. Electron Devices*, vol. 54, no. 1, pp. 131-140, 2007.

7. Y. Taur, X. Liang, W. Wang and H. Lu, "Continuous, Analytic Drain-Current Model for DG MOSFETs", *Electron Device Lett.*, vol. 25, no. 2, pp. 107-109, 2004.
8. Teja Pulakhandam, K.SinhaSanjeet, Durgesh Addala "Comparison of MOSFET, MESFET, FinFET and SOIFinFET" *JETIR*, Volume 7, Issue 11, November 2020.
9. D. Singh; S. K. Mohapatra; K. P. Pradhan; P. K. Sahu, IEEE, "Variation study of process parameters in Trigate SOIFinFET," *2014 Annual IEEE India Conference (INDICON)*, Pages: 1 - 4, 2014.
10. Anterpreet Gill; Charu Madhu; Pardeep Kaur, IEEE, "Investigation of short channel effects in Bulk MOSFET and SOI FinFET at 20nm node technology," *2015 Annual IEEE India Conference (INDICON)*, pp. 1-4, 2015.
11. A. Bhardwaj, P. Kumar, B. Raj, and S. Anand, "Design and Performance Optimization of Doping-Less Vertical Nanowire TFET Using Gate Stack Technique," *J. Electron. Mater. (JEMS)*, Springer, Vol. 41, No. 7, pp. 4005–4013, 2022.
12. J.Singh, and B.Raj, "Tunnel Current Model of Asymmetric MIM Structure Levying Various Image Forces to Analyze the Characteristics of Filamentary Memristor", *Appl. Phys. A*, Springer, Vol. 125, No. 3, pp. 203.1 to 203.11, 2019.
13. C.Goyal, J. S.Ubhi, and B.Raj, "Low Leakage Zero Ground Noise Nanoscale Full Adder Using Source Biasing Technique", *J. Nanoelectron. Optoelectron.*, American Scientifc Publishers,Vol. 14, pp. 360–370, 2019.
14. G.Singh, R. K.Sarin, and B.Raj, "A Novel Robust Exclusive-OR Function Implementation in QCA Nanotechnology with Energy Dissipation Analysis", *J. Comput. Electron.*, Springer, Vol. 15, No. 2, pp. 455–465, 2016.
15. T. Wadhera, D. Kakkar, G. Wadhwa, and B. Raj, " Recent Advances and Progress in Development of the Field Effect Transistor Biosensor: A Review", *J. Electron. Mater.*, Springer,Vol. 48, No. 12, pp. 7635–7646, 2019.
16. G.Wadhwa, P.Kamboj, and B.Raj, "Design Optimisation of Junctionless TFET Biosensor for High Sensitivity", *Adv. Nat. Sci.: Nanosci. Nanotechnol.*, Vol. 10, No. 7, p. 045001, 2019.
17. P.Bansal, and B.Raj, "Memristor Modeling and Analysis for Linear Dopant Drift Kinetics", *J. Nanoeng. Nanomanuf.*, American Scientifc Publishers, Vol. 6, pp. 1–7, 2016.
18. A.Singh, M.Khosla, and B.Raj, "Circuit Compatible Model for Electrostatic Doped Schottky Barrier CNTFET," *J. Electron. Mater.*, Springer, Vol. 45, No. 12, pp. 4825–4835, 2016.
19. Ashima, D.Vaithiyanathan, and B.Raj, "Performance Analysis of Charge Plasma induced Graded Channel Si Nanotube", *J. Eng. Res. (JER)*, EMSME Special Issue, Vol. 9, pp. 146–154, 2021.
20. A. S.Tomar, V. K.Magraiya, and B.Raj, "Scaling of Access and Data Transistor for High Performance DRAM Cell Design", *Quantum Matter*, Vol. 2, pp. 412–416, 2013.
21. N.Jain, and B.Raj, "Parasitic Capacitance and Resistance Model Development and Optimization of Raised Source/Drain SOI FinFET Structure for Analog

Circuit Applications", *J. Nanoelectron. Optoelectron.*, ASP, USA, Vol. 13, pp. 531–539, 2018.

22. S.Singh, S. K.Vishvakarma, and B.Raj, "Analytical Modeling of Split-Gate Junction-Less Transistor for a Biosensor Application", *Sens. Bio-Sens., Elsevier*, Vol. 18, pp. 31–36, 2018.
23. M.Gopal, and B.Raj, "Low Power 8T SRAM Cell Design for High Stability Video Applications", *ITSI Trans. Electr. Electron. Eng.*, Vol. 1, No. 5, pp. 91–97, 2013.
24. B.Raj, J.Mitra, D. K.Bihani, V.Rangharajan, A. K. Saxena, and S. Dasgupta, "Analysis of Noise Margin, Power and Process Variation for 32 nm FinFET Based 6T SRAM Cell", *J. Comput. (JCP), Academy Publisher, FINLAND*, Vol. 5, No. 6, PP. 1-6, 2010.
25. D.Sharma, R.Mehra, and B.Raj, "Comparative Analysis of Photovoltaic Technologies for High Efficiency Solar Cell Design", *Superlattices Microstruct.*, Elsevier, Vol. 153, pp. 106861, 2021.
26. P.Kaur, A. S.Buttar, and B.Raj, "A Comprehensive Analysis of Nanoscale Transistor Based Biosensor: A Review", *Indian J. Pure Appl. Phys.*, Vol. 59, pp. 304–318, 2021.
27. D.Yadav, B.Raj, and B.Raj, "Design and Simulation of Low Power Micro-controller for IoT Applications", *J. Sens. Lett.*, ASP USA, Vol. 18, pp. 401–409, 2020.
28. S.Singh, and B.Raj, "A 2-D Analytical Surface Potential and Drain Current Modeling of Double-Gate Vertical t-Shaped Tunnel FET", *J. Comput. Electron.*, Springer, Vol. 19, pp. 1154–1163, 2020.
29. J.Singh, and B.Raj, "An Accurate and Generic Window function for Nonlinear Memristor Model" *J. Comput. Electron., Springer*, Vol. 18, No. 2, pp. 640–647, 2019.
30. M.Kaur, N.Gupta, S.Kumar, B.Raj, and A. K.Singh, "Comparative RF and Crosstalk Analysis of Carbon Based Nano Interconnects", *IET Circuits, Devices Syst.*, Vol. 15, No. 6, pp. 493–503, 2021.
31. N.Kandasamy, F.Ahmad, D.Ajitha, B.Raj, and N.Telagam, "Quantum Dot Cellular Automata Based Scan Flip Flop and Boundary Scan Register", *IETE J. Res.*, Vol. 69, PP. 535-548, 2023.
32. S. K.Sharma, B.Raj, and M.Khosla, "Enhanced Photosensitivity of Highly Spectrum Selective Cylindrical Gate In1-xGaxAs Nanowire MOSFET Photodetector", *Mod. Phys. Lett.-B*, Vol. 33, No. 12, p. 1950144, 2019.
33. J.Singh, and B.Raj, "Design and Investigation of 7T2M NVSARM with Enhanced Stability and Temperature Impact on Store/Restore Energy", *IEEE Trans. Very Large Scale Integr. Syst.*, Vol. 27, No. 6, pp. 1322–1328, 2019.
34. A. K.Bhardwaj, S.Gupta, B.Raj, and A.Singh,"Impact of Double Gate Geometry on the Performance of Carbon Nanotube Field Effect Transistor Structures for Low Power Digital Design", *J. Comput. Theor. Nanosci.*, ASP USA, Vol. 16, pp. 1813–1820, 2019.

35. N.Jain, and B.Raj, "Thermal Stability Analysis and Performance Exploration of Asymmetrical Dual-k Underlap Spacer (ADKUS) SOI FinFET for Security and Privacy Applications", *Indian J. Pure Appl. Phys. (IJPAP)*, Vol. 57, pp. 352-360, 2019.
36. A.Singh, M.Khosla, and B.Raj, "Design and Analysis of Dynamically Configurable Electrostatic Doped Carbon Nanotube Tunnel FET", *Microelectronics J.*, Elsevier, Vol. 85, pp. 17–24, 2019.
37. N.Jain, and B.Raj, "Dual-k Spacer Region Variation at the Drain Side of Asymmetric SOI FinFET Structure: Performance Analysis towards the Analog/RF Design Applications", *J. Nanoelectron. Optoelectron.*, American Scientifc Publishers, Vol. 14, pp. 349–359, 2019.
38. J.Singh, S.Sharma, B.Raj, and M.Khosla, "Analysis of Barrier Layer Thickness on Performance of In1-xGaxAs Based Gate Stack Cylindrical Gate Nanowire MOSFET", JNO, ASP USA, Vol. 13, pp. 1473–1477, 2018.
39. N.Jain, and B.Raj, "Analysis and Performance Exploration of High-k SOI FinFETs over the Conventional Low-k SOI FinFET toward Analog/RF Design", *J. Semicond. (JoS), IOP Science*, Vol. 39, No. 12, pp. 124002-1-7, 2018.
40. C.Goyal, J. S.Ubhi, and B.Raj, "A Reliable Leakage Reduction Technique for Approximate Full Adder with Reduced Ground Bounce Noise", *J. Math. Probl. Eng., Hindawi*, Vol. 2018, Article ID 3501041, PP.1-16, 2018.
41. Anuradha, J. Singh, B.Raj, and M.Khosla, "Design and Performance Analysis of Nano-Scale Memristor-Based Nonvolatile SRAM", *J. Sens. Lett.*, American Scientific Publishers USA, Vol. 16, pp. 798–805, 2018.
42. G.Wadhwa, and B.Raj, "Parametric Variation Analysis of Charge-Plasma-Based Dielectric Modulated JLTFET for Biosensor Application", *IEEE Sens. J.*, Vol. 18, No. 15, PP. 6070 - 6077, 2018.
43. J.Singh, and B.Raj, "Comparative Analysis of Memristor Models for Memories Design", *JoS, IoP*, Vol. 39, No. 7, PP. 074006-1-12, 2018.
44. D.Yadav, S. S.Chouhan, S. K.Vishvakarma, and B.Raj, "Application Specific Microcontroller Design for IoT Based WSN", *Sens. Lett.*, ASP, Vol. 16, pp. 374–385, 2018.
45. G.Singh, R. K.Sarin, and B.Raj, "Fault-Tolerant Design and Analysis of Quantum-Dot Cellular Automata Based Circuits", *IEEE/IET Circuits, Devices Syst.*, Vol. 12, pp. 638–664, 2018.
46. J.Singh, and B.Raj, "Modeling of Mean Barrier Height Levying Various Image Forces of Metal Insulator Metal Structure to Enhance the Performance of Conductive Filament Based Memristor Model", *IEEE Nanotechnology*, Vol. 17, No. 2, pp. 268–275, 2018.
47. A.Jain, S.Sharma, and B.Raj, "Analysis of Triple Metal Surrounding Gate (TM-SG) III-V Nanowire MOSFET for Photosensing Application", *Opto-Electron. J.*, Elsevier, Vol. 26, No. 2, PP. 141–148, 2018.
48. A.Jain, S.Sharma, and B.Raj, "Design and Analysis of High Sensitivity Photosensor Using Cylindrical Surrounding Gate MOSFET for Low Power

Sensor Applications", *Eng. Sci. Technol. Int. J.*, Elsevier's, Vol. 19, No. 4, pp. 1864–1870, 2016.

49. A.Singh, M.Khosla, and B.Raj, "Analysis of Electrostatic Doped Schottky Barrier Carbon Nanotube FET for Low Power Applications," *J. Mater. Sci.: Mater. Electron.*, Springer, Vol. 28, PP. 1762–1768, 2017.
50. G.Saiphani Kumar, A.Singh, B.Raj, "Design and Analysis of Gate All around CNTFET Based SRAM Cell Design", *J. Comput. Electron.*, Springer, Vol. 17, No.1, PP. 138–145, 2018.
51. G. P.Singh, B. S.Sohi, and B.Raj, "Material Properties Analysis of Graphene Base Transistor (GBT) for VLSI Analog Circuits", *Indian J. Pure Appl. Phys. (IJPAP)*, Vol. 55, pp. 896–902, 2017.
52. S.Kumar and B.Raj, "Estimation of Stability and Performance Metric for Inward Access Transistor Based 6T SRAM Cell Design Using n-type/p-type DMDG-GDOV TFET", *IEEE VLSI Circuits Syst. Lett.*, Vol. 3, No. 2, PP. 25–39, 2017.
53. S.Sharma, A.Kumar, M.Pattanaik, and B.Raj, "Forward Body Biased Multimode Multi-Threshold CMOS Technique for Ground Bounce Noise Reduction in Static CMOS Adders", *International Journal of Information and Electronics Engineering*, Vol. 3, No. 3, PP. 567–572, 2013.
54. H.Singh, P.Kumar, and B.Raj, "Performance Analysis of Majority Gate SET Based 1-bit Full Adder", *Int. J. Inf. Electron. Eng. (IJCCE)*, IACSIT Press Singapore, ISSN: 2010-3743, Vol. 2, No. 4, PP.1-8, 2013.
55. A. K.Bhardwaj, S.Gupta, and B.Raj, "Investigation of Parameters for Schottky Barrier (SB) Height for Schottky Barrier Based Carbon Nanotube Field Effect Transistor Device", *J. Nanoelectron. Optoelectron.*, ASP, Vol. 15, PP. 783–791, 2020.
56. P.Bansal, and B.Raj, "Memristor: A Versatile Nonlinear Model for Dopant Drif and Boundary Issues", *JCTN, American Scientifc Publishers*, Vol. 14, No. 5, P P. 2319–2325, 2017.
57. N.Jain, and B.Raj, "An Analog and Digital Design Perspective Comprehensive Approach on Fin-FET (Fin-Field Effect Transistor) Technology - A Review", *Rev. Adv. Sci. Eng. (RASE)*, ASP, Vol. 5, PP. 1–14, 2016.
58. S.Sharma, B.Raj, and M.Khosla, "Subthreshold Performance of In1-xGaxAs Based Dual Metal with Gate Stack Cylindrical/Surrounding Gate Nanowire MOSFET for Low Power Analog Applications", *J. Nanoelectron. Optoelectron.*, American Scientific Publishers, USA, Vol. 12, PP. 171– 176, 2017.
59. B.Raj, A. K.Saxena, and S.Dasgupta, "Analytical Modeling for the Estimation of Leakage Current and Subthreshold Swing Factor of Nanoscale Double Gate FinFET Device" *Microelectron. Int.*, UK, Vol. 26, PP. 53–63, 2009.
60. S. S.Soniya, G.Wadhwa, and B.Raj, "A n Analytical Modeling for Dual Source Vertical Tunnel Field Effect Transistor", *Int. J. Recent Technol. Eng, (IJRTE)*, Vol. 8, No. 2, 2019.

61. S.Singh, and B.Raj, "Design and Analysis of Hetrojunction Vertical T-Shaped Tunnel Field Effect Transistor", *J. Electron. Mater.*, Springer, Vol. 48, No. 10, PP. 6253–6260, 2019.
62. C.Goyal, J. S.Ubhi, and B.Raj, "A Low Leakage CNTFET Based Inexact Full Adder for Low Power Image Processing Applications", *Int. J. Circuit Theory Appl.*, Wiley, Vol. 47, No. 9, PP. 1446–1458, 2019.
63. B.Raj, A. K.Saxena, and S.Dasgupta, "A Compact Drain Current and Threshold Voltage Quantum Mechanical Analytical Modeling for FinFETs", *J. Nanoelectron. Optoelectron. (JNO)*, USA, Vol. 3, No. 2, PP. 163–170, 2008.
64. G.Wadhwa, and B.Raj, "A n Analytical Modeling of Charge Plasma Based Tunnel Field Effect Transistor with Impacts of Gate Underlap Region", *Superlattices Microstruct.*, Elsevier, Vol. 142, PP. 106512-22, 2020.
65. S.Singh, and B.Raj, "Modeling and Simulation Analysis of SiGe Hetero-junction Double GateVertical t-Shaped Tunnel FET", *Superlattices Microstruct., Elsevier*, Vol. 142, PP. 106496-104, 2020.
66. A.Singh, D. K.Saini, D.Agarwal, S.Aggarwal, M.Khosla, and B.Raj, "Modeling and Simulation of Carbon Nanotube Field Effect Transistor and Its Circuit Application", *J. Semicond. (JoS)*, IOP Science, Vol. 37, PP. 074001–074006, 2016.
67. N.Jain, and B.Raj, "Device and Circuit Co-Design Perspective Comprehensive Approach on FinFET Technology - A Review", *J. Electron Devices*, Vol. 23, No. 1, PP. 1890–1901, 2016.
68. S.Kumar and B.Raj, "Analysis of ION and Ambipolar Current for Dual-Material Gate-Drain Overlapped DG-TFET", *J. Nanoelectron. Optoelectron.*, American Scientific Publishers, USA, Vol. 11, PP. 323–333, 2016.
69. N.Anjum, T.Bali, and B.Raj, "Design and Simulation of Handwritten Multiscript Character Recognition", *Int. J. Adv. Res. Comput. Commun. Eng.*, Vol. 2, No. 7, PP. 2544–2549, 2013.
70. S.Sharma, B.Raj, and M.Khosla, "A Gaussian Approach for Analytical Subthreshold Current Model of Cylindrical Nanowire FET with Quantum Mechanical Effects", *Microelectronics J.*, Elsevier, Vol. 53, PP. 65–72, 2016.
71. K.Singh, and B.Raj, "Performance and Analysis of Temperature Dependent Multi-Walled Carbon Nanotubes as Global Interconnects at Different Technology Nodes", *J. Comput. Electron.*, Springer, Vol. 14, No. 2, PP. 469–476, 2015.
72. S.Kumar and B.Raj, "Compact Channel Potential Analytical Modeling of DG-TFET Based on Evanescent–Mode Approach", *J. Comput. Electron.*, Springer, Vol. 14, No. 2, PP. 820–827, 2015.
73. K.Singh, and B.Raj, "Temperature Dependent Modeling and Performance Evaluation of Multi-Walled CNT and Single-Walled CNT as Global Interconnects", *J. Electron. Mater.*, Springer,Vol. 44, No. 12, PP. 4825–4835, 2015.

74. V. K. Sharma, M. Pattanaik, and B. Raj, "INDEP Approach for Leakage Reduction in Nanoscale CMOS Circuits", *Int. J. Electron.*, Taylor & Francis, Vol. 102, No. 2, PP. 200–215, 2014.
75. K.Singh, and B.Raj, "Influence of Temperature on MWCNT Bundle, SWCNT Bundle and Copper Interconnects for Nanoscaled Technology Nodes", *J. Mater. Sci.: Mater. Electron.* Springer, Vol. 26, No. 8, PP. 6134–6142, 2015.
76. N.Anjum, T.Bali, and B.Raj, "Design and Simulation of Handwritten Gurumukhi and Devanagri Numerical Recognition", *Int. J. Comput. Appl., Published by Foundation of Computer Science*, New York, USA,Vol. 73, No. 12, PP. 16–21, 2013.
77. S.Khandelwal, V.Gupta, B.Raj, and R. D.Gupta, "Process Variability Aware Low Leakage Reliable Nano Scale DG-FinFET SRAM Cell Design Technique", *J. Nanoelectron. Optoelectron.*, Vol. 10, No. 6, PP. 810–817, 2015.
78. V. K. Sharma, M. Pattanaik, and B.Raj, "ONOFIC Approach: Low Power High Speed Nanoscale VLSI Circuits Design", *Int. J. Electron.*, Taylor & Francis, Vol. 101, No. 1, PP. 61–73, 2014.
79. S.Khandelwal, B.Raj, and R. D.Gupta, "FinFET Based 6T SRAM Cell Design: Analysis of Performance Metric, Process Variation and Temperature Effect", *J. Comput. Theor. Nanosci.*, ASP, USA, Vol. 12, PP. 2500–2506, 2015.
80. S.Singh, Y.Shekhar, R.Jagdeep, S.Anurag, and B.Raj, "Impact of HfO2 in Graded Channel Dual Insulator Double Gate MOSFET", *J. Comput. Theor. Nanosci.*, American Scientific Publishers, Vol. 12, No. 6, PP. 950–953, 2015.
81. V. K.Sharma, M.Pattanaik, and B.Raj, "PVT Variations Aware Low Leakage INDEP Approach for Nanoscale CMOS Circuits", *Microelectron. Reliab.*, Elsevier, Vol. 54, PP. 90–99, 2014.
82. B.Raj, A. K.Saxena and S.Dasgupta, "Quantum Mechanical Analytical Modeling of Nanoscale DG FinFET: Evaluation of Potential, Threshold Voltage and Source/Drain Resistance", Elsevier's *J. Mater. Sci. Semicond. Process.*, Elsevier, Vol. 16, No. 4, PP. 1131–1137, 2013.
83. M.Gopal, S. S. D.Prasad, and B.Raj, "8T SRAM Cell Design for Dynamic and Leakage Power Reduction", *Int. J. Comput. Appl.*, Published by Foundation of Computer Science, New York, USA,Vol. 71, No. 9, PP. 43–48, 2013.
84. M.Pattanaik, B.Raj, S.Sharma, and A.Kumar, "Diode Based Trimode Multi-Treshold CMOS Technique for Ground Bounce Noise Reduction in Static CMOS Adders", *Adv. Mat. Res., Trans Tech Publications*, Switzerland, Vol. 548, PP. 885–889, 2012.
85. B.Raj, A. K.Saxena, and S.Dasgupta, "Nanoscale FinFET Based SRAM Cell Design: Analysis of Performance Metric, Process Variation, Underlapped FinFET and Temperature Effect", ***IEEE Circuits and Syst. Mag.***, Vol. 11, No. 2, PP. 38–50, 2011.
86. V. K. Sharma, M. Pattanaik, and B.Raj, "Leakage Current ONOFIC Approach for Deep Submicron VLSI Circuit Design", *International Journal of Electrical, Computer, Electronics and Communication Engineering*, World Academy of Sciences, Engineering and Technology, Vol. 7, No. 4, PP. 239–244, 2013.

87. T.Chawla, M.Khosla, and B.Raj, "Design and Simulation of Triple Metal Double-Gate Germanium on Insulator Vertical Tunnel Field Effect Transistor", *Microelectronics J.*, Elsevier, Vol. 114, PP. 105125-32, 2021.
88. P.Kaur, S. S.Gill, and B.Raj, "Comparative Analysis of OFETs Materials and Devices for Sensor Applications", *J. Silicon*, Springer, Vol. 14, PP. 4463–4471, 2022.
89. S. K.Sharma, P.Kumar, B.Raj, and Balwinder Raj, "In1-xGaxAs Double Metal Gate-Stacking Cylindrical Nanowire MOSFET for Highly Sensitive Photo Detector", *J. Silicon*, Springer, Vol. 14, PP. 3535–3541,2022.
90. B.Raj, A. K.Saxena, and S.Dasgupta, "Analytical Modeling of Quasi Planar Nanoscale Double Gate FinFET with Source/Drain Resistance and Field Dependent Carrier Mobility: A Quantum Mechanical Study", *J. Comput. (JCP)*, Academy Publisher, FINLAND, Vol. 4, No. 9, PP.1-8, 2009.
91. S.Bhushan, S.Khandelwal, and B.Raj, "Analyzing Diferent Mode FinFET Based Memory Cell at Different Power Supply for Leakage Reduction", *Seventh International Conference on Bio-Inspired Computing: Theories and Application, (BIC-TA 2012)*, organized IIITM Gwalor, PP. 89–100, 14-16 Dec 2012.
92. J.Singh, and B.Raj, "Temperature Dependent Analytical Modeling and Simulations of Nanoscale Memristor", *Eng. Sci. Technol. Int. J.*, Elsevier's, Vol. 21, PP. 862–868, 2018.
93. S.Singh, S.Bala, B.Raj, and Balwinder Raj, "Improved Sensitivity of Dielectric Modulated Junctionless Transistor for Nanoscale Biosensor Design", *Sens. Lett.*, ASP, Vol. 18, PP. 328–333, 2020.
94. V.Kumar, S. K.Vishvakarma, and B.Raj, "Design and Performance Analysis of ASIC for IoT Applications", *Sens. Lett., ASP*, Vol. 18, PP. 31–38, 2020.
95. A.Jaiswal, R. K.Sarin, B.Raj, and S.Sukhija, "A Novel Circular Slotted Microstrip-fed Patch Antenna with Three Triangle Shape DEffected Ground Structure for Multiband Applications", *Adv. Electromagn. (AEM)*, Vol. 7, No. 3, PP. 56–63, 2018.
96. G.Wadhwa, and BRaj, "Label Free Detection of Biomolecules Using Charge-Plasma-Based Gate Underlap Dielectric Modulated Junctionless TFET", *J. Electron. Mater. (JEMS)*, Springer, Vol. 47, No. 8, PP. 4683– 4693, 2018.
97. G.Singh, R. K.Sarin, and B.Raj, "Design and Performance Analysis of a New Efcient Coplanar Quantum-Dot Cellular Automata Adder", *Indian J. Pure Appl. Phys. (IJPAP)*, Vol. 55, PP. 97–103, 2017.
98. A.Singh, M.Khosla, and B.Raj, "Design and Analysis of Electrostatic Doped Schottky Barrier CNTFET Based Low Power SRAM", *Int. J. Electron. Commun.*, (AEÜ), Elsevier, Vol. 80, PP. 67–72, 2017.
99. P.Kaur, V.Pandey, and B.Raj, "Comparative Study of Efficient Design, Control and Monitoring of Solar Power Using IoT", *Sens. Lett.*, ASP, Vol. 18, PP. 419–426, 2020.
100. G.Wadhwa, P.Kamboj, J.Singh, B.Raj, "Design and Investigation of Junctionless DGTFET for Biological Molecule Recognition", *Trans. Electr. Electron. Mater.*, Springer, Vol. 22, PP. 282–289, 2020.

101. T.Chawla, M.Khosla, and B.Raj, "Optimization of Double-Gate Dual Material GeOI-Vertical TFET for VLSI Circuit Design", *IEEE VLSI Circuits Syst. Lett.*, Vol. 6, No. 2, PP. 13–25, 2020.
102. S. K.Verma, S.Singh, G.Wadhwa and B.Raj, "Detection of Biomolecules Using Charge-Plasma Based Gate Underlap Dielectric Modulated Dopingless TFET", *Trans. Electr. Electron. Mater. (TEEM)*, Springer, Vol. 21, PP. 528–535, 2020.
103. N.Jain, and B.Raj, "Impact of Underlap Spacer Region Variation on Electrostatic and Analog/RF Performance of Symmetrical High-k SOI FinFET at 20 nm Channel Length", *J. Semicond. (JoS), IOP Science*, Vol. 38, No. 12, PP. 122002-10, 2017.
104. S.Singh, and B.Raj, "Analytical Modeling and Simulation Analysis of T-Shaped III-V Heterojunction Vertical T-FET", *Superlattices Microstruct.*, Elsevier, Vol. 147, PP. 106717-25, 2020.
105. G.Singh, R. K.Sarin, and B.Raj, "Design and Analysis of Area Efcient QCA Based Reversible Logic Gates", *J. Microprocess. Microsyst.*, Elsevier, Vol. 52, PP. 59–68, 2017.
106. A.Singh, M.Khosla, and B.Raj, "Compact Model for Ballistic Single Wall CNTFET under Quantum Capacitance Limit", *J. Semicond. (JoS), IOP Science*, Vol. 37, PP. 104001–104008, 2016.
107. S.Singh, M.Khosla, G.Wadhwa, and B.Raj, "Design and Analysis of Double-Gate Junctionless Vertical TFET for Gas Sensing Applications", *Appl. Phys. A*, Springer, Vol. 127, No. 16, PP. 883-890, 2021.
108. I.Singh, B.Raj, M.Khosla, and B. K.Kaushik, "Potential MRAM Technologies for Low Power SoCs", *SPIN World Scientific Publisher, SCIE*; Vol. 10, No. 04, PP. 2050027-35, 2020.
109. S.Singh, and B.Raj, "Parametric Variation Analysis on Hetero-Junction Vertical t-Shape TFET for Suppressing Ambipolar Conduction", *Indian J. Pure Appl. Phys*., Vol. 58, PP. 478–485, 2020.
110. S. S.Soniya, G.Wadhwa, and B.Raj, "Design and Analysis of Dual Source Vertical Tunnel Field Effect Transistor for High Performance", *Trans. Electr. Electron. Mater.*, Springer, Vol. 21, PP. 74–82, 2019.
111. M.Kaur, N.Gupta, S.Kumar, B.Raj, and A. K.Singh, "RF Performance Analysis of Intercalated Graphene Nanoribbon Based Global Level Interconnects", *J. Comput. Electron.*, Springer, Vol. 19, PP. 1002–1013, 2020.
112. G.Wadhwa, B.Raj, "Design and Performance Analysis of Junctionless TFET Biosensor for High Sensitivity" *IEEE Nanotechnol.*, Vol.18, PP. 567–574, 2019.
113. J.Singh, and B.Raj, "Enhanced Nonlinear Memristor Model Encapsulating Stochastic Dopant Drift", *JNO, ASP*, Vol. 14, PP. 958–963, 2019.

2

Nanosheet FET for Future Technology Scaling

Aruru Sai Kumar[1], V. Bharath Sreenivasulu[2], M. Deekshana[1], G. Shanthi[1] and K. Srinivasa Rao[3]*

[1]Department of ECE, VNR Vignana Jyothi Institute of Engineering and Technology, Hyderabad, India
[2]Department of Electrical Engineering, IIT Patna, Patna, India
[3]Department of Electronics and Communication Engineering, MEMS Research Center, Koneru Lakshmaiah Education Foundation, Vaddeswaram, Guntur, India

Abstract

Improvements in the VLSI industry have always been striving to justify the Moore's law by implanting, twice count transistors from the existing one. This law has made a significant improvement in the trends of FET family starting with the straightforward MOSFET with one gate controlling the channel, then next followed by FET with Dual gates, later Tri-gate/Fin FET and now persisting Gate-All-Around Field Effect Transistor. Each era of technology contributed its own advantages and disadvantages. Presently the Gate-All-Around FETs are ruling the FET industry because of its major advantage of improved gate electrostatic integrity over the channel from all the directions, with reduced overall size of the FET. In this research work, the vertically stacked GAA Nanosheet FET is simulated at the device-level using Visual TCAD – 3D Cogenda tool. Research is conducted to substantiate the influence of geometrical variations with respect to thickness and width on the performance of the FET. Parameters analyzed to in the research are I_{ON}, I_{OFF}, Switching ratio (I_{ON}/I_{OFF}), Subthreshold swing (SS), DIBL, and Threshold voltage (V_{th}). Device is optimal if it offers better ON current, minimum OFF current. To evaluate this optimal performance the thickness of the nanosheet (N_T) is varied from 5 nm to 9 nm, and the width is varied from 10 nm to 50 nm. Visual TCAD - 3D Cogenda tool is used in analyzing the outputs generated and conclude the best geometric dimension for optimum characteristics.

**Corresponding author*: srinivasakarumuri@gmail.com

Balwinder Raj, Suman Lata Tripathi, Tarun Chaudhary, K. Srinivasa Rao and Mandeep Singh (eds.) Integrated Devices for Artificial Intelligence and VLSI, (25–48) © 2024 Scrivener Publishing LLC

***Keywords*:** FinFET, nanowire FET, nanosheet FET, subthreshold swing (SS), DIBL

2.1 Introduction

The semiconductor industry has always been striving to make new advancements in the technology trends to meet the ITRS standards. The major motivation for all the advancements is to abide the law stated by Moore which states the count of transistors on the chip must be doubled up for every 18 months. Most optimum way to double the count of transistors is to reduce the size of each transistor. Transistor size diminishing is done by channel length reduction which initially was the effective solution to improve the technology nodes [1, 2]. Conventional MOSFET built in 1959 ruled the initial era of FET family in various fields of applications. MOSFETs prevailed until the SCEs become prominent due to which performance declined. When the channel length of a MOSFET is like the depletion layer widths at the source and drain junctions, short-channel effects occur. Various SCE'S such as velocity saturation, drain-induced barrier lowering (DIBL), hot carrier degradation, and quantum confinement. To address these SCE's various FET structures – Double gate FETs, FinFETs, Nanowire FET, Nanosheet FET, and Tunnel FET have come into existence [3, 4]. Double gate FETs are constructed with gates on the top and bottom which served a superior gate electrostatic integrity but, major drawback is the alignment between the top and bottom gate. Misalignment in the process of fabrication of Double gate FET may cause degradation in the conduction current in the ON state. To address the problem, next generation of FETs were developed with Tri-gate structures known as FinFET. FinFET is created with the gate surrounding the channel from tri planes. Family of FinFET ruled the Semiconductor industry providing the advantage of enormous reduction in SCEs having lower DIBL and SS characteristics for sub-22 nm technology nodes. Progressing towards instrument measuring of FinFET, the Fins of the design are erected taller and finer to drive best currents. But making FinFETs accompanying finer, and more exaggerated fins is a troublesome movement, and their small bases can bring about design questions [5].

The GAA structures of nanosheet (NS) and nanowire (NW) are introduced to address all these issues with conventional FETs. GAA structures are designed with the channel enclosed with the gates from all surrounding planes. A nanowire has a same width and height, however a nanosheet is created when the height of the channel is slightly increased.

Although the lower SCEs of nanowires are their strongest suit, their performance has suffered because of their rough surface [6–10]. The next generation of FinFET devices for sub-7 nm and beyond is called nanosheets. NSFETs exhibit less SCEs because of great electrostatic control over the channel area [26–32].

They are also capable of driving large capacitive loads. Nanosheets are the following generation devices of FinFET for sub-7 nm and beyond. Because of their excellent electrostatic control over the channel region, NSFETs have fewer SCEs. It has been demonstrated that single stack nanosheets operate demonstrably better than FinFETs or stacked nanowires [33–38]. They can also drive substantial capacitive loads. To drive the large current, nanosheets are stacked together, but the disadvantage of this approach is the larger height of the device may build parasitic capacitance. Nanosheets are piled together to drive a huge current, however this method has the drawback that the device's greater height may cause parasitic capacitance to develop. NSFETs associated with these advantages replace the traditional MOSFET. Combining the advantages of GAA Nanosheet FET with the vertical stacking enhances the overall performance of the device by providing robust leakage control, increased effective width, good electrostatic performance over the channel, and increase in the driver current. Count of the nanosheet stacking vertically, also influences the performance of the device constructed which can have negative effect on parasitic capacitance. Hence this paper limits the stacking of nanosheet to two [39–43].

Sresta *et al.*, [8] investigates the various dimensions of the Junctionless NS Field effect Transistor by varying the gate length, Nanosheet thickness, and width. Analysis of results exposed that the scaling down of gate length results in a rise the SS, DIBL, and decrease the ION/IOFF ratio. The influence of the substrate material on the performance of three channels Stacked NS Field Effect Transistor is investigated, and the Super-Steep-Retrograde silicon substrate is bestowed as a capable option for optimal operation [9]. After a comparative analysis of FinFET, NWs, NSs based on the doping concentrations of source and drain concludes NS is the most favorable device [10–15].

The proposed NSFET examines the influence of geometrical variations – width of the NS, thickness of the NS by keeping the length of the gate, doping concentration, work function constant throughout the simulation. Paper highlights the device dimensions technology to evaluate the performance of the device. Thickness is varied from 5 nm ranging up to 9 nm and width is varied from 10 nm to 50 nm, and the performance of the two vertically stacked NS FETs is analyzed throughout this work using DC metrics [44–50]. The lower technology node functionality of two stacks of GAA MOSFETs is anticipated in this research, which also includes different DC characteristics than ITRS assumptions. The gate stack uses the high-k dielectric HfO_2 to further minimize gate leakage current. Nitride Spacers along the source to gate and gate to drain can increase the switching ratio due to field coupling through fringing effect. These characteristics are carefully evaluated in this study [51–56].

This research paper is structured as follows, section 2 covers the mechanisms and characteristics of the devices, section 3 describes results and analysis, section 4 describes conclusions.

2.2 Device Description and Simulation Parameters

Device is structured and simulated is using Visual TCAD -3D Cogenda tool. It provides Genius, a device simulator, which has a graphical user interface called Visual TCAD. GAA NSFET is constructed with the gate length of 16 nm. Terminals of the NSFET, channel are formed by doping Si with N-type impurities of concentration 10^{20} cm^{-3} and 10^{15} cm^{-3}. To avoid the gate tunneling effect the EOT of 0.78 nm is maintained while includes interfacial oxide of 0.5 nm and HfO_2 of 1.5 nm [57–62]. A constant metal gate's work function of 4.6 eV is maintained through the research. To overcome the hurdle of direct tunneling the spacer of 5 nm nitride with the dielectric constant of 7.5 is used. Figure 2.1 shows the 3D view of GAA NSFET. Device is constructed using the coding method in Visual TCAD. Thickness of the device is varied between 5 nm to 9 nm. Width of the device is varied between the 10–50 nm. SOI technology is been used in the device construction in which the Buried Oxide layer of SiO_2 material is used to enhance the device performance by providing diminished parasitic capacitance, enhanced switching, low power dissipation.

Figure 2.2 depicts the cross sectional view of the device across the X-Z plane. As depicted in the below figure the gate stack architecture is used to avoid the gate tunneling effect. In this architecture design the high dielectric materials are used to reduce the gate leakage current. Here the HfO_2 with the dielectric constant approximately equal to 20 with the thickness of 1.5 nm is used along with the interfacial oxide, providing the total EOT of 0.78 nm [63–70]. In parallel, the advantage of spacer is also being used in the application in which the nitride spacer of 5 nm is used on the either sides of the gate. Spacer technology in the sub-nm devices is important to reduce SCE's. But imposing the spacer can pose a hazard of increase in the series resistance due to which the drain current degrades. Hence, the

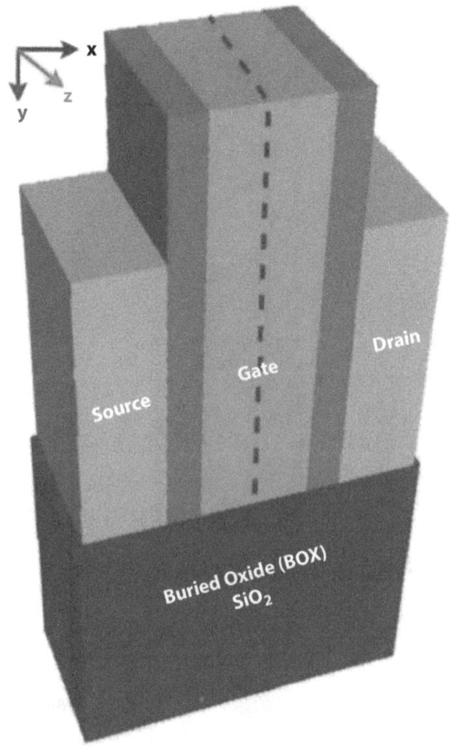

Figure 2.1 GAA NSFET 3D view.

effect can be dominated by using the high dielectric spacer with the fringing effect induces field coupling and increases switching ratio[16].

Figure 2.3 depicts the Cross-sectional view of GAA NSFET. In simulation and device construction numerous models and methods [71–73]. The simulation uses Shockley-Read-Hall (SRH), Schenk's bandgap narrowing model, Lombardi mobility model. Lombardi mobility model is used when due to surface roughness, the carrier degradation phenomena occur. Shockley-Read-Hall (SRH), model manages the generation and recombination of the carriers.

The simulation parameters used in the research are understood in the Table 2.1.

2.2.1 Analysis of the Results Obtained

This section discusses the DC Figure of Merits such as ION, IOFF, threshold voltage, switching ratio, Subthreshold swing (SS), Drain induced

Figure 2.2 2-D cross-sectional view across XZ plane.

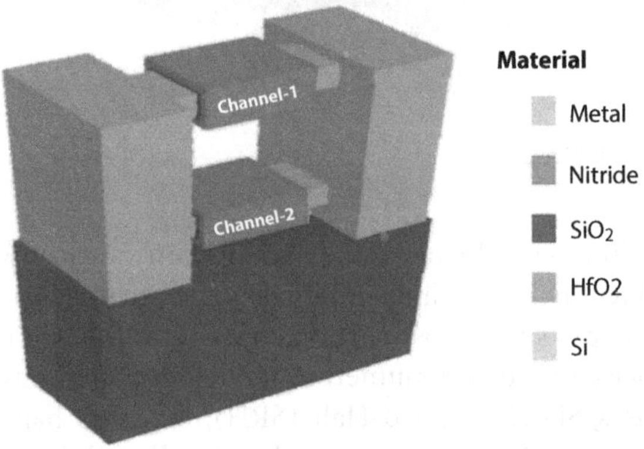

Figure 2.3 Cross-sectional view of GAA NSFET.

barrier lowering (DIBL) by varying the device geometry in the incremental format [74–78]. The NS width is varied from 10 nm to 50 nm, NS thickness is varied from 5 nm to 10 nm. Research deals with the dimensional engineering based on width and thickness of the nanosheet. Through analysis is performed on the various device parameters based on the simulation results obtained from constructed device [79–83].

Table 2.1 List of parameters used in simulating GAA NSFET

Parameter	Value used
NS width (nm)	10–50 nm
NS thickness (nm)	5–9 nm
Doping of terminals (cm^{-3})	10^{20} cm^{-3}
Gate Length (nm)	16 nm
Gate height (nm)	60 nm
Effective oxide thickness (nm)	0.78 nm
Terminal pad length (nm)	12 nm
Underlap material	Nitride
Terminal underlap length (nm)	5 nm
Channel doping (cm^{-3})	10^{15} cm^{-3}
Gate work function (eV)	4.6 eV

2.2.2 Impact of Variation in Width Across Various Thickness Values on Device Characteristics

The relationship between electron mobility and the channel's on-state is depicted in Figure 2.4. The device's enhanced gate electrostatics boost the

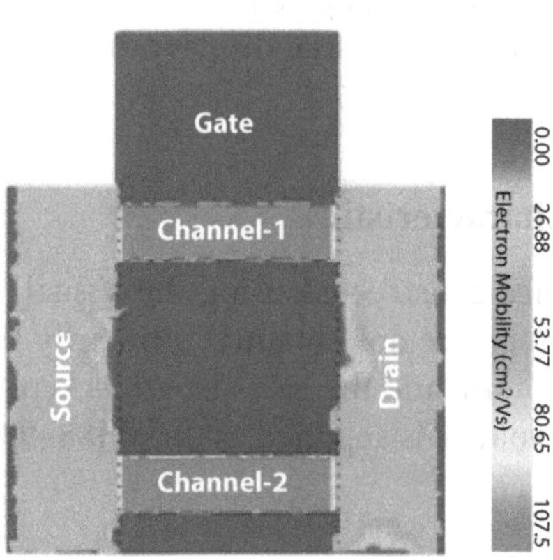

Figure 2.4 Electron mobility.

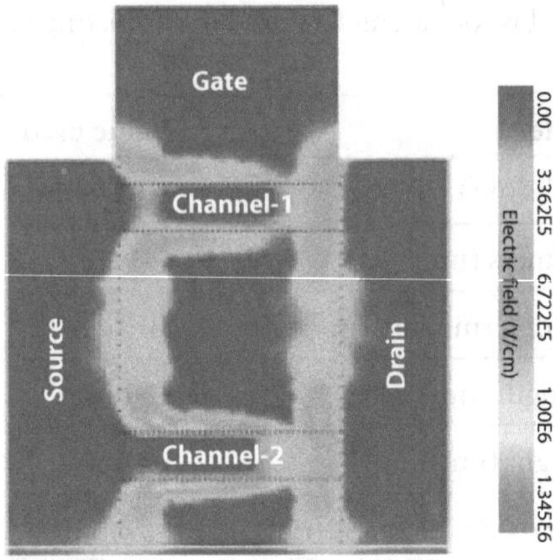

Figure 2.5 Electric field distribution.

mobility of the channel area [84–87]. The device's spacer area exhibits a greater electric field. The impact is less obvious on the channel.

The electric field distribution of a Nanosheet FET is shown in Figure 2.5. The increased current flowing in this circuit and the use of large dielectric spacers to reduce delay across the circuit appear to be at odds. Improved spacer designs will be required to increase device and circuit performance by reducing parasitic capacitances while maintaining a high driving current [17–20]. Due to its poor power efficiency degradation between Ie_{ff} and Ce_{ff} [21], a nitride spacer with a thickness of 5 nm is utilized in our work. The switching ratio is increased and field coupling is produced using high-dielectric (k) spacers [22–25]. Nitride spacers, as seen in Figure 2.5, widen the gap between the channel and the drain, lowering the drain potential across the channel and minimizing SCE.

The Figure 2.6 shows the potential displays the potential distribution of a Nanosheet FET at $V_{DS}= V_{GS}= 0.7$ V.

2.2.3 Transfer Characteristics

To examine the device constructed using the Visual-TCAD the transfer characteristics of the device are obtained. Primarily the thickness of the NS is maintained at 5 nm, and the gate's length is kept constant at 16 nm. The drain to source voltage is supplied as 0.7 V [88–93]. Width of the NS

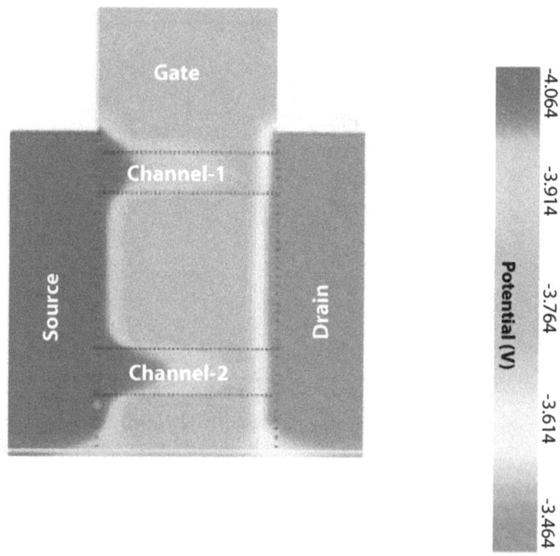

Figure 2.6 Potential of NSFET.

is varied between 10 nm to 50 nm. Figure 2.7 indicates the transfer characteristics of the device simulated. The result depicts the drain current in the linear scale and logarithmic scale, across the various gate voltage (V_{GS}) varying between 0 to 0.7V.

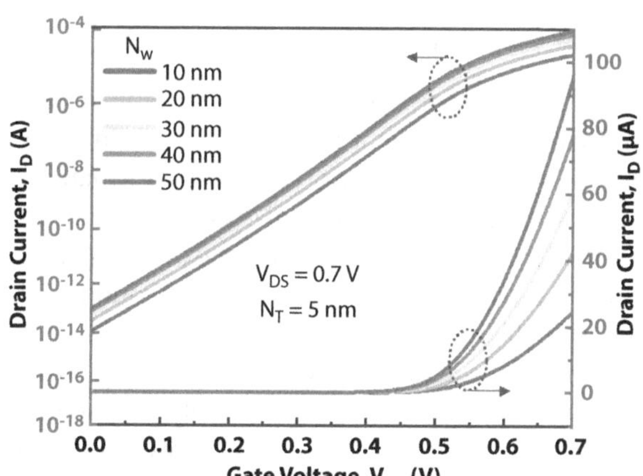

Figure 2.7 I_D vs. V_{GS} of nanosheet FET in log and a linear scale for various N_W values.

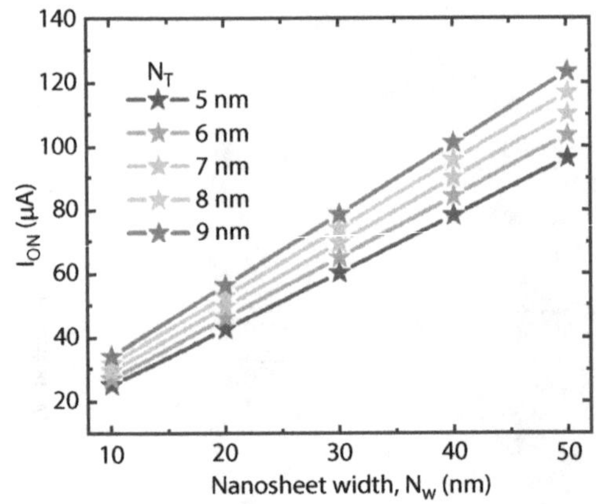

Figure 2.8 ON-state current.

2.2.4 Impact of Geometrical Variations on ON Current

Figure 2.8 depicts the current across the device in the ON state. To bias the device to the on-state, on-state drain current is ID with the designated V_{GS}. The NS thickness and width are varied in incremental fashion. Initially the I_{ON} showed up less but by varying geometry the I_{ON} also increased which is desirable and considered to be prominent factor for an optimum device [94–98].

2.2.5 Impact of Geometrical Variations on OFF-Current

Figure 2.9 shows the OFF-state current. Apart from the increment in the I_{ON} there is also a drastic improvement in the I_{OFF} which increases the leakage currents across the device and deteriorates the performance of the device. This issue arises from degraded gate control over the channel, which results in greater leaks in the device as NW's width increases [99–102]. It was obvious that using thinner nanosheets reduced the I_{OFF}. The conduction band energy and potential barrier height of the channel diminish when the N_T rises in the OFF condition, increasing leakage currents. The I_{ON}, however, decreases when the thickness decreases due to the mobility degradation brought on by an increase in the perpendicular electric field. Moreover, during an N_T of 9 nm, its NW increases between 10 and 50 nm, and a 2-order rise in I_{OFF} is observed. The I_{OFF} increases significantly because the electrostatic integrity of the channel declines as N_T increases.

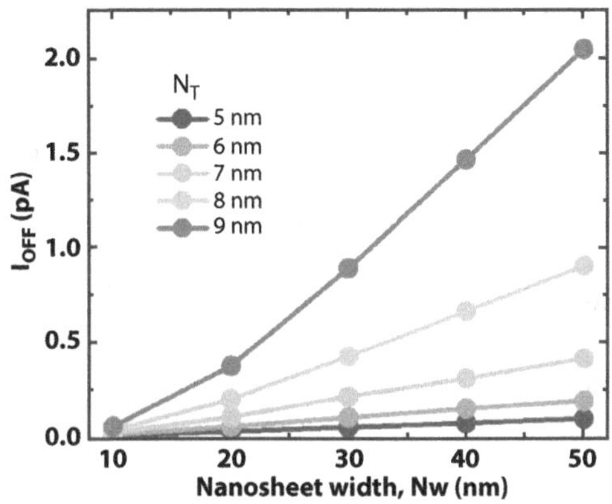

Figure 2.9 OFF-state current for various N_T.

2.2.6 Impact of Geometrical Variations on Switching Ratio

Switching ratio is defined as the ratio of ON-sate current to OFF-state current. For better logic applications, the switching ratio of 10^6 is workable. According to Figure 2.10, as the N_W increases, the switching ratio performance declines because of a considerable rise in IOFF. The ratio tends to decrease as N_W expands due to a significant rise in the IOFF and a slight rise in the ION. Also, as N_T rises, the switching ratio decreases, which is brought on by a sharp rise in I_{OFF} brought on by the gate's deteriorated electrostatic integrity over the channel.

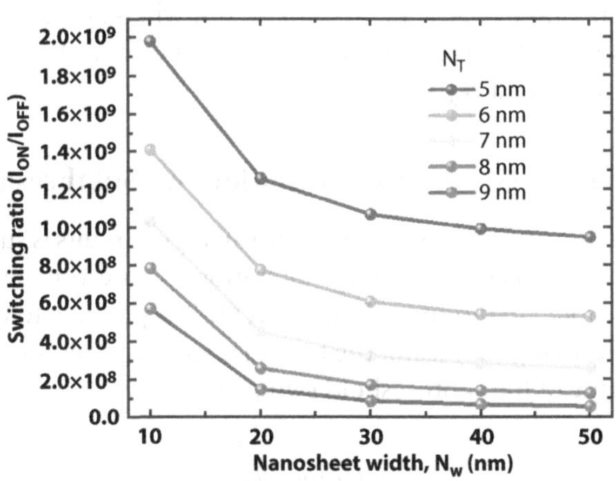

Figure 2.10 Switching ratio variation due to geometrical variations.

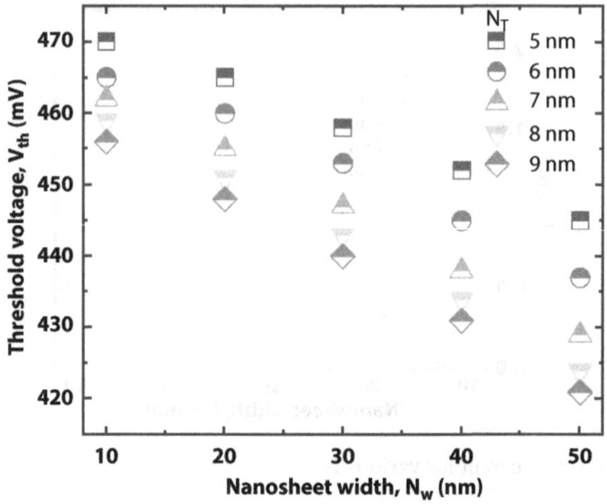

Figure 2.11 Threshold voltage variations due to geometrical variations.

2.2.7 Impact of Geometrical Variations on Threshold Voltage

The threshold voltage of a field-effect transistor (FET) is the lowest gate-to-source voltage (V_{GS}) required to establish a conducting route between the source and drain terminals. It is frequently abbreviated as Vth or V_{GS}. To maintain power efficiency, it is a crucial scaling factor [103–105]. Threshold voltage variations due to geometrical variations are depicted in the Figure 2.11. Due to increased leakage current, V_{th} drops as N_W increases. When N_T is raised, the similar pattern emerges. V_{th} decreases by 2.9% for a N_W of 10 nm and by 5.4% for a N_W of 50 nm when N_T varies from 5 nm to 9 nm, respectively. Due to static gate degradation in the channel area, $V_{throll\text{-}off}$ rises as N_T and N_W values rise.

2.2.8 Impact of Geometrical Variations on Subthreshold Swing

A crucial parameter for device applicability in logic circuits is sub-threshold swing (SS). The SS is defined as the alteration in gate voltage required to produce a tenfold alteration in drain current. It is possible to calculate the Sub-threshold swing (SS) using the formula. Subthreshold slope is also the reciprocal value of the subthreshold swing (SS).

$$SS = \left(\frac{\partial log_{10} I_D}{\partial V_{GS}} \right)^{-1} \tag{2.1}$$

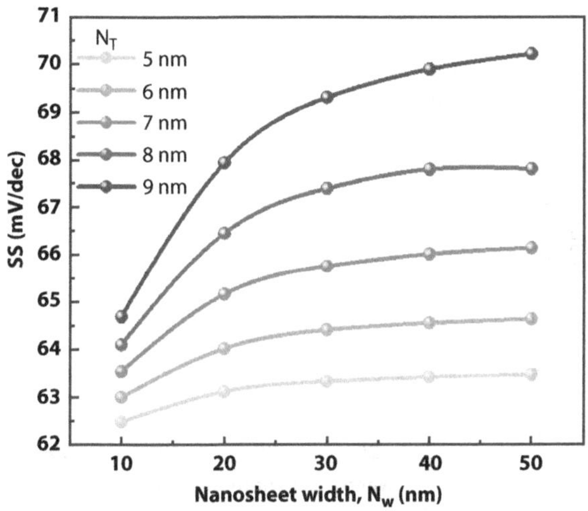

Figure 2.12 Subthreshold swing variation due to geometrical variations.

Figure 2.12 depicts the SS variations due to variation in the geometry. With a near-ideal SS of 62.5 mV/dec and N_W and N_T geometries of 10 nm and 5 nm, the NSFET performs the best. When N_W grew, a maximum SS increment of 8.6% was discovered from 9 nm to N_T. With increasing N_T, a cumulative 10.6% rise in SS is seen at a N_W of 50 nm. Because SS is high, larger values of N_W and N_T diminish subliminal power.

2.2.9 Impact of Geometrical Variations on DIBL

One of the most important sub-threshold performance indicators is drain induced barrier lowering (DIBL), which should be as low as feasible for the device to operate well. Equation is used to calculate the DIBL.

$$DIBL = \left(\frac{V_{tlin} - V_{tsat}}{V_{Dsat} - V_{Dlin}} \right) \tag{2.2}$$

Here $VD_{lin} = 0.04V$, $VD_{sat} = 0.7V$ and the V_{tlin} and V_{tsat} are the threshold voltages obtained at respective V_{Dlin}, V_{Dsat}. As seen, the DIBL grows in proportion to the size of the nanosheets. Due to the low $V_{throll\text{-}off}$, it is possible to use wider, thinner nanosheets to reduce the DIBL. The DIBL rise for N_W 50 nm is the largest, at 30.6%. Additionally, at 9 nm N_T, the biggest increase

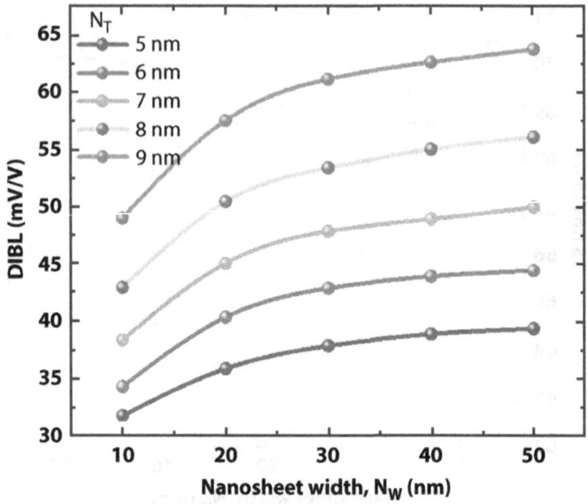

Figure 2.13 DIBL variations due to geometrical variations.

Table 2.2 Comparison of proposed work with the performance of existing devices.

Reference	I_{ON} (A)	I_{OFF} (A)	I_{ON}/I_{OFF}	SS (mV/dec)	DIBL (mV/V)
Ref [13]	6.8×10^{-5} A	6.7×10^{-12} A	2.3×10^{7}	71	22.8
Ref [17]	16.6×10^{-5} A	-	-	-	32
Ref [18]	2.37×10^{-5} A/μm	1.15×10^{-10} A/μm	2×10^{5}	70.89	-
Ref [19]	832 μA/μm	13.3 nA/μm	6.3×10^{4}	74	-
Proposed work	2.45×10^{-5} A	1.24×10^{-14} A	1.98×10^{9}	62.48	31.81

in DIBL, 62%, is seen. The DIBL may be susceptible to significant changes in width and thickness, according to Figure 2.13.

2.2.10 Comparison with Previous Works

Many previous research has been performed in the same domain of nanosheets. Below is the comparison of proposed work with existing work. The design parameters used in the below mentioned papers are slightly different from the design approach used in this work clearly depicted in Table 2.2.

2.3 Conclusions

Nanosheet the next generation of FET family providing the advantages of less SCEs because of their great electrostatic control over the channel area, driving large capacitive loads stood to be the prominent device for the next generation FET. This proposed work discusses the nanosheet's structure at the 7 nm technological node to test GAA NSFET's DC performance. When the thickness of the nanosheet N_T and the width of the nanosheet N_W are increased, the on-state current I_{ON} and the off-state current I_{OFF} both increases. Ideal (I_{ON}/I_{OFF}) switching ratio, SS, and DIBL values are produced by lower N_W and N_T values. Hence based in the research conducted the Nanosheet FET with the minimal width and Thickness of the nanosheet are the most optimal geometric variation.

References

[1] V. B. Sreenivasulu and V. Narendar, "Junctionless SOI FinFET with advanced spacer techniques for sub-3 nm technology nodes," *AEU - Int. J. Electron. Commun.*, vol. 145, no. August, p. 154069, 2022, doi: 10.1016/j.aeue.2021.154069.

[2] A. Sai Kumar, M. Deekshana, V. Bharath Sreenivasulu, R. Prasad Somineni, and D. Kanthi Sudha, "Characterization for Sub-5nm Technology Nodes of Junctionless Gate-All-Around Nanowire FETs," *2022 13th Int. Conf. Comput. Commun. Netw. Technol. ICCCNT 2022*, 2022, doi: 10.1109/ICCCNT54827.2022.9984269.

[3] V. Narula and M. Agarwal, "Study of analog performance of common source amplifier using rectangular core-shell based double gate junctionless transistor," *Semicond. Sci. Technol.*, vol. 35, no. 10, 2020, doi: 10.1088/1361-6641/abaaed.

[4] R. Vaddi, R. P. Agarwal, and S. Dasgupta, "Compact modeling of a generic double-gate MOSFET with gate-S/D underlap for subthreshold operation," *IEEE Trans. Electron Devices*, vol. 59, no. 10, pp. 2846–2849, 2012, doi: 10.1109/TED.2012.2208464.

[5] N. Loubet *et al.*, "Stacked nanosheet gate-all-around transistor to enable scaling beyond FinFET," *Dig. Tech. Pap. - Symp. VLSI Technol.*, pp. T230–T231, 2017, doi: 10.23919/VLSIT.2017.7998183.

[6] Lin, Yu-Ru and Lin, Yu-Hsien and Chen, Yu-Fang and Hsu, Ya-Ting and Chen, Ya-Han and Huang, Yu-Hsien and Wu, Yung-Chun, "Performance of Junctionless and Inversion- Mode Thin-Film Transistors with Stacked Nanosheet Channels," *IEEE Trans. Nanotechnol.* 19, 84-88 (2020). doi: 10.1109/TNANO.2019.2960836.

[7] N. Aruna Kumari *et al.* 2023 impact of scaling on nanosheet FET and CMOS circuit applications ECS *J. Solid State Sci. Technol.* 12 053009.

[8] Sresta Valasa, Shubham Tayal, and Laxman Raju Thoutam, "Optimization of Design Space for Vertically Stacked Junctionless Nanosheet FET for Analog/RF Applications", *Silicon*. doi: 10.1007/s12633-022-01793-6.

[9] V. B. Sreenivasulu and V. Narendar, "Junctionless SOI FinFET with advanced spacer techniques for sub-3 nm technology nodes," AEU - *Int. J. Electron. Commun.*, vol. 145, 2022, doi: 10.1016/j.aeue.2021.154069.

[10] N. A. Kumari and P. Prithvi, "Device and circuit-level performance comparison of GAA nanosheet FET with varied geometrical parameters," *Microelectronics J.*, vol. 125, 2022, doi: 10.1016/j.mejo.2022.105432.

[12] V. B. Sreenivasulu and V. Narendar, "Design insights into RF/analog and linearity/distortion of spacer engineered multi-fin SOI FET for terahertz applications," *Int. J. RF Microw. Comput. Eng.*, vol. 31, no. 12, 2021, doi: 10.1002/mmce.22875.

[13] V. Jegadheesan, K. Sivasankaran, and A. Konar, "Impact of geometrical parameters and substrate on analog/RF performance of stacked nanosheet feld effect transistor," *Mater. Sci. Semicond. Process.*, vol. 93, pp. 188–195, 2019, doi: 10.1016/j.mssp.2019.01.003.

[14] V. B. Sreenivasulu *et al*, "Common Source Amplifier and Ring Oscillator Performance Optimization Using Multi-Bridge/Nanosheet FETs" *ECS J. Solid State Sci. Technol.*, 2023, doi: 10.1149/2162-8777/ acbb9e.

[15] N. A Kumari, V. B. Sreenivasulu, P. Prithvi, "Impact of Scaling on Nanosheet FET and CMOS Circuit Applications", *ECS J. Solid State Sci. Technol.*, 12(3), 2023, doi: 10.1149/2162-8777/acbcf2.

[16] Sreenivasulu, V.B., Narendar, V. "Circuit Analysis and Optimization of GAA Nanowire FET Towards Low Power and High Switching", *Silicon*, 2022.

[17] V. Jegadheesan and K. Sivasankaran, "A source/drain-on-insulator structure to improve the performance of stacked nanosheet field-effect transistors," *J. Comput. Electron.*, vol. 19, no. 3, pp. 1136–1143, 2020, doi: 10.1007/s10825-020-01502-9.

[18] B. Kumar and R. Chaujar, "Analog and RF Performance Evaluation of Junctionless Accumulation Mode (JAM) Gate Stack Gate All Around (GS-GAA) FinFET," *Silicon*, vol. 13, no. 3, pp. 919–927, 2021, doi: 10.1007/s12633-020-00910-7.

[19] D. Nagy, G. Espineira, G. Indalecio, A. J. Garcia-Loureiro, K. Kalna, and N. Seoane, "Benchmarking of FinFET, Nanosheet, and Nanowire FET Architectures for Future Technology Nodes," *IEEE Access*, vol. 8, pp. 53196–53202, 2020, doi: 10.1109/ACCESS.2020.2980925.

[20] S. Kaya and A. Kulkarni, "A novel voltage-controlled ring oscillator based on nanoscale DG-MOSFETs," *Proc. Int. Conf. Microelectron.* ICM, pp. 417–420, 2008, doi: 10.1109/ICM.2008.5393792.

[21] R. Islam, A. N. K. Suprotik, S. M. Z. Uddin, and M. T. Amin, "Design and analysis of 3 stage ring oscillator based on MOS capacitance for wireless applications," ECCE 2017 - *Int. Conf. Electr. Comput. Commun. Eng.*, pp. 723–727, 2017.

[22] N. Aruna Kumari *et al.*, "Spacer engineering on Nanosheet FET device and circuit perspective" , *ECS J. Solid State Sci. Technol.*, Vol. 12, 2023.

[23] Aruru Sai Kumar *et al.*, "Device Analysis of Vertically Stacked GAA Nanosheet FET at Advanced Technology node", *IEEE 3rd Int. Conf. on Advances In Computing, Communication*, Embedded and Secure Systems (ACCESS-2023), 2023.

[24] A. S. Kumar *et al.*, "A Novel RRAM-based FPGA architecture with Improved Performance and Optimization Parameters," *2022 IEEE 19th India Council International Conference (INDICON)*, Kochi, India, pp. 1-5, 2022.

[25] J. Jalil, M. B. I. Reaz, M. A. M. Ali, and T. G. Chang, "A low power 3-stage voltage- controlled ring oscillator in 0.18 μm CMOS process for active RFID transponder," *Elektron. ir Elektrotechnika*, vol. 19, no. 8, pp. 69–72, 2013, doi: 10.5755/j01.eee.19.8.2524.

[26] A.Bhardwaj, P.Kumar, B.Raj, and S.Anand, "Design and Performance Optimization of Doping-Less Vertical Nanowire TFET Using Gate Stack Technique," *J. Electron. Mater. (JEMS)*, Springer, Vol. 41, No. 7, pp. 4005–4013, 2022.

[27] J.Singh, and B.Raj, "Tunnel Current Model of Asymmetric MIM Structure Levying Various Image Forces to Analyze the Characteristics of Filamentary Memristor", *Appl. Phys. A*, Springer, Vol. 125, No. 3, pp. 203.1 to 203.11, 2019.

[28] C.Goyal, J. S.Ubhi, and B.Raj, "Low Leakage Zero Ground Noise Nanoscale Full Adder Using Source Biasing Technique", *J. Nanoelectron. Optoelectron.*, American Scientific Publishers,Vol. 14, pp. 360–370, 2019.

[29] G.Singh, R. K.Sarin, and B.Raj, "A Novel Robust Exclusive-OR Function Implementation in QCA Nanotechnology with Energy Dissipation Analysis", *J. Comput. Electron.*, Springer, Vol. 15,No. 2, pp. 455–465, 2016.

[30] T.Wadhera, D.Kakkar, G.Wadhwa, and B.Raj, "Recent Advances and Progress in Development of the Field Effect Transistor Biosensor: A Review", *J. Electron. Mater.*, Springer,Vol. 48, No. 12, pp. 7635–7646, 2019.

[31] G.Wadhwa, P.Kamboj, and B.Raj, "Design Optimisation of Junctionless TFET Biosensor for High Sensitivity", *Adv. Nat. Sci.: Nanosci. Nanotechnol.*, Vol. 10, No. 7, p. 045001, 2019.

[32] P.Bansal, and B.Raj, "Memristor Modeling and Analysis for Linear Dopant Drif Kinetics", *J. Nanoeng. Nanomanuf., American Scientifc Publishers*, Vol. 6, pp. 1–7, 2016.

[33] A.Singh, M.Khosla, and B.Raj, "Circuit Compatible Model for Electrostatic Doped Schottky Barrier CNTFET," *J. Electron. Mater.*, Springer, Vol. 45, No. 12, pp. 4825–4835, 2016.

[34] Ashima, D.Vaithiyanathan, and B.Raj, "Performance Analysis of Charge Plasma induced Graded Channel Si Nanotube", *J. Eng. Res. (JER)*, EMSME Special Issue, Vol. 9, pp. 146–154, 2021.

[35] A. S.Tomar, V. K.Magraiya, and B.Raj, "Scaling of Access and DataTransistor for High Performance DRAM Cell Design", *Quantum Matter*, Vol. 2, spp. 412–416, 2013.

[36] N.Jain, and B.Raj, "Parasitic Capacitance and Resistance Model Development and Optimization of Raised Source/Drain SOI FinFET Structure for Analog Circuit Applications *J. Nanoelectron. Optoelectron.*, ASP, USA, Vol. 13, pp. 531–539, 2018.

[37] S.Singh, S. K.Vishvakarma, and B.Raj, "Analytical Modeling of Split-Gate Junction-Less Transistor for a Biosensor Application", *Sens. Bio-Sens.*, Elsevier, Vol. 18, pp. 31–36, 2018.

[38] M.Gopal, and B.Raj, "Low Power 8T SRAM Cell Design for High Stability Video Applications", *ITSI Trans. Electr. Electron. Eng.*., Vol. 1, No. 5, pp. 91–97, 2013.

[39] B.Raj, J.Mitra, D. K.Bihani, V.Rangharajan, A. K. Saxena, and S. Dasgupta, "Analysis of Noise Margin, Power and Process Variation for 32 nm FinFET Based 6T SRAM Cell", *J. Comput. (JCP)*, Academy Publisher, FINLAND, Vol. 5, No. 6, PP.1-6, 2010.

[40] D.Sharma, R.Mehra, and B.Raj, "Comparative Analysis of Photovoltaic Technologies for High Efciency Solar Cell Design", *Superlattices and Microstruct.*, Elsevier, Vol. 153, pp. 106861, 2021.

[41] P.Kaur, A. S.Buttar, and B.Raj, "A Comprehensive Analysis of Nanoscale Transistor Based Biosensor: A Review", *Indian J. Pure Appl. Phys.*, Vol. 59, pp. 304–318, 2021.

[42] D.Yadav, B.Raj, and B.Raj, "Design and Simulation of Low Power Microcontroller for IoT Applications", *J. Sens. Lett.*, ASP USA, Vol. 18, pp. 401–409, 2020.

[43] S.Singh, and B.Raj, "A 2-D Analytical Surface Potential and Drain Current Modeling of Double-Gate Vertical t-Shaped Tunnel FET", *J. Comput. Electron.*, Springer, Vol. 19, pp. 1154–1163, 2020.

[44] J.Singh, and B.Raj, "An Accurate and Generic Window function for Nonlinear Memristor Model". *J. Comput. Electron.*, Springer, Vol. 18, No. 2, pp. 640–647, 2019.

[45] M.Kaur, N.Gupta, S.Kumar, B.Raj, and A. K.Singh, "Comparative RF and Crosstalk Analysis of Carbon Based Nano Interconnects", *IET Circuits, Devices Syst.*, Vol. 15, No. 6, pp. 493–503, 2021.

[46] N.Kandasamy, F.Ahmad, D.Ajitha, B.Raj, and N.Telagam, "Quantum Dot Cellular Automata Based Scan Flip Flop and Boundary Scan Register", *IETE J. Res.*, Vol. 69, P P. 535-548, 2023.

[47] S. K.Sharma, B.Raj, and M.Khosla, "Enhanced Photosensitivity of Highly Spectrum Selective Cylindrical Gate In1-xGaxAs Nanowire MOSFET Photodetector", *Mod. Phys. Lett. B*, Vol. 33, No. 12, p. 1950144, 2019.

[48] J.Singh, and B.Raj, "Design and Investigation of 7T2M NVSARM with Enhanced Stability and Temperature Impact on Store/Restore Energy", *IEEE Transactions on Very Large Scale Integration Systems*, Vol. 27, No. 6, pp. 1322–1328, 2019.

[49] A. K.Bhardwaj, S.Gupta, B.Raj, and A.Singh,"Impact of Double Gate Geometry on the Performance of Carbon Nanotube Field Effect Transistor Structures for Low Power Digital Design", *Computational and Theoretical Nanoscience*, ASP USA, Vol. 16, pp. 1813–1820, 2019.

[50] N.Jain, and B.Raj, "Thermal Stability Analysis and Performance Exploration of Asymmetrical Dual-k Underlap Spacer (ADKUS) SOI FinFET for Security and Privacy Applications", *Indian Journal of Pure & Applied Physics (IJPAP)*, Vol. 57, pp. 352–360, 2019.

[51] A.Singh, M.Khosla, and B.Raj, "Design and Analysis of Dynamically Configurable Electrostatic Doped Carbon Nanotube Tunnel FET", *Microelectronics J.* Elsevier, Vol. 85, pp. 17–24, 2019.

[52] N.Jain, and B.Raj, "Dual-k Spacer Region Variation at the Drain Side of Asymmetric SOI FinFET Structure: Performance Analysis towards the Analog/RF Design Applications", *J. Nanoelectron. Optoelectron., American Scientifc Publishers*, Vol. 14, pp. 349–359, 2019.

[53] J.Singh, S.Sharma, B.Raj, and M.Khosla, "Analysis of Barrier Layer Thickness on Performance of In1-xGaxAs Based Gate Stack Cylindrical Gate Nanowire MOSFET", JNO, ASP USA, Vol. 13, pp. 1473–1477, 2018.

[54] N.Jain, and B.Raj, "Analysis and Performance Exploration of High-k SOI FinFETs over the Conventional Low-k SOI FinFET toward Analog/ RF Design", *J. Semicond. (JoS)*, IOP Science, Vol. 39, No. 12, pp. 124002-1-7, 2018.

[55] C.Goyal, J. S.Ubhi, and B.Raj, "A Reliable Leakage Reduction Technique for Approximate Full Adder with Reduced Ground Bounce Noise", *J. Math. Probl. Eng.*, Hindawi, Vol. 2018, Article ID 3501041, P P. 1-16, 2018.

[56] Anuradha, J. Singh, B.Raj, and M.Khosla, "Design and Performance Analysis of Nano-Scale Memristor-Based Nonvolatile SRAM", *J. Sens. Lett.*, American Scientifc Publishers USA, Vol. 16, pp. 798–805, 2018.

[57] G.Wadhwa, and B.Raj, "Parametric Variation Analysis of Charge-Plasma-Based Dielectric Modulated JLTFET for Biosensor Application", *IEEE Sens. J.*, Vol. 18, No. 15, PP. 6070 – 6077, 2018.

[58] J.Singh, and B.Raj, "Comparative Analysis of Memristor Models for Memories Design", *JoS, IoP*, Vol. 39, No. 7, P P. 074006-1-12, 2018.

[59] D.Yadav, S. S.Chouhan, S. K.Vishvakarma, and B.Raj, "Application Specific Microcontroller Design for IoT Based WSN", *Sensor Letter, ASP*, Vol. 16, pp. 374–385, 2018.

[60] G.Singh, R. K.Sarin, and B.Raj, "Fault-Tolerant Design and Analysis of Quantum-Dot Cellular Automata Based Circuits", *IEEE/IET Circuits, Devices & Systems*, Vol. 12, pp. 638–664, 2018.

[61] J.Singh, and B.Raj, "Modeling of Mean Barrier Height Levying Various Image Forces of Metal Insulator Metal Structure to Enhance the Performance of Conductive Filament Based Memristor Model", *IEEE Nanotechnol.*, Vol. 17, No. 2, pp. 268–275, 2018.

[62] A.Jain, S.Sharma, and B.Raj, "Analysis of Triple Metal Surrounding Gate (TM-SG) III-V Nanowire MOSFET for Photosensing Application", *OptoElectron. J.*, Elsevier, Vol. 26, No. 2, P P. 141–148, 2018.

[63] A.Jain, S.Sharma, and B.Raj, "Design and Analysis of High Sensitivity Photosensor Using Cylindrical Surrounding Gate MOSFET for Low Power Sensor Applications", *Eng. Sci. Technol. Int. J.*, Elsevier's, Vol. 19, No. 4, pp. 1864–1870, 2016.

[64] A.Singh, M.Khosla, and B.Raj, "Analysis of Electrostatic Doped Schottky Barrier Carbon Nanotube FET for Low Power Applications," *J. Mater. Sci. Mater. Electron,*, Springer,Vol. 28, PP. 1762–1768, 2017.

[65] G.Saiphani Kumar, A.Singh, B.Raj, "Design and Analysis of Gate All around CNTFET Based SRAM Cell Design", *J. Comput. Electron.*, Springer, Vol. 17, No. 1, P P. 138–145, 2018.

[66] G. P.Singh, B. S.Sohi, and B.Raj, "Material Properties Analysis of Graphene Base Transistor (GBT) for VLSI Analog Circuits", *Indian J. Pure Appl. Phys. (IJPAP)*, Vol. 55, pp. 896–902, 2017.

[67] S.Kumar and B.Raj, "Estimation of Stability and Performance Metric for Inward Access Transistor Based 6T SRAM Cell Design Using n-type/p-type DMDG-GDOV TFET", *IEEE VLSI Circuits Syst. Lett.*, Vol. 3, No. 2, PP. 25–39, 2017.

[68] S.Sharma, A.Kumar, M.Pattanaik, and B.Raj, "Forward Body Biased Multimode Multi-Threshold CMOS Technique for Ground Bounce Noise Reduction in Static CMOS Adders", *Int. J. Inf. Electron. Eng.*, Vol. 3, No. 3, PP. 567–572, 2013.

[69] H.Singh, P.Kumar, and B.Raj, "Performance Analysis of Majority Gate SET Based 1-bit Full Adder", *Int. J. Comput. Commun. Eng. (IJCCE)*, IACSIT Press Singapore, ISSN: 2010-3743, Vol. 2, No. 4, PP.1-8, 2013.

[70] A. K.Bhardwaj, S.Gupta, and B.Raj, "Investigation of Parameters for Schottky Barrier (SB) Height for Schottky Barrier Based Carbon Nanotube Field Effect Transistor Device", *J. Nanoelectron. Optoelectron.*, ASP, Vol. 15, P P. 783–791, 2020.

[71] P.Bansal, and B.Raj, "Memristor: A Versatile Nonlinear Model for Dopant Drif and Boundary Issues", *JCTN, American Scientifc Publishers*, Vol. 14, No. 5, P P. 2319–2325, 2017.

[72] N.Jain, and B.Raj, "An Analog and Digital Design Perspective Comprehensive Approach on Fin-FET (Fin-Field Effect Transistor) Technology - A Review", *Reviews in Advanced Sciences and Engineering (RASE)*, ASP, Vol. 5, P P. 1–14, 2016.

[73] S.Sharma, B.Raj, and M.Khosla, "Subthreshold Performance of In1-xGaxAs Based Dual Metal with Gate Stack Cylindrical/Surrounding Gate Nanowire MOSFET for Low Power Analog Applications", *J. Nanoelectron. Optoelectron.*, American Scientifc Publishers, USA, Vol. 12, P P. 171– 176, 2017.

[74] B.Raj, A. K.Saxena, and S.Dasgupta, "Analytical Modeling for the Estimation of Leakage Current and Subthreshold Swing Factor of Nanoscale Double Gate FinFET Device" *Microelectron. Int.*, UK, Vol. 26, PP. 53–63, 2009.

[75] S. S.Soniya, G.Wadhwa, and B.Raj, "An Analytical Modeling for Dual Source Vertical Tunnel Field Effect Transistor", *Int J. Recent Technol. Eng. (IJRTE)*, Vol. 8, No. 2, 2019.

[76] S.Singh, and B.Raj, "Design and Analysis of Hetrojunction Vertical T-Shaped Tunnel Field Efect Transistor", *J. Electron. Mater., Springer*, Vol. 48, No. 10, PP. 6253–6260, 2019.

[77] C.Goyal, J. S.Ubhi, and B.Raj, "A Low Leakage CNTFET Based Inexact Full Adder for Low Power Image Processing Applications", *Int. J. Circuit Theory Appl.*, Wiley, Vol. 47, No. 9, PP. 1446–1458, 2019.

[78] B.Raj, A. K.Saxena, and S.Dasgupta, "A Compact Drain Current and Treshold Voltage Quantum Mechanical Analytical Modeling for FinFETs", *J. Nanoelectron. Optoelectron. (JNO)*, USA, Vol. 3, No. 2, PP. 163–170, 2008.

[79] G.Wadhwa, and B.Raj, "An Analytical Modeling of Charge Plasma Based Tunnel Field Effect Transistor with Impacts of Gate Underlap Region", *Superlattices and Microstruct.*, Elsevier, Vol. 142, P P. 106512-22, 2020.

[80] S.Singh, and B.Raj, "Modeling and Simulation Analysis of SiGe Heterojunction Double GateVertical t-Shaped Tunnel FET", *Superlattices and Microstruct.*, Elsevier,Vol. 142, P P. 106496-104, 2020.

[81] A.Singh, D. K.Saini, D.Agarwal, S.Aggarwal, M.Khosla, and B.Raj, "Modeling and Simulation of Carbon Nanotube Field Effect Transistor and Its Circuit Application", *J. Semicond. (JoS), IOP Science*, Vol. 37, PP. 074001–074006, 2016.

[82] N.Jain, and B.Raj, "Device and Circuit Co-Design Perspective Comprehensive Approach on FinFET Technology - A Review", *J. Electron Devices*, Vol. 23, No. 1, P P. 1890–1901, 2016.

[83] S.Kumar and B.Raj, "Analysis of ION and Ambipolar Current for Dual-Material Gate-Drain Overlapped DG-TFET", *J. Nanoelectron. Optoelectron., American Scientific Publishers*, USA, Vol. 11, PP. 323–333, 2016.

[84] N.Anjum, T.Bali, and B.Raj, "Design and Simulation of Handwritten Multiscript Character Recognition", *Int. J. Adv. Res. Comput. Commun. Eng.*, Vol. 2, No. 7, PP. 2544–2549, 2013.

[85] S.Sharma, B.Raj, and M.Khosla, "A Gaussian Approach for Analytical Subthreshold Current Model of Cylindrical Nanowire FET with Quantum Mechanical Effects", *Microelectronics J.*, Elsevier, Vol. 53, PP. 65–72, 2016.

[86] K.Singh, and B.Raj, "Performance and Analysis of Temperature Dependent Multi-Walled Carbon Nanotubes as Global Interconnects at Different Technology Nodes", *J. Comput. Electron.*, Springer, Vol. 14, No. 2, P P. 469–476, 2015.

[87] S.Kumar and B.Raj, "Compact Channel Potential Analytical Modeling of DG-TFET Based on Evanescent-Mode Approach", *J. Comput. Electron.*, Springer, Vol. 14, No. 2, P P. 820–827, 2015.

[88] K.Singh, and B.Raj, "Temperature Dependent Modeling and Performance Evaluation of Multi-Walled CNT and Single-Walled CNT as Global Interconnects", *J. Electron. Mater.*, Springer, Vol. 44, No. 12, PP. 4825–4835, 2015.

[89] V. K. Sharma, M. Pattanaik, and B. Raj, "INDEP Approach for Leakage Reduction in Nanoscale CMOS Circuits", *Int. J. Electron.*, Taylor & Francis, Vol. 102, No. 2, P P. 200–215, 2014.

[90] K.Singh, and B.Raj, "Influence of Temperature on MWCNT Bundle, SWCNT Bundle and Copper Interconnects for Nanoscaled Technology Nodes", *Journal of Materials Science: Materials in Electronics*, Springer, Vol. 26, No. 8, PP. 6134–6142, 2015.

[91] N.Anjum, T.Bali, and B.Raj, "Design and Simulation of Handwritten Gurumukhi and Devanagri Numerical Recognition", *Int. J. Comput. Appl., Published by Foundation of Computer Science*, New York, USA, Vol. 73, No. 12, P P. 16–21, 2013.

[92] S.Khandelwal, V.Gupta, B.Raj, and R. D.Gupta, "Process Variability Aware Low Leakage Reliable Nano Scale DG-FinFET SRAM Cell Design Technique", *J. Nanoelectron. Optoelectron.*, Vol. 10, No. 6, P P. 810–817, 2015.

[93] V. K. Sharma, M. Pattanaik, and B.Raj, "ONOFIC Approach: Low Power High Speed Nanoscale VLSI Circuits Design", *Int. J. Electron.*, Taylor & Francis, Vol. 101, No. 1, P P. 61–73, 2014.

[94] S.Khandelwal, B.Raj, and R. D.Gupta, "FinFET Based 6T SRAM Cell Design: Analysis of Performance Metric, Process Variation and Temperature Effect", *J. Comput. Theor. Nanosci.*, ASP, USA, Vol. 12, PP. 2500–2506, 2015.

[95] S.Singh, Y.Shekhar, R.Jagdeep, S.Anurag, and B.Raj, "Impact of HfO2 in Graded Channel Dual Insulator Double Gate MOSFET", *J. Comput. Theor. Nanosci.*, American Scientifc Publishers, Vol. 12, No. 6, P P. 950–953, 2015.

[96] V. K.Sharma, M.Pattanaik, and B.Raj, "PVT Variations Aware Low Leakage INDEP Approach for Nanoscale CMOS Circuits", *Microelectron. Reliab.*, Elsevier, Vol. 54, PP. 90–99, 2014.

[97] B.Raj, A. K.Saxena and S.Dasgupta, "Quantum Mechanical Analytical Modeling of Nanoscale DG FinFET: Evaluation of Potential, Treshold Voltage and Source/Drain Resistance", *Elsevier's J. Mat Sci Semicon. Proc*, Elsevier, Vol. 16, No. 4, P P. 1131–1137, 2013.

[98] M.Gopal, S. S. D.Prasad, and B.Raj, "8T SRAM Cell Design for Dynamic and Leakage Power Reduction", *Int. J. Comput. Appl., Published by Foundation of Computer Science*, New York, USA,Vol. 71, No. 9, P P. 43–48, 2013.

[99] M.Pattanaik, B.Raj, S.Sharma, and A.Kumar, "Diode Based Trimode Multi-Threshold CMOS Technique for Ground Bounce Noise Reduction in Static CMOS Adders", *Adv. Mat. Res., Trans Tech Publications*, Switzerland, Vol. 548, P P. 885–889, 2012.

[100] B.Raj, A. K.Saxena, and S.Dasgupta, "Nanoscale FinFET Based SRAM Cell Design: Analysis of Performance Metric, Process Variation, Underlapped FinFET and Temperature Effect", *IEEE Circuits Syst. Mag.*, Vol. 11, No. 2, PP. 38–50, 2011.

[101] V. K. Sharma, M. Pattanaik, and B.Raj, "Leakage Current ONOFIC Approach for Deep Submicron VLSI Circuit Design", *Int. J. Electr., Comput., Electron. Commun. Eng.*, World Academy of Sciences, Engineering and Technology, Vol. 7, No. 4, PP. 239–244, 2013.

[102] T.Chawla, M.Khosla, and B.Raj, "Design and Simulation of Triple Metal Double-Gate Germanium on Insulator Vertical Tunnel Field Effect Transistor", *Microelectronics J.*, Elsevier, Vol. 114, PP. 105125-32, 2021.

[103] P.Kaur, S. S.Gill, and B.Raj, "Comparative Analysis of OFETs Materials and Devices for Sensor Applications", *J. Silicon*, Springer, Vol. 14, PP. 4463–4471, 2022.

[104] S. K.Sharma, P.Kumar, B.Raj, and Balwinder Raj, "In1-xGaxAs Double Metal Gate-Stacking Cylindrical Nanowire MOSFET for Highly Sensitive Photo Detector", *Journal of Silicon*, Springer, Vol. 14, P P. 3535–3541,2022.

[105] B.Raj, A. K.Saxena, and S.Dasgupta, "Analytical Modeling of Quasi Planar Nanoscale Double Gate FinFET with Source/Drain Resistance and Field Dependent Carrier Mobility: A Quantum Mechanical Study", *J. Comput. (JCP)*, Academy Publisher, FINLAND, Vol. 4, No. 9, PP.1-8, 2009.

3

Comparison of Different TFETs: An Overview

Rama Satya[1], Nageswara Rao[1,2] and K. Srinivasa Rao[1*]

[1]*MEMS Research Center, Department of ECE, Koneru Lakshmaiah Educational Foundation (Deemed to be University), Green Fields, Vaddeswaram, Guntur, Andhra Pradesh, India*

[2]*Department of ECE, B V C Engineering College, Odalarevu, Andhra Pradesh, India*

Abstract

The standard MOSFET used in modern IC technology is expected to be replaced by new tunnel field effect transistor (TFET), which is seen to be the utmost viable contender. Researchers have been very interested in it as a result of its capacity to attain a steep sub-threshold slope, a stronger resilience to short channel effect's, and a reduced standby power dissipation. The current transport technique, band to band tunneling (BTBT), causes TFET's two main bottlenecks, lower Ion current and ambipolar conductivity, despite the fact that it promises several benefits over other MOSFET competitors. This article provides a thorough overview of the many methods researchers have proposed to increase the ON-State current and subthreshold swing in Tunnel FETs.

***Keywords*:** Tunnel field-effect-transistor, band to band tunneling, sub-threshold slope, ambipolar current, drain current, surface potential

3.1 Introduction

There has been constant development in CMOS technology afterward presentation of initial Field-Effect Transistor in 1960 and CMOS (Complementary Metal Oxide Semiconductor) disposition in 1963.

**Corresponding author*: srinivasakarumuri@gmail.com

Balwinder Raj, Suman Lata Tripathi, Tarun Chaudhary, K. Srinivasa Rao and Mandeep Singh (eds.) Integrated Devices for Artificial Intelligence and VLSI, (49–74) © 2024 Scrivener Publishing LLC

The Moore's law-inspired exponential trend of continuous CMOS scaling, or the downsizing of transistors, is what propels and sustains these developments. It is amazing that the trend of shrinking transistor size and the ensuing aggressive development of semiconductor industry have persisted unabated for more than five decades despite a number of challenging technological obstacles. Before the momentum could be reversed, the challenges related to the impending downsizing of transistors that once seemed insurmountable were finally removed adopting inventive methods. The difficulty of a transistor to go from the OFF-state to ON-state outwardly increasing power-dissipation puts the further scalability of transistors at genuine risk when they get smaller than 10 nm. The design of transistors and the technologies to create them have changed significantly over time because to CMOS scaling, yet the fundamental workings of transistors have stayed the same. Further reduction of supply voltages in an integrated circuit is not possible due to limitations imposed by current transport physics in a typical transistor (IC). Because of this, the present trend of miniaturizing transistors while maintaining a constant supply voltage prevents increasing the efficiency of integrated circuits (ICs), which might have been accomplished by lowering the input supply voltage [1–3]. In order to address the shortcomings of conventional transistors, it is necessary to investigate transistors based on unique non-silicon-based material systems, better designs, and new operating principles. In this regard, it had been shown that the Tunnel-Field-Effect Transistors (TFET's), based on quantum mechanical tunneling, have exceptional switching qualities that go beyond what is theoretically possible for a regular transistor. Because of this, TFETs may be used in integrated circuits (IC's) with low input supply voltages, producing extremely energy competent circuits. Due to their intriguing nature, TFETs have recently been the subject of extensive experimental and theoretical investigation.

The organization of this paper is followed with discussion on the Tunnel FET in Section 3.2. Different types of Gate engineering's have been discussed in the Section 3.3. Section 3.4 explains the different tunneling junction engineering to boost the drain current. Material engineering explained in the Section 3.5 for improvement in the performance of the TFET device and the conclusion concluded in Section 3.6.

3.2 Tunnel FET

A traditional MOSFET's current is determined by the carrier thermionic emission through the potential barrier. The outcome is, one of thee largest problems for traditional MOSFETs at advanced technological nodes is maintaining the power consumption within an acceptable range. Due to its unique current transport mechanism, Tunnel Field-Effect Transistors (TFETs) display a number of intriguing electrical properties and may provide a viable alternative to MOSFETs as we know them today.

A gated, reverse-biased p-i-n structure is what the TFET is, to put it simply. Figure 3.1 illustrates an adaptation of ntype TFET. The sort of doping used in the drain and source is the structural characteristic of a TFET that makes it stand out the most. Drain and source of a TFET have dopants are of different kinds. However, in a traditional MOSFET, both source and the drain doping are of the same kind. The source and drain of an n-type TFET are both doped with n+ and p+, respectively. A p-type or n-type semiconductor having little doping or inherent properties makes up the channel. Comparable to a traditional MOSFET, a dielectric separates the channel from the gate electrode.

Figure 3.1 displays the biasing schemes of an ntype TFET. An elevated voltage is provided pertaining to the drain and gate electrodes of an n-type TFET while the source electrode is grounded. An adverse voltage is provided to the electrodes of drain and gate of a p-type TFET while the source is grounded. When the TFET is turned on, the dominant carrier in the channel produced under the gate changes, a TFET is either referred to as an n-type TFET or a p-type TFET. The TFET is referred to as an n-type TFET when electrons are the channel's predominant carriers, and a p-type TFET when holes are the channel's predominant carriers. According to whether the majority carriers enter or exit the channel wound-up a particular terminal, the terminals are referred to as source or drain. In an n-type TFET, the source and drain are the points at which electrons enter and exit the channel, respectively as shown in Figures 3.2 to 3.4.

Finely scaled MOSFET technology is found to be more susceptible to the aforementioned problems than any of the other candidates [34–39]. The restriction of Boltzmann inebriated by MOSFETs, for charge carriers through thermionic emission generates the current conduction, is

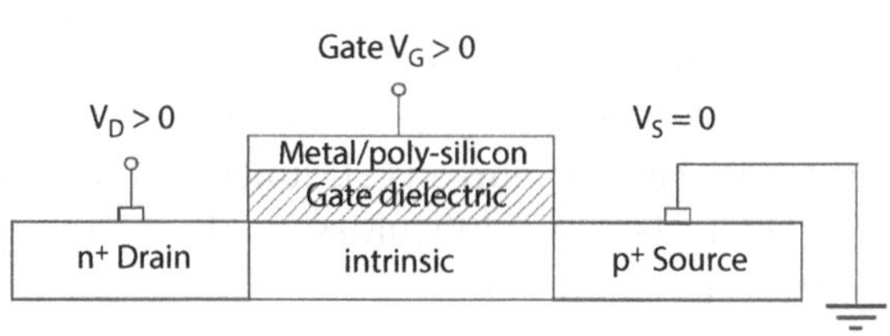

Figure 3.1 TFET and its biasing scheme n-type TFET [4].

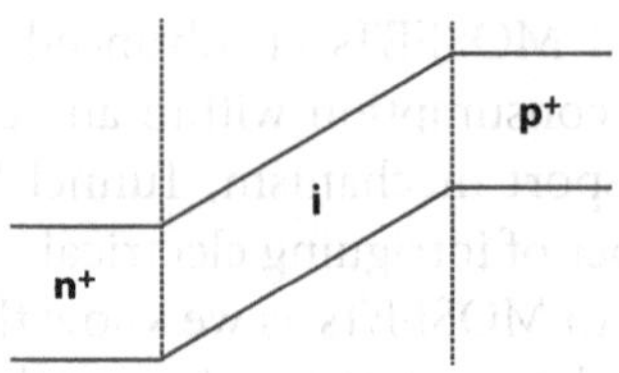

Figure 3.2 Under thermal equilibrium band-diagram [5].

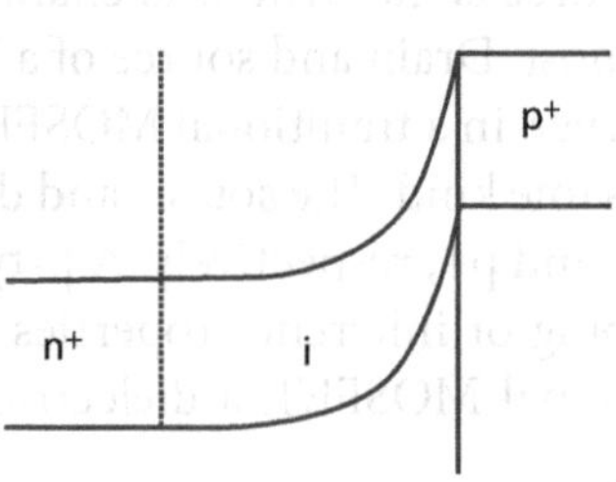

Figure 3.3 Under positive gate bias band-diagram [5].

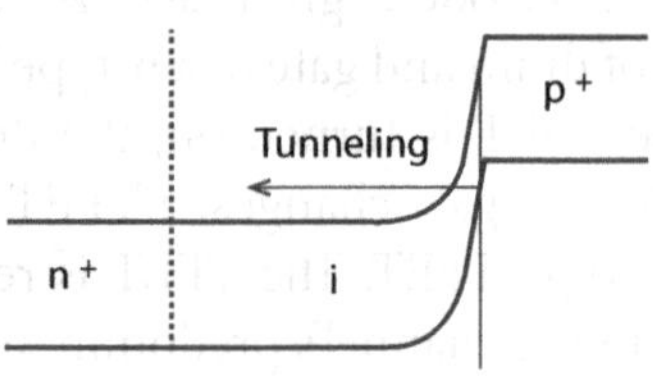

Figure 3.4 Under larger positive gate bias band-diagram [5].

supposedly avoided by TFETs since their current conduction primarily depends on inter-band tunneling process astride interface between source and channel. Furthermore, ambipolarity and reduced conducting current during the on-state are TFETs' two main drawbacks. Numerous researchers have developed numerous methods to deal with the problem the essential component of ambipolar-conduction in TFET's is including the drain-channel interface with dielectric pocket [6, 7], drain overlapped gate [8], and placing pocket near channel-drain interface with high doping [9]. Researchers have identified a variety of methods to enhance ON-state current, including Hetero-gate structure [10], source with minimal bandgap material [11], gate overlapped source [12], Junction-less HG TFET [13], SG Junction-less TFET [14]. The difficulties in TFET and some potential solutions are depicted in Figure 3.5.

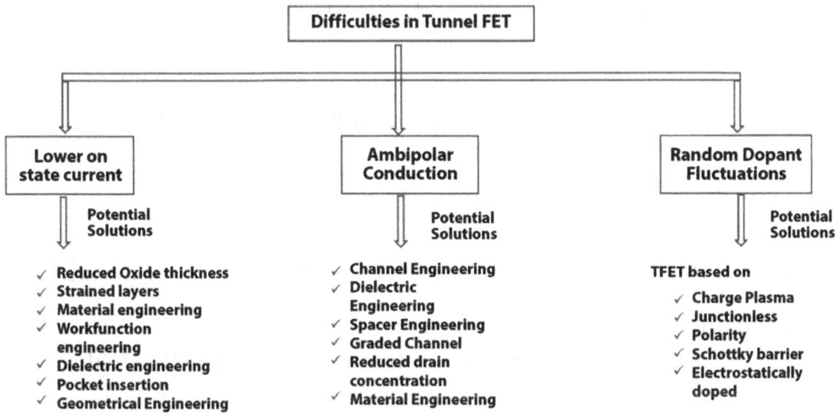

Figure 3.5 Difficulties in TFET and potential solutions.

Current conduction in tunnel FETs is primarily determined by the tunneling probability, which also determines the rate at which charge carriers tunnel at the channel-source contact [40–44]. The one that has a lower effective mass material, a steeper doping profile with in area, materials with a smaller bandgap energy (Eg), and an strong electro-static command of the gate on the input-tunneling interfaces can all be used to reduce the tunneling probability T(E) and increase ON-state current. We have categorized the numerous research studies into the following categories basing on the various methods used to increase tunneling probability and subsequently ON-state current:

1. Gate engineering
2. Tunneling-junction engineering
3. Materials engineering

3.3 Gate Engineering

In a TFET, the gate modifies the channel's carrier density. Valence band and conduction band alignment there at junction of source-channel is caused by the modulation of carrier density as in channel, which starts the tunneling of the carriers from the source through into channel. As a result, the ION of TFET is significantly impacted by channel potential and gate voltage is connected. This has inspired researchers to use the methods outlined below to engineer the gate design of a TFET in order to enhance the ION [45–48].

- Oxide-thickness and Dielectric-constant of gateoxide
- Multiple Gates
- Spacer Engineering

3.3.1 Oxide-Thickness and Dielectric-Constant of Gateoxide

The fundamental static operation and the features of a DG Tunnel FET in Figure 3.6 [15] as a current switch that is better than 60 mV/dec were inspected through simulation. With the suggested design changes—a double-gate, a high-k gate dielectric, and an optimized silicone body thickness the examined Tunnel FET had superior characteristics are shown in the Figure 3.7 and Figure 3.8 which includes the variations in dielectric material and thickness of the oxide layer respectively, including a larger on-current and a reduced subthreshold-swing. By increasing on state current up to 0.23 mA for Vg = 1.8 V and offering an comparable a reduction in the minimum point swing of 11 mV/dec and a 57 mV/dec minimal subthreshold swing over an average, the Double Gate and high-k-dielectric are able to achieve this.

3.3.2 Multiple Gates

Several gate configurations can be used in a TFET to enhance the ION, the effective tunneling area, and the manipulation of the channel's potential via the gate as presented in Figures 3.7 and 3.8. The tunneling zone of a TFET is typically found close to the source region, right beyond the channel's border. Channels occur beneath every gate in multiple gate designs [49–54].

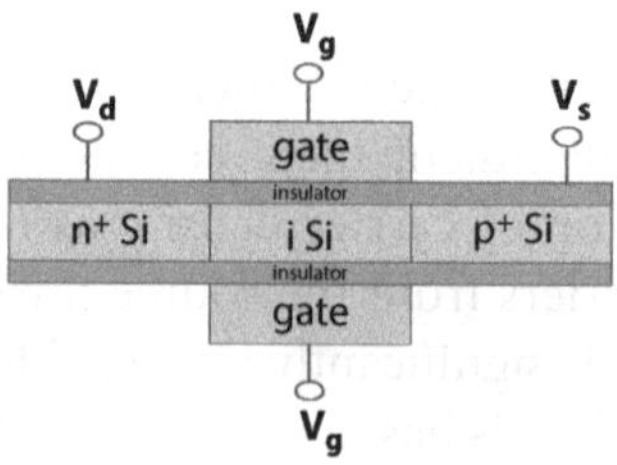

Figure 3.6 Simulated tunnel n-FET DG. SiO2 and high-κ gate dielectrics were studied.

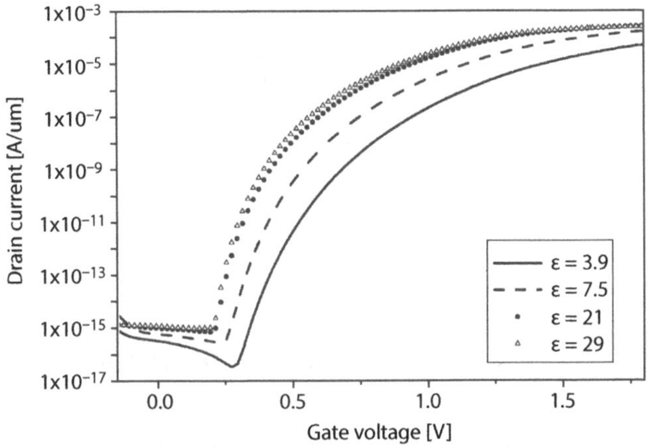

Figure 3.7 DG-Tunnel-FET characteristics for various gate dielectrics [15].

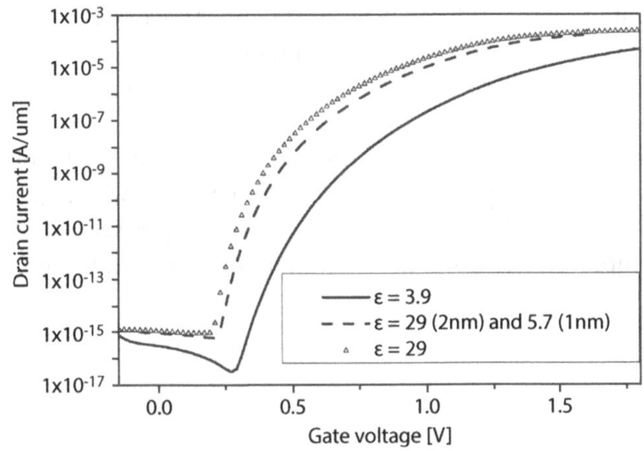

Figure 3.8 Double-Gate-Tunnel-FET structure with 2 nm of a high-κ dielectric [15].

Sneh Saurabh *et al.* [16] suggested a Tunnel FET configuration with Dual-Material Gate Field Effect Transistor. Figure 3.9 displays a cross section image of a DGTFET. They discovered that when work-function of the gate-metal closest to the 4.0 eV is assumed to break down when the source, and a metal possessing a greater work-function is utilized on the drain region side, the tunneling the likelihood that the charge carriers across the channel and source intersection reap a maximal value. The channel area conduction band gets closer to the valency band of the source region when the work function difference between the Tunnel Gate (TG) is minimized, increasing the ON-state current [55–60].

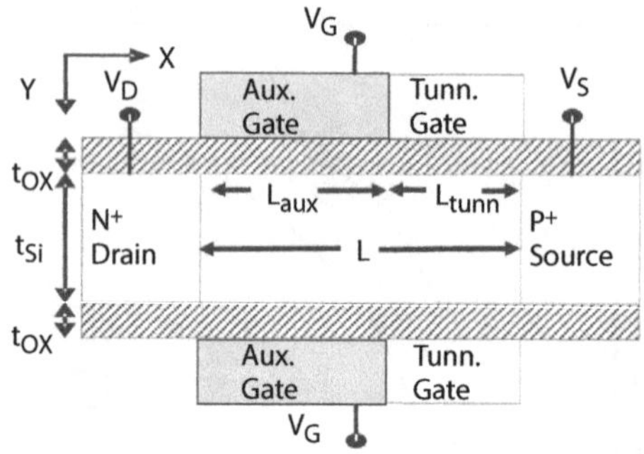

Figure 3.9 Schematic cross-sectional view of the DMG-DGTFET [16].

They also noticed that the performance of the device is reduced if the work-function of the source side gate metal is selected to be greater than 4.4 eV (in Figure 3.10). It primarily occurs because the energy (potential) barrier created by the difference with TG and AGwork-functions at their interface acts as a restraint to the movement of charge carriers in the state of OFF. This barrier is reduced as TG's work-function is elapsed farther a certain point, or 4.40 eV.

The concept of tri-material to the terminal of gate in Double Gated-TFET was developed by Bagga *et al.* [17]. A metal gate with a greater work-function is interceded between two metals with depressed work-functions that are employed on both the terminals of drain and source in their

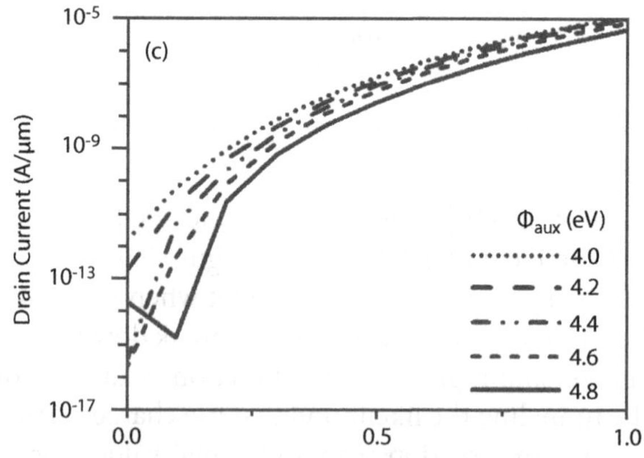

Figure 3.10 I–V_{gs} characteristics for different work function of metals [16].

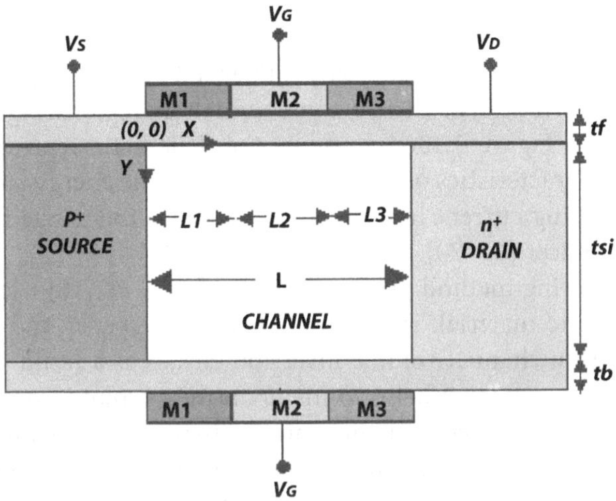

Figure 3.11 Triple metal double gate TFET (n-type) [17].

suggested construction in Figure 3.11. The band-bending at the input tunneling interface is enhanced by a metal with low work-function that is located closer to the source terminal. This ultimately increases the likelihood of BTBT for the charge carriers [61–64].

As seen in Figure 3.12, increased tunneling probability further improves current conduction when in the on-state.

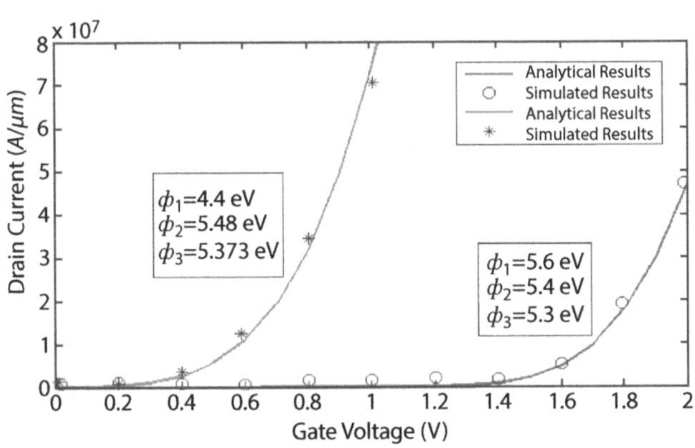

Figure 3.12 I–V_{gs} characteristics for different work function of three metals [17].

3.3.3 Spacer Engineering

The electrostatics of a TFET is greatly influenced by the fringing fields that go from the electrode of gate to the source and drain regions. Figure 3.13(a); Figure 3.13(b); Figure 3.13(c) presents the Schematic representation of TFETs. The characteristics of the spacer influence the energy band profiles and the tunneling current in Tunnel FET because the fringe fields travel through the spacer [65–70].

The engineering method presented by Anghel *et al.* [18] uses low- and high-k dielectric material, respectively, being the spacer and gate oxide layer (a). Within channel volume inversion ensues as a result of the double gate's existence, which helps to maintain the current in on state. They contrasted their suggested structure with the two other ones already in use, one of which lacks a spacer but has a thin oxide coating across device entire region, and the other of which has a high- spacer.

As delineated in Figure 3.14, the current in om state is not much increased in these two existing structures because tunneling occurs at the depth of the body, where the likelihood of tunneling is quite low since there is larger tunneling width [71–74]. The suggested architecture lessens

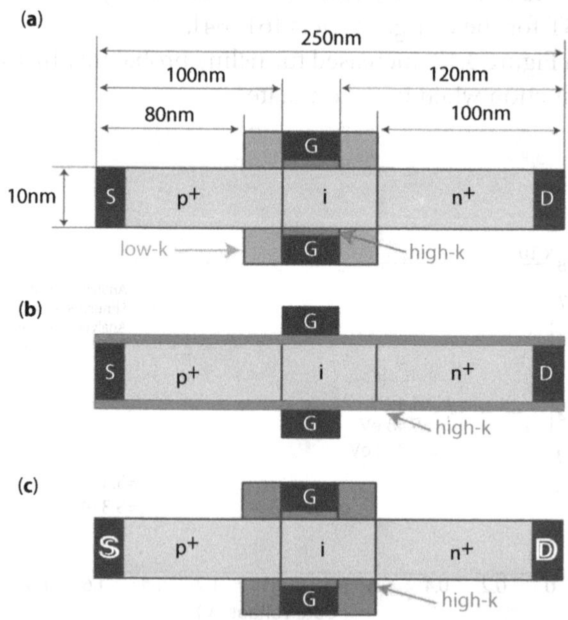

Figure 3.13 Schematic representation of TFET [18]. (a) TFET with low-k dielectric materials. (b) TFET with high-k dielectric materials. (c) TFET with dual gate high-k dielectric materials.

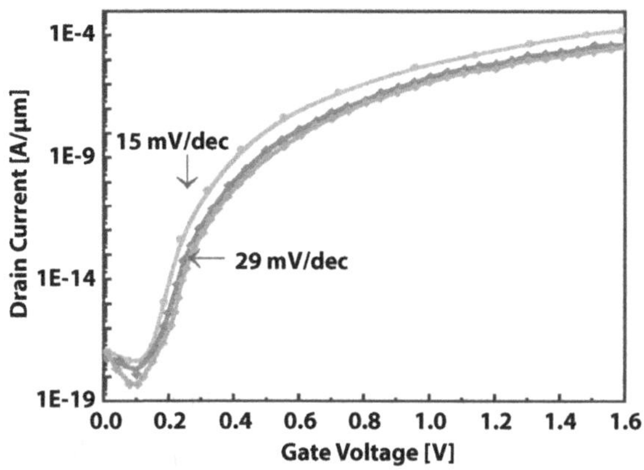

Figure 3.14 Comparison of transfer characteristics[18].

the fringing electric fields, which prevents depletion from occurring at the source-channel tunneling interface.

3.4 Tunneling-Junction Engineering

The tunneling feature has a significant impact on ION in TFET. Especially on source side of the intersection [75–78]. The aforementioned parameters, which may be constructed to generate a high on current I_{on}, are what essentially define the specification of the tunneling junction on side of source:

- Doping of Source
- Heterojunctions

3.4.1 Doping of Source

We initially start with in metal-oxide-semiconductor field-effect transistor like structure in which the source and drain are both N+ doped as drawn in Figure 3.15 in order to provide the built-in N+ pocket P-N-P-N TFET (a). Now we employ the concept of charge plasma with a metal as electrode with the proper work function to change a portion of the "N+" doped source into a "P+" region, as seen in Figure 3.15, a pocket of N+-doped area should be left intermediate the source and the channel in order the identical doping amounts strategy as in an PNPN TFET is to be achieved (b).

The Figure 3.16 displays the suggested in-built N+ pocket PNPN TFET's transfer properties as well as the traditional p-i-n TFET for various drain to source voltages (VDS) [79–84]. As compared to the traditional pin structured TFET, the built-in pocket N+ PNPN Tunnel FET has much

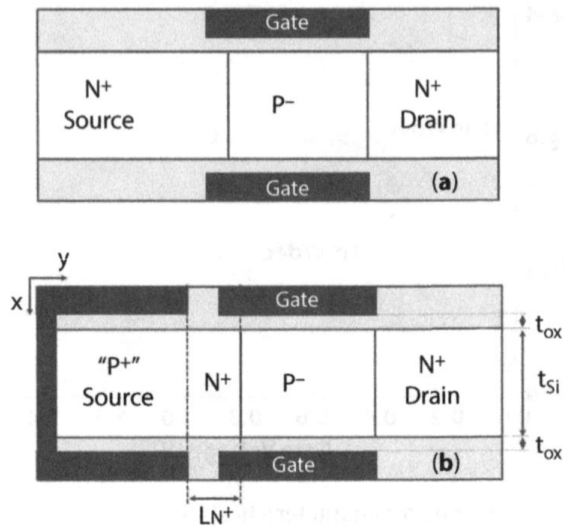

Figure 3.15 Schematic of (a) MOSFET structure with N+ pocket PNPN TFET. (b) N+ pocket in built PNPN TFET [19].

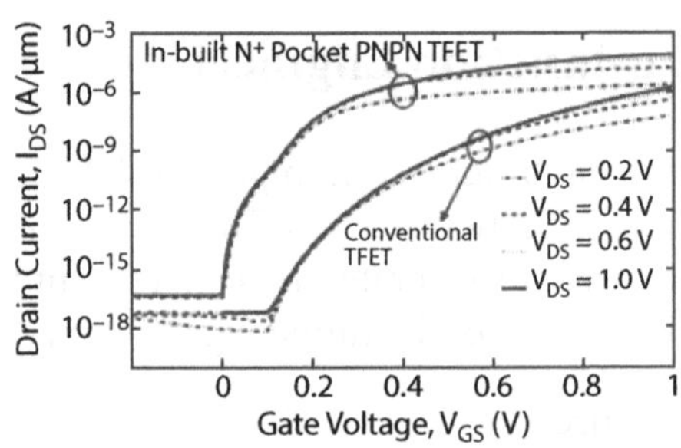

Figure 3.16 N+ pocket PNPN TFET characteristics [19].

better subthreshold slope and drain current. For the suggested structure and the traditional pin structured Tunnel FET, the subthreshold average slope, extracted as provided in [19], is settled to be 25 mV/dec and 65 mV/dec, respectively.

3.4.2 Heterojunctions

An Silicon on insulator Tunnel FET with a dual-dielectric pocket was suggested by Debika Das *et al.* [20] in paper [20]. The findings imply that this device is less likely to degrade its ON current when interface traps are present than ambipolar behavior. Reverse leakage issues are avoided and competent ION is provided via dielectric gate oxide. The HDSilicon

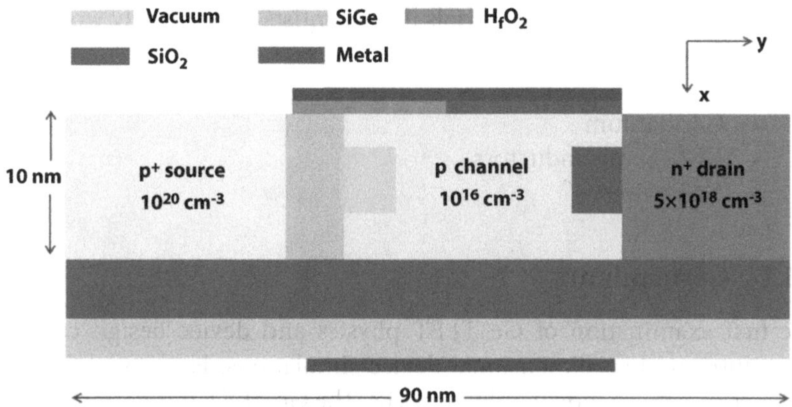

Figure 3.17 HD gate SOI Tunnel FET 2D structure [20].

on insulator Tunnel FET gate, which employs the 'P+' $Si_{1-x}Ge_x$ layer in between source and channel, is part of the structure. The smaller bandgap material utilized at the interface between source to channel boosts ION due to the valence band's position in relation to the conductivity band at the tunnel contact [85–93]. SiO2 serves as a dielectric pocket in the drain channel junction and a vacuum pocket dielectric material in the source channel junction. This pocket configuration allows reliable to performance for less power applications with steeper SS, lower threshold voltage, high ON current (ION), low OFF current (IOFF), and high ON current (IOFF) (Vth). The hetero-dielectric gate SOI TFET 2D construction is seen in Figure 3.17. In account of the increase in tSiGe, the ON current also rises.

3.5 Materials Engineering

TFETs built on silicon are indeed very desirable since silicon dependent manufacturing technologies had advanced and developed over the past proportionate decades. This will make it easier to integrate TFETs into the current CMOS process flows and will allow TFETs to benefit from the silicon processing technology that has already been proved. However, because to its wide bandgap and indirect tunneling process, silicon supports a modest tunneling current. Because of this, silicon-only TFETs are unable to satisfy the advanced technology nodes' ION requirement. So that a greater ION may be attained, researchers are now looking into other materials for TFETs. To achieve TFETs, for instance, germanium, III-V semiconductors, nanowires, carbon-based nanomaterials, etc. Under this section, we outline a few of the other materials that scientists are currently working to develop for use in TFETs.

- Germanium
- III-V Semiconductors
- Nanowires

3.5.1 Germanium

The first examination of Ge TFET physics and device design concepts using intricate two dimensional device simulations. N-channel Si or Ge TFET schematic is shown in Figure 3.18. The Ge TFET was designed using source and drain engineering to fulfil ITRS expectations. Decrease of off-state tunneling current. Controlling the concentration of drain doping in Ge results in TFET. It has been demonstrated it as though Ge has a lesser Eg bandgap than Silicon and may cause TFET Ioff to be significantly greater, the concentration of dopants in the drain has an impact on electric-field and tunneling off-state distance at the drain side and may being altered to reduce off current. Germanium based TFET needs source region dopants profile engineerings to provide a small tunneling barrier-width in the on state in order to obtain noticeably high Ion.

Due to the smaller width at the drain side in the off state caused by Ge's shorter bandgap, there is a lot of unwanted electron tunneling between the channel and the drain, which results in high Ioff. The transfer characteristics of Ids vs. Vgs for the Si and Ge TFETs are simulated in Figure 3.19. To better understand how a smaller bandgap material affects the device performance of TFETs. Both ND and NS are determined, Ge TFET has an exceptional Ion in comparison to Si TFETs, this is more than an order of magnitude higher [21] and surpasses the present ITRS criterion due to a smaller T in the on state. Regrettably, Ge TFET produces high Ioff due to small width in the off state.

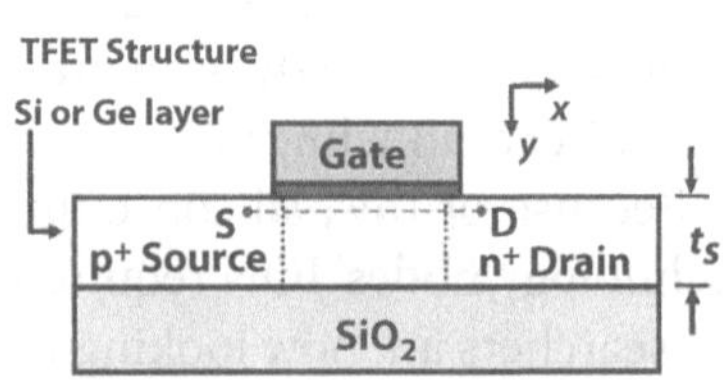

Figure 3.18 N-channel Si or Ge TFET schematic [21].

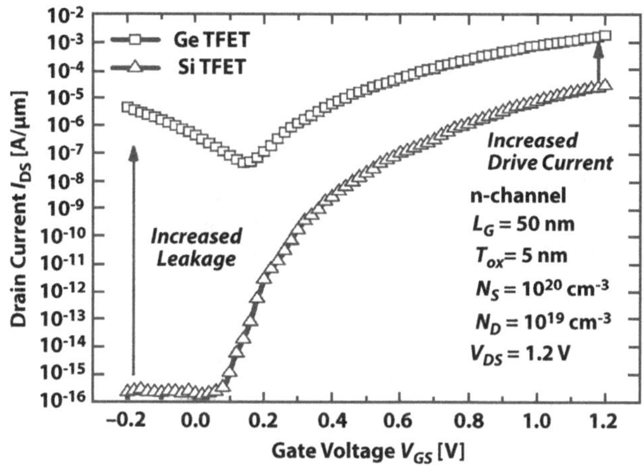

Figure 3.19 Ids vs. Vgs characteristics for Si and Ge TFET [21].

3.5.2 III-V Semiconductors

Higher ION can be provided by III-V material-based TFETs than by traditional Silicon-based TFETs [22]. Indirect transitions enabled by phonons play an important role in BTBT at channel-source tunneling junctions mechanism, typical Silicon TFETs have relatively poor tunneling efficiency. When compared to vertical line TFETs, T shaped TFETs (TTFET) had a much lower ambipolar current [22].

Figure 3.20. depicts the device architectures of well-known GaAs or GaAs alloy-based channel TFETs. Ge/GaAs heterojunction nanowire TFETs with 30 nm channel lengths were used to demonstrate a common source (CS) amplifier in 2013, according to Seongjae Cho *et al.* [23].

With a rise in Na, the staggered TFET's driving current gets better. Figure 3.21. shows how the source doping concentration (Na) affects the $GaAs_{0.9}Bi_{0.1}/GaAs_{0.9}N_{0.1}$ heterojunction Type-II staggered TFET's transfer characteristics.

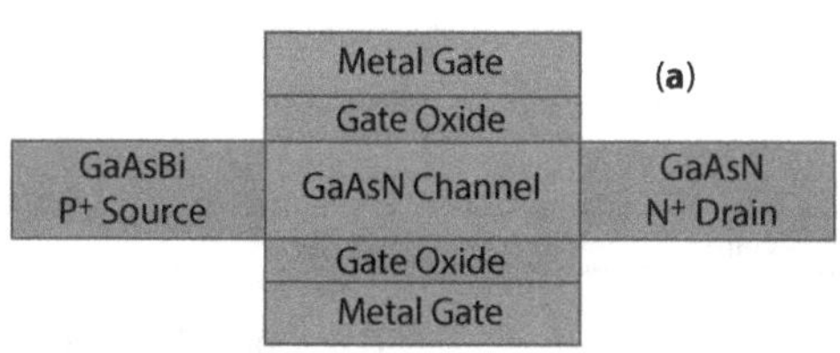

Figure 3.20 Device structure of GaAsBi/GaAsN TFET.

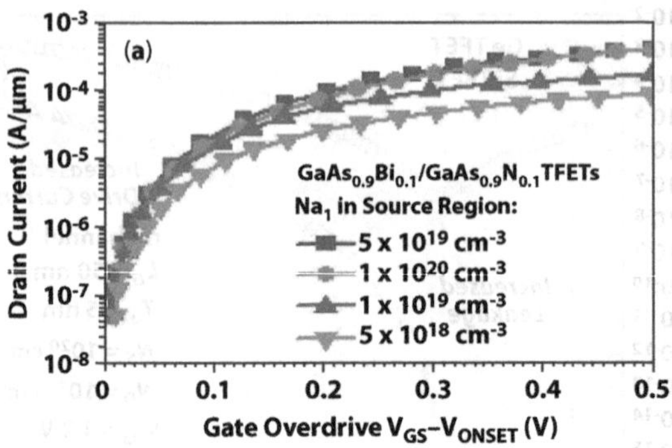

Figure 3.21 $GaAs_{0.9}Bi_{01}/GaAs_{0.9}N_{0.1}$ TFET drain characteristics.

3.5.3 Nanowires

Research on nanowires provided great strides past the initial 2000s. Many organizations are looking at unique semiconductor nanowire properties may be exploited for nanoelectronics. Nanowires are a logical alternative for progressively getting smaller transistors due to their inherent smallness. Short channel effects are especially helped by the precise gate control that nanowire transistors can provide. Nanowires' bottom-up synthesis methods make it feasible to realize device structures those are exceedingly challenging to make using conventional top to down approaches. Due to that, device researchers started looking at nanowire FETs (NWFET), and significant progress has been achieved recently. TFETs can further benefit Nanowires are geometrically small, the gate is electrostatically controlled over the channel potential is excellent, and complex architectures that can be realized utilizing nanowires as shown in Figure 3.22.

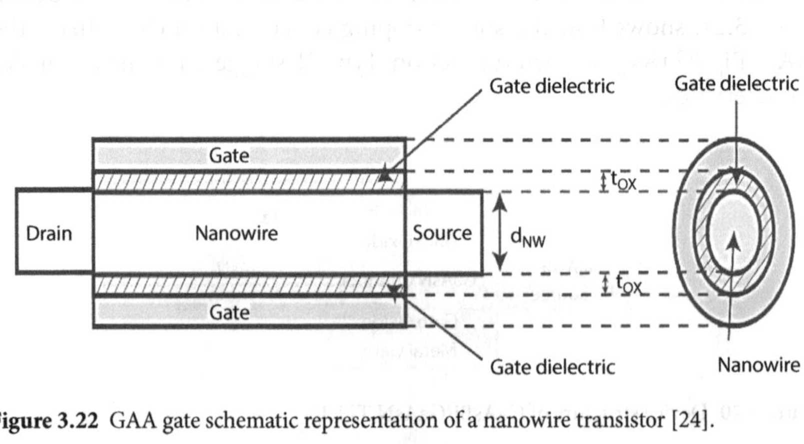

Figure 3.22 GAA gate schematic representation of a nanowire transistor [24].

A comparison is done to demonstrate how well various device topologies work [24]. The simulation results are displayed in Figure 3.23 for verification of the device's veracity. Comparative parametric values of nanowire TEFT structures is given in Table 3.1.

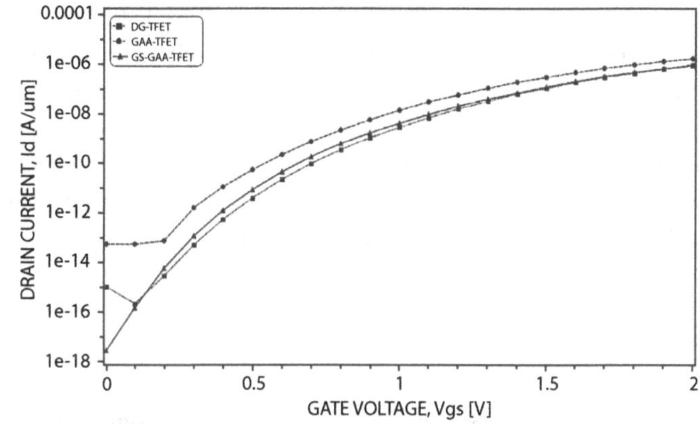

Figure 3.23 Id-Vg curves for DG, GAA, GS GAA TFET [25].

Table 3.1 Comparative parametric values of nanowire TEFT structures.

Ref no	Structure	Ion (μA)	Ioff (A)	SS (mv/ dec)	Ion/Ioff
[26]	N type channel J-L Nanowire TFET	0.1	1×10^{-15}	60	1×10^{6}
[27]	G-A-A Nanowire p-TFET	0.14	1×10^{-9}	45	----
[28]	Hetero-Gate Dielectric Nanowire TFET	----	----	----	2×10^{12}
[29]	Charge Plasma Nanowire TFET	5.54	0.1a	46	5×10^{6}
[30]	DM DG HTFET	116	3.3×10^{-17}	20	3.84×10^{9}
[31]	HV-TFET	10	1×10^{-14}	48.2	3.49×10^{12}
[32]	DG HTFET	166.6	7.55×10^{-9}	50	22.06×10^{3}
[33]	TG TFET	----	-----	24.4	$\sim 10^{10}$

3.6 Conclusion

An overview in regard to tunnel FET's is provided in this review study. There are many different types of hybrid designs, such as DG Tunnel FET, hetero material based junction TFET, doping less TFET, PNPN TFET, tunnel FET with pocket dielectrics, and Tunnel FET with elevated Ge Source region. These studies have been conducted on architectures reveals several benefits, including enhanced performance compared to traditional MOSFETs and larger on state currents, reduced Subthreshold-Swing, and lower off state current. Tunnel FET devices had provided use in the field of bio medical Sensing, analog and digital applications.

References

1. Chokkara SP, Gaur A, Sravani KG *et al.* (2021) "Design, Simulation and Analysis of a Slotted RF MEMS Switch." *Trans. Electr. Electron. Mater.* https:// doi.org/10.1007/s42341-021-00363-8,24.09.21
2. Srinivasa Rao K, Ashok Kumar P, Guha K, Sailaja BV S, Vineetha KV, Baishnab KL, Girija Sravani K. — "Design and simulation of ixed–ixedlexure type RF MEMS switch for reconfgurable antenna||." *Microsyst. Technol.*, to be published. https:// doi.org/ 10. 1007/ s00542- 018- 3983-2.
3. Alluri S, Balaji B, Ch.cury (2021) "Low power, high speed VLSI circuits in 16nm technology." ISSN: 0094–243X, Volume-2358 Issue-1 July. 030001–1–16. https://doi.org/10.1063/5.0060101
4. Balaji B, Aditya M, Adithya G, Ayyappa Vijay VVSSK, Chandu K (2019) "Implementation of low-power 1-bit hybrid full adder with reduced area." *IJITEE* 8(7):61–64 (ISSN: 2278-3075)
5. Sravani Girija K, Guha K, Srinivasa Rao K (2018) "Analysis on selection of beam material for novel step structured RF-MEMS switch used for satellite communication applications." *Trans. Electr. Electron. Mater.* 19(6):467–474
6. Rao, I. R. S. N., *et al.* "Wireless secured data transmission using cryptographic techniques through fpga." *Int. J. Eng. Technol. (IJET)*, e-ISSN (2016): 0975-4024.
7. Kumar, S. T., *et al.* "Design of Various Shaped MEMS Based Cantilevers Executed in COMSOL Multiphysics." *Eur. J. Mol. Clin. Med.* 7.10 (2020): 1972-1979.
8. Pandey CK, Singh A, Chaudhury S (2020) "Effect of asymmetric gate–drain overlap on ambipolar behavior of double-gate TFET and its impact on HF performances." *Appl. Phys. A Mater. Sci. Process.* 126(3):1–12.
9. ahay S, Kumar MJ (2015) "Controlling the drain side tunnelling width to reduce ambipolar current in tunnel FETs using heterodielectric box." *IEEE Trans. Electron Devices* 62(11):3882–3886

10. Garg S, Saurabh S (2018) "Suppression of ambipolar current in tunnel FETs using drain-pocket: Proposal and analysis." *Superlattices Microstruct.* 113:261–270
11. Choi WY, Lee W (2010) "Hetero-gate-dielectric tunneling field effect transistors." *IEEE Trans. Electron Devices* 57(9):2317–2319.
12. Kim SH, Kam H, Hu C, Liu T-JK (2009) "Germanium-source tunnel field effect transistors with record high ion/iof." *In: 2009 Symposium on VLSI Technology*, pp. 178–179 .
13. Abdi DB, Kumar MJ (2014) "In-built n+ pocket p-n-n tunnel feild-effect transistor." *IEEE Electron. Device Lett.* 35(12):1170–1172.
14. Kao K-H *et al.* (2012) "Optimization of gate-on-source-only tunnel FETs with counter-doped pockets." *IEEE Trans. Electron Devices* 59(8):2070–2077.
15. K. Boucart and A. M. Ionescu, "Double-Gate Tunnel FET With High-κ Gate Dielectric," in *IEEE Trans. Electron Devices*, vol. 54, no. 7, pp. 1725-1733, July 2007, doi: 10.1109/TED.2007.899389.
16. S. Saurabh and M. J. Kumar, "Novel Attributes of a Dual Material Gate Nanoscale Tunnel Field-Effect Transistor," in *IEEE Trans. Electron Devices*, vol. 58, no. 2, pp. 404-410, Feb. 2011, doi: 10.1109/TED.2010.2093142.
17. N. Bagga and S. K. Sarkar, "An Analytical Model for Tunnel Barrier Modulation in Triple Metal Double Gate TFET," in *IEEE Trans. Electron Devices*, vol. 62, no. 7, pp. 2136-2142, July 2015, doi: 10.1109/ TED.2015.2434276
18. Chauhan A, Saini G, Yerur PK (2018) Improving the performance of dual-k spacer underlap double gate TFET. *Superlattices Microstruct.* 124:79–91.
19. D. B. Abdi and M. J. Kumar, "In-Built N+ Pocket p-n-p-n Tunnel Field-Effect Transistor," in *IEEE Electron Device Lett.*, vol. 35, no. 12, pp. 1170-1172, Dec. 2014, doi: 10.1109/LED.2014.2362926.
20. Das, D., Chakraborty, U. "A Study on Dual Dielectric Pocket Heterojunction SOI Tunnel FET Performance and Flicker Noise Analysis in Presence of Interface Traps." *Silicon* 13, 787–798 (2021). https://doi.org/10.1007/s12633-020-00488-0
21. Eng-HuatToh, Grace Huiqi Wang, Ganesh Samudra, and Yee-Chia Yeo, "Device physics and design of germanium tunneling field-effect transistor with source and drain engineering for low power and high performance applications", *J. Appl. Phys.* 103, 104504 (2008).
22. M. Saravanan, Eswaran Parthasarathy, "A review of III-V Tunnel Field Effect Transistors for future ultra low power digital / analog applications," *Microelectronics J.*,Volume 114, 2021, 105102, ISSN 0026-2692.
23. Sungjin Cho, Jeon-Young Kang, Ansar-Ul-Haque Yasar, LukKnapen, Tom Bellemans, Davy Janssens, Geert Wets, Chul-Sue Hwang, "An Activity-based Carpooling Microsimulation Using Ontology", *Procedia Comput. Sci.*, Volume 19, 2013, Pages 48-55, ISSN 1877-0509.
24. Saurabh, S., & Kumar, M.J. (2016). Fundamentals of Tunnel Field-Effect Transistors (1st ed.). *CRC Press.* https://doi.org/10.1201/9781315367354

25. Anne S. Verhulst, Bart Sorée, Daniele Leonelli, William G. Vandenberghe, and Guido Groeseneken , "Modeling the single-gate, double-gate, and gate-all-around tunnel field-effect transistor", *J. Appl. Phys.* 107, 024518 (2010) https:/doi.org/10.1063/1.3277044.
26. Ravindran A, George A, Praveen CS, Kuruvilla N (2017) Gate all around nanowire TFET with high ON/OFF current ratio. *Mater. Today Proc.* 4:10637–10642
27. Vishnoi R, Kumar MJ (2015) A compact analytical model for the drain current of gate-all-around nanowire tunnel FET accurate from sub-threshold to ON-state. *IEEE Trans. Nanotechnol.* 14:358–362
28. Lazarov VK, Lari L, Lytvyn PM, Kholevchuk V V, Mitin VF (2012) Structural study of Ge/GaAs thin films. In: *J. Phys.: Conference Series*. IOP Publishing, p 12040
29. Kurniawan ED, Yang S-Y, Thirunavukkarasu V, Wu Y-C (2017) Analysis of Ge-Si heterojunction nanowire tunnel FET: impact of tunneling window of band-to-band tunneling model. *J. Electrochem. Soc.* 164:E3354
30. Das D, Chakraborty U (2020) A study on dual dielectric pocket Heterojunction SOI tunnel FET performance and ficker noise analysis in presence of Interface traps. *Silicon*. 13:787–798.
31. Vanlalawmpuia K, Bhowmick B (2020) Optimization of a Heterostructure vertical tunnel FET for enhanced electrical performance and effects of temperature variation on RF/linearityparameters. *Silicon*.
32. Ahish S, Sharma D, Kumar YBN, Vasantha MH (2016) Performance enhancement of novel InAs/Si hetero double-gate tunnel FET using Gaussian doping. *IEEE Tran. Electron Devices* 63:288–295.
33. Dubey PK, Kaushik BK (2017) T-shaped iii–v heterojunction tunneling-field-effect transistor. *IEEE Trans. Electron Devices* 64(8):3120–3125
34. Anjana Bhardwaj, Pradeep Kumar, B. Raj, Sunny Anand, "Design and Performance Optimization of doping-less Vertical Nanowire TFET using Gate Stack Technique", *J. Electron. Mater. (JEMS)* Springer, Vol. 41, Issuse 7, PP. 4005-4013, 2022.
35. Jeetendra Singh, BRaj, "Tunnel Current Model of Asymmetric MIM Structure Levying Various Image Forces to Analyze the Characteristics of Filamentary Memristor", *Appl. Phys. A*, Springer, Vol. 125, No. 3, pp. 203.1 to 203.11 Feb 2019.
36. Candy Goyal, Jagpal Singh Ubhi and B Raj, Low Leakage Zero Ground Noise Nanoscale Full Adder using Source Biasing Technique, "*J. Nanoelectron. Opto electron.*", American Scientifc Publishers, Vol. 14, PP. 360–370, March 2019.
37. Gurmohan Singh, R. K. Sarin and B Raj, "A Novel Robust Exclusive-OR Function Implementation in QCA Nanotechnology with Energy Dissipation Analysis", *J. Comput. Electron.*, Springer, Volume 15, Issue 2, PP. 455-465, June 2016.
38. Tanu Wadhera, Deepti Kakkar, Girish wadhwa, B Raj, "Recent Advances and Progress in Development of the Field Effect Transistor Biosensor: A Review"

J. Electron. Mater., Springer, Volume 48, Issue 12, pp 7635–7646, December 2019.

39. Girish Wadhwa, Priyanka Kamboj, Balwinder Raj" Design optimisation of junctionless TFET biosensor for high sensitivity" *Advances in Natural Sciences: Nanoscience and Nanotechnology*, Vol. 10, PP. 045001 (7), (2019).
40. Priya Bansal, BRaj, "Memristor Modeling and Analysis for Linear Dopant Drift Kinetics," *J. Nanoeng. Nanomanuf.*, American Scientific Publishers, Vol. 6, pp. 1–7, 2016.
41. Amandeep Singh, Mamta Khosla, BRaj, "Circuit Compatible Model for Electrostatic Doped Schottky Barrier *CNTFET*," *J. Electron. Mater.*, Springer, Vol. 45, Issue 12, pp 4825-4835, 2016.
42. Ashima, D Vaithiyanathan, B Raj," Performance Analysis of Charge Plasma induced Graded Channel Si nanotube", *J. Eng. Res. (JER)*, EMSME Special Issue, Vol. 9, pp. 146-154 Aug 2021.
43. Abhishek Singh Tomar, Vijay Kumar Magraiya, and B Raj, Scaling of Access and Data Transistor for High Performance DRAM Cell Design, *Quantum Matter*, Vol. 2, PP. 412-416 Oct. 2013.
44. Neeraj Jain, BRaj, "Parasitic Capacitance and Resistance Model Development and Optimization of Raised Source/Drain SOI FinFET Structure for Analog Circuit Applications, *J. Nanoelectron. Optoelectron.*, ASP, USA, Vol. 13, PP. 531–539, Apl 2018.
45. Shradhya Singh, S. K. Vishvakarma and B Raj, "Analytical Modeling of Split-Gate Junction-Less Transistor for a Biosensor Application", *Sens. Biosens.*, Elsevier, Vol. 18, Pages 31-36, April 2018.
46. Maisagalla Gopal, and Balwinder Raj, Low Power 8T SRAM Cell Design for High Stability Video Applications, *ITSI Transaction on Electrical and Electronics Engineering*, Vol. 1(5), pp. 91-97, 2013.
47. Balwinder Raj, Jatin Mitra, Deepak Kumar Bihani, Rangharajan V, A.K. Saxena and S. Dasgupta, "Analysis of Noise Margin, Power and Process Variation for 32 nm FinFET Based 6T SRAM Cell" *J. Comput. (JCP)*, Academy Publisher, FINLAND, vol. 5, issue 6, PP. 1-8, 2010.
48. Divya Sharma, Rajesh Mehra and B Raj, "Comparative Analysis of Photovoltaic Technologies for High Efficiency Solar Cell Design", *Superlattices Microstruct.*, Elsevier, Volume 153, PP. 106861, May 2021.
49. Pawandeep Kaur, Avtar Singh Buttar, B Raj, "A comprehensive Analysis of Nanoscale Transistor based Biosensor: A Review, *Indian Journal of Pure and Applied Physics (IJPAP)*, Vol. 59, pp. 304-318, April 2021.
50. Divya Yadav, Balwant Raj and B Raj, "Design and Simulation of Low Power Microcontroller for IoT Applications", *J. Sens. Lett.*, ASP, Vol. 18, PP. 401–409, May 2020.
51. Shailendra Singh, B Raj," A 2-D Analytical Surface Potential and Drain current Modeling of Double-Gate Vertical t-shaped Tunnel FET", *J. Comput. Electron.*, Springer, Vol. 19, PP.1154–1163, Apl 2020.

52. Jeetendra Singh, B Raj, "An Accurate and Generic Window function for Nonlinear Memristor Model" *J. Comput. Electron.*, Springer, Volume 18 Issue 2, PP. 640-647 June 2019.
53. Manjit Kaur; Neena Gupta; Sanjeev Kumar; BRaj; Arun Kumar Singh, "Comparative RF and Crosstalk Analysis of Carbon Based Nano Interconnects" *IET Circuits, Devices Syst.*, Vol.15(6), PP. 493-503, Feb 2021.
54. Nehru Kandasamy, Firdous Ahmad, D. Ajitha, B Raj, and Nagarjuna Telagam "Quantum Dot Cellular Automata based Scan Flip Flop and Boundary Scan Register, *IETE J. Res.*, Volume 66, PP. 535-548, 2020.
55. Sharma, S. K.., Raj, B., Khosla, M., "Enhanced Photosensivity of Highly Spectrum Selective Cylindrical Gate In1-xGaxAs Nanowire MOSFET Photodetector, *Mod. Phys. Lett.-B*, Vol. 33, No. 12, PP. 1950144 (2019).
56. Jeetendra Singh, BRaj, "Design and Investigation of 7T2M NVSARM with Enhanced Stability and Temperature Impact on Store/Restore Energy", *IEEE Trans. Very Large Scale Integr. Syst.*, Vol. 27, Issue 6, PP. 1322 - 1328 June 2019.
57. Anil Kumar Bhardwaj, Sumeet Gupta, BRaj, Amandeep Singh, "Impact of Double Gate Geometry on the Performance of Carbon Nanotube Field Effect Transistor Structures for Low Power Digital Design", *J. Comput. Theor. Nanosci.*, ASP, Vol. 16, PP. 1813–1820, 2019.
58. Neeraj Jain, BRaj, "Thermal Stability Analysis and Performance Exploration of Asymmetrical Dual-k underlap Spacer (ADKUS) SOI FinFET for Security and Privacy Applications", *Indian J. Pure Appl. Phys. (IJPAP)*, Vol. 57, pp. 352-360, May 2019.
59. Amandeep Singh, Mamta Khosla, BRaj, "Design and Analysis of Dynamically Confgurable Electrostatic Doped Carbon Nanotube Tunnel FET", *Microelectronics J.*, Elesvier, Volume 85, Pages 17-24, March 2019.
60. Neeraj Jain and Balwinder Raj, "Dual-k Spacer Region Variation at the Drain Side of Asymmetric SOI FinFET Structure: Performance Analysis towards the Analog/RF Design Applications", communicated in "*J. Nanoelectron. Optoelectron.*", American Scientifc Publishers, Vol. 14, PP. 349–359, March 2019.
61. Jeetendra Singh, Sanjeev Sharma, B Raj, Mamta Khosla Analysis of barrier layer thickness on performance of In1-xGaxAs based Gate Stack Cylindrical Gate Nanowire MOSFET, " *JNO, ASP*, Vol. 13, PP 1473-1477, Oct 2018.
62. Neeraj Jain, B Raj, "Analysis and Performance Exploration of High-k SOI FinFETs Over the Conventional Low-k SOI FinFET toward Analog/ RF Design", *Journal of Semicond. (JoS)*, IOP Science, Vol.39, Issue 12, PP- 124002-1-7, Dec 2018.
63. Candy Goyal, Jagpal Singh Ubhi and BRaj, A reliable leakage reduction technique for approximate full adder with reduced ground bounce noise, *J. Math. Probl. Eng.*, Hindawi, Volume 2018, Article ID 3501041, 16 pages, 15 Oct 2018.
64. Anuradha, Jeetendra Singh, B Raj and Mamta Khosla, "Design and Performance Analysis of Nano-scale Memristor-based Nonvolatile SRAM" *J. Sens. Lett.*", American Scientifc Publishers, Vol. 16, PP 798– 805, Oct 2018.

65. Girish Wadhwa, BRaj, "Parametric Variation Analysis of Charge-Plasma-based Dielectric Modulated JLTFET for Biosensor Application" *IEEE Sens. J.*, VOL. 18, NO. 15, AUGUST 1, PP. 6070 - 6077, 2018.
66. Jeetendra Singh, BRaj, "Comparative Analysis of Memristor Models for Memories Design", *JoS, IoP*, Vol. 39, No. 7, PP. 074006-1-12, July 2018.
67. Divya Yadav, Shailesh Singh Chouhan, Santosh Kumar Vishvakarma and BRaj, "Application Specific Microcontroller Design for IoT based WSN", *Sens. Lett., ASP*, Vol. 16, P P. 374–385, May 2018.
68. Gurmohan Singh, R. K. Sarin and BRaj, "Fault-Tolerant Design and Analysis of Quantum-Dot Cellular Automata Based Circuits", *IEEE/IET Circuits, Devices & Systems*, Vol. 12, PP. 638 - 64, 2018.
69. Jeetendra Singh, BRaj, "Modeling of Mean Barrier Height Levying Various Image Forces of Metal Insulator Metal Structure to Enhance the Performance of Conductive Filament Based Memristor Model", *IEEE Nanotechnol.*, Vol. 17, No. 2, PP. 268-267, March 2018.
70. Aakash Jain, Sanjeev Sharma, BRaj, "Analysis of Triple Metal Surrounding Gate (TM-SG) III-V Nanowire MOSFET for Photosensing Application", *Opto-electronics Journal*, Elsevier, Volume 26, Issue 2, Pages 141-148, May 2018.
71. Aakash Jain, Sanjeev Sharma, B Raj, "Design and Analysis of High Sensitivity Photosensor Using Cylindrical Surrounding Gate MOSFET For Low Power Sensor Applications", *Eng. Sci. Technol. Int. J.*, Elsevier's, Volume 19, Issue 4, Pages 1864–1870, December 2016.
72. Amandeep Singh, Mamta Khosla, B Raj, "Analysis of Electrostatic Doped Schottky Barrier Carbon Nanotube FET for Low Power Applications," *J. Mater. Sci.: Mater. Electron.*, Springer, Vol. 28, PP. 1762–1768, 2017.
73. G. Saiphani Kumar, Amandeep Singh, BRaj, "Design and Analysis of Gate All Around CNTFET based SRAM cell Design", *J. Comput. Electron.*, Springer, Volume 17, Issue 1, pp 138–145, March 2018.
74. Gurinder pal Singh, B. S. Sohi, Balwinder Raj, "Material Properties Analysis of Graphene Base Transistor (GBT) for VLSI Analog Circuits, *Indian J. Pure Appl. Phys. (IJPAP)*, Vol. 55, pp. 896-902, Dec-2017.
75. Amandeep Singh, Mamta Khosla, B Raj, "Comparative Analysis of Carbon Nanotube Field Effect Transistor and Nanowire Transistor for Low Power Circuit Design," *J. Nanoelectron. Optoelectron.*, American Scientifc Publishers, USA, Vol. 11, PP. 388-393, June 2016.
76. Sunil Kumar and B Raj, "Estimation of Stability and Performance metric for Inward Access Transistor based 6T SRAM Cell Design using n-type/p-type DMDG-GDOV TFET" *IEEE VLSI Circuits Syst. Lett.*, Volume 3, Issue 2, PP 25-39, June 2017.
77. Shashikant Sharma, Anjan Kumar, Manisha Pattanaik and B Raj, "Forward Body Biased Multimode Multi-Threshold CMOS Technique for Ground Bounce Noise Reduction in Static CMOS Adders" *Int. J. Inf. Electron. Eng.* , pp. 567-572, Vol. 3, issue 3, 2013.

78. Hamendra Singh, Pankaj Kumar, Balwinder Raj, "Performance Analysis of Majority Gate SET Based 1-bit Full Adder", *Int. J. Comput. Commun. Eng.* (IJCCE), IACSIT Press Singapore, ISSN: 2010-3743, Volume 2, No. 4, PP.1-8, 2013.
79. Anil Kumar Bhardwaj, Sumeet Gupta and B Raj, "Investigation of Parameters for Schottky Barrier (SB) Height for Schottky Barrier Based Carbon Nanotube Field Effect Transistor Device", *J. Nanoelectron. Optoelectron.*, ASP, Vol. 15, PP. 783–791, July 2020.
80. Priya Bansal, BRaj, "Memristor: A Versatile Nonlinear Model for Dopant Drift and Boundary Issues," *JCTN*, American Scientifc Publishers, Vol. 14 (5), P P. 2319-2325, May 2017.
81. Neeraj Jain, B. Raj, "An Analog and Digital Design Perspective Comprehensive Approach on Fin-FET (Fin-Field Effect transistor) Technology - A Review", *Rev. Adv. Sci. Eng. (RASE)*, ASP, Vol. 5, PP. 1-14, 2016.
82. Sanjeev Sharma, B Raj, Mamta Khosla, "Subthreshold Performance of In1-xGaxAs based Dual Metal with Gate Stack Cylindrical/ Surrounding Gate Nanowire MOSFET for Low Power Analog Applications, *J. Nanoelectron. Optoelectron.*, American Scientifc Publishers, USA, Vol.12, PP. 171–176, 2017.
83. Balwinder Raj, A. K. Saxena and S. Dasgupta, "Analytical Modeling for the Estimation of Leakage Current and Subthreshold Swing Factor of Nanoscale Double Gate FinFET Device" *Microelectron. Int.*, UK, vol. 26, pp. 53-63, 2009.
84. Soniya, Shailendra Singh, Girish Wadhwa and Balwinder Raj, "An Analytical Modeling for Dual Source Vertical Tunnel Field Effect Transistor" submitted to '*Int. J. Recent Technol. Eng. (IJRTE)*' at Volume-8 Issue-2, PP. 603-608, July 2019.
85. Shailendra Singh, B Raj, "Design and analysis of hetrojunction Vertical T-shaped Tunnel Field Effect Transistor", *J. Electron. Mater.*, Springer, Volume 48, Issue 10, pp 6253–6260, October 2019.
86. Candy Goyal, Jagpal Singh Ubhi and B Raj, "A Low Leakage CNTFET based Inexact Full Adder for Low Power Image Processing Applications", *Int. J. Circuit Theory Appl.*, Wiley, Volume 47, Issue 9, Pages 1446-1458, September 2019.
87. B Raj, A. K. Saxena and S. Dasgupta, "A Compact Drain Current and Threshold Voltage Quantum Mechanical Analytical Modeling for FinFETs" *J. Nanoelectron. Optoelectron. (JNO)*, USA, vol. 3, no. 2, pp. 163-170, 2008.
88. Girish Wadhwa, B Raj, "An Analytical Modeling of Charge Plasma based Tunnel Field Effect Transistor with Impacts of Gate underlap Region" *Superlattices Microstruct.*, Elsevier, Vol. 142, PP.106512, June 2020.
89. Shailendra Singh, B Raj, "Modeling and Simulation analysis of SiGe heterojunction Double GateVertical t-shaped Tunnel FET", *Superlattices Microstruct.*, Elsevier Volume 142, PP. 106496, June 2020.
90. Amandeep Singh, Dinesh Kumar Saini, Dinesh Agarwal, Sajal Aggarwal, Mamta Khosla, BRaj, "Modeling and Simulation of Carbon Nanotube Field Effect Transistor and its Circuit Application," *J. Semicond. (JoS)*, IOP Science, Vol.37, PP. 074001-6, July 2016.

91. Neeraj Jain, Balwinder Raj, "Device and Circuit Co-Design Perspective Comprehensive Approach on FinFET Technology - A Review, *J. Electron Devices*, Vol. 23, Num. 1, pp 1890-1901, 2016.
92. Sunil Kumar and BRaj, "Analysis of I_{ON} and Ambipolar Current for Dual-Material Gate-drain Overlapped DG-TFET," *J. Nanoelectron. Optoelectron.*, American Scientifc Publishers, USA, Vol.11, PP. 323-333, June 2016.
93. Naveed Anjum, Tarun Bali and B Raj, Design And Simulation Of Handwritten Multiscript Character Recognition, *Int. J. Adv. Res. Comput. Commun. Eng.*, Vol. 2 (7), PP. 2544-2549, July 2013.

4

GaAs Nanowire Field Effect Transistor

Shailendra Yadav[1], Mandeep Singh[2], Tarun Chaudhary[2], Balwinder Raj[2*], Alok Kumar Shukla[1] and Brajesh Kumar Kaushik[1]

[1]*Department of Electronics and Communication Engineering, Indian Institute of Technology Roorkee, Roorkee, India*
[2]*Department of Electronics and Communication Engineering, Dr. B. R. Ambedkar National Institute of Technology Jalandhar, Punjab, India*

Abstract
The unique characteristics of nanowire field effect transistors (NW FETs), such as high electron mobility, low power consumption, and scalability, have made them a promising technology for upcoming electronic devices. An overview of the underlying concepts, methods of fabrication, and performance traits of NW FETs are given in this chapter. The chapter discusses the benefits and drawbacks of NW FETs, including how contact resistance, defect density, and surface effects affect device performance. The chapter also highlights the potential uses of NW FETs in several industries, including sensing, computing, and energy harvesting. Overall, this chapter offers a thorough reference for researchers and engineers interested in this innovative technology since NW FETs are anticipated to play a significant role in the creation of next-generation electronic devices.

Keywords: FET, NW FET, NEMS

4.1 Introduction

Nanowires are small, cylindrical constructions that range in size from a few micrometers to several millimeters. Their diameter is generally less than 100 nm. They may be produced using a range of substances, including metals, semiconductors, and oxides. Nanowires are attractive materials for a variety of applications, including electronics, photonics, sensors, and

**Corresponding author*: rajb@nitj.ac.in

Balwinder Raj, Suman Lata Tripathi, Tarun Chaudhary, K. Srinivasa Rao and Mandeep Singh (eds.) Integrated Devices for Artificial Intelligence and VLSI, (75–100) © 2024 Scrivener Publishing LLC

energy conversion due to their distinctive physical and chemical characteristics. Several techniques, like chemical vapor deposition, electrochemical deposition, and template-assisted growth, can be used to create nanowires [1]. The high surface-to-volume ratio of nanowires, which makes them extremely responsive to environmental changes, is one of their main benefits. They may be used for sensing applications like identifying trace levels of chemicals or biomolecules in a sample because of this feature. Due to their distinct electrical characteristics, nanowires may also be employed as the foundation for nanoscale electronic devices, including transistors and diodes [2].

Due to their tiny size, high surface-to-volume ratio, and quantum confinement effects, nanowires have unusual electrical, optical, and mechanical capabilities. They are strong prospects for a variety of applications in electronics, photonics, energy conversion, and biological sensing and imaging because of their characteristics. High-performance transistors, sensors, and memory components may all be made in electronics using nanowires [3–5]. Nanowires can be utilized in photonics to make effective solar cells, photodetectors, and light-emitting diodes. Nanowires can be utilized to make extremely effective fuel cells and batteries for energy conversion [6–10]. Nanowires can be utilized for high-resolution imaging of biological tissues and the detection of biomolecules in biomedical sensing and imaging. Researchers in many domains have focused heavily on the special characteristics of nanowires, and work is still being done to create novel techniques for creating and modifying these tiny structures as well as to investigate their potential uses in a variety of sectors. Overall, due to their distinct physical and chemical features, which might enable novel technologies and applications, nanowires provide a viable route for research and development in a range of sectors. Based on their composition, structure, and characteristics, nanowires can be categorized [11–15]. Here are a few typical nanowire kinds. Figure 4.1 depicts the numerous kinds of nanowires.

4.1.1 Semiconductor Nanowires

Nanostructures known as semiconductor nanowires are created from semiconductor substances like silicon, germanium, or gallium arsenide. They can be many micrometers long and often have a diameter of just a few nanometers. These structures have special electrical and optical characteristics that set them apart from bulk semiconductors, which makes them very interesting for nanoelectronics and nanophotonics. For instance, highly sensitive sensors, effective solar cells, and fast transistors may all be

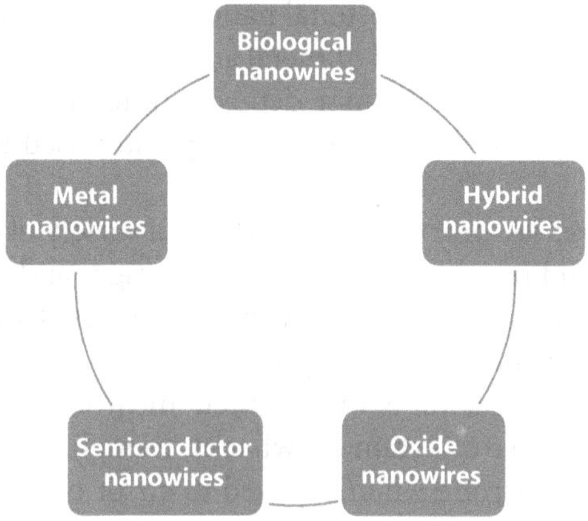

Figure 4.1 Classification of nanowires.

made with nanowires. Semiconductor nanowires [16–20] have mechanical qualities in addition to electrical and optical ones that make them desirable for application in nanoelectromechanical systems (NEMS) and flexible electronics. For instance, nanowires are perfect for use in electronics that need flexibility and can be bent, stretched, and twisted without breaking [21–26].

4.1.2 Metal Nanowires

Si gate all around nanowire-FET is presented in Figure 4.2. Metals such as gold, silver, copper, and platinum are used to create these nanowires.

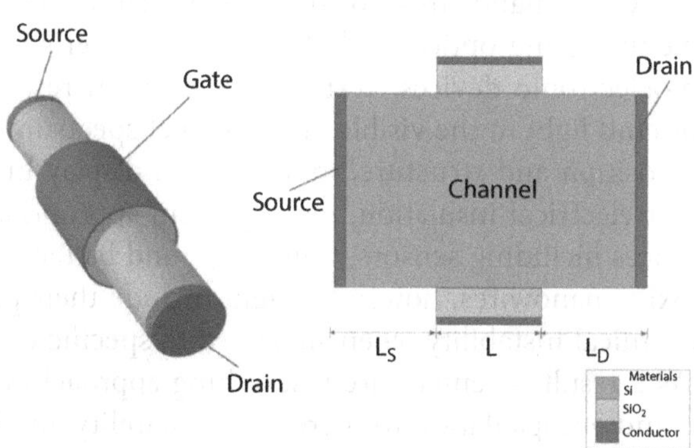

Figure 4.2 Si gate all around nanowire-FET [10].

Metal nanowires are one-dimensional nanostructures with a diameter of a few nanometers that are formed of metal atoms or metal alloys. Numerous techniques, such as chemical vapor deposition, template-assisted synthesis, and electrochemical deposition, are frequently used to produce them [27–31].

Metal nanowires are potential prospects for a variety of nanotechnology applications, including electronics, energy storage, and catalysis, due to their distinctive physical and chemical characteristics. Their high electrical conductivity, which makes them appropriate for use in electronic devices like transistors and sensors, is one of their most crucial characteristics. Metal nanowires do have certain drawbacks, though, namely the propensity to oxidize and corrode when exposed to air or other reactive species. Their mechanical and electrical qualities may deteriorate as a result over time. As a result, scientists are researching solutions like protective coatings and encapsulation to enhance the stability and longevity of metal nanowires [32–36].

4.1.3 Oxide Nanowires

Oxide nanowires are one-dimensional structures formed of oxide substances with a diameter in the range of a few nanometers, such as zinc oxide, titanium dioxide, and iron oxide. Numerous techniques, such as chemical vapor deposition, hydrothermal synthesis, and sol-gel procedures are frequently used to produce them. Oxide nanowires are desirable for a range of nanotechnology applications, including electrical devices, energy conversion, and sensing, due to their distinctive features [37–43]. Their high surface area to volume ratio, which boosts their reactivity and makes them excellent for use in catalytic applications, is one of their most crucial characteristics. Oxide nanowires not only have a large surface area, but they also have intriguing optical and electrical characteristics. For application in optoelectronic devices, certain oxide nanowires, for instance, can absorb or emit light in the visible or ultraviolet spectrum. Depending on their composition and structure, they can also display high electrical conductivity or electrical insulation, making them appropriate for use in electronic devices including sensors, transistors, and batteries. The disadvantages of oxide nanowires, however, might include their propensity to degrade or chemical instability when subjected to specific environmental conditions. As a result, scientists are researching approaches like surface modification and encapsulation to increase the stability and longevity of oxide nanowires [44–48].

4.1.4 Hybrid Nanowires

Hybrid nanowires have been researched for their potential use in plasmonics, catalysis, and sensing because they can display special qualities that are not found in individual materials. One-dimensional hybrid nanowires are formed of a variety of substances, including metals, semiconductors, and oxides. They are frequently produced using a variety of processes, including electrochemical deposition, solution-based approaches, and chemical vapor deposition. Due to their special qualities that come from the blending of several materials, hybrid nanowires are appealing for several nanotechnology applications, including sensing, energy conversion, and biological ones [49–53]. For instance, due to their distinct electrical and optical characteristics, the combination of metallic and semiconducting materials in hybrid nanowires can result in improved sensing capabilities. Hybrid nanowires can display intriguing structural characteristics in addition to their improved properties. As an illustration, the fusion of several materials can produce core-shell structures or heterostructures, which can further improve their characteristics for certain purposes. However, due to the complexity of the materials used and the requirement for exact control over the growth conditions, the synthesis and characterization of hybrid nanowires can be difficult. To create hybrid nanowires with specialized features for certain applications, researchers are attempting to develop novel methodologies [54–59].

4.1.5 Biological Nanowires

The bacterial pili and fungi that create these nanowires are living things. Biological nanowires have been investigated for their possible use in bioelectronics and biotechnology because of their distinctive electrical and mechanical characteristics. There are other kinds of nanowires as well, such as one-dimensional carbon structures called carbon nanotubes and graphene nanoribbons, which have special electrical and mechanical characteristics. Living things include one-dimensional structures called biological nanowires that are made of biological molecules like DNA or proteins [60–64]. These nanowires play a crucial role in many biological systems, including viruses, muscles, and neurons. Actin filaments, which are long, thin protein fibers essential for muscle contraction and cell movement, are one type of biological nanowire. Microtubules are another illustration. These cylinder-shaped tubulin protein structures play a role in cell division and the movement of organelles inside cells. DNA nanowires are another form of biological nanowire that may be created in the lab by arranging

DNA molecules into certain configurations [65–70]. There are several uses for these DNA nanowires, including medication delivery and biosensing. The capacity of biological nanowires to self-assemble into complicated patterns, which may be challenging with synthetic materials, is one of its advantages. In addition, biological nanowires can have special qualities, such as certain proteins' capacity to conduct electricity, which makes them desirable for use in bioelectronics devices. However, the application of biological nanowires in nanotechnology is still in its infancy, and issues like repeatability and stability need to be resolved. However, scientists

Table 4.1 Comparative analysis of various types of nanowires.

Nanowire type	Composition	Properties	Applications
Semiconductor Nanowire	Inorganic semi-conducting material (e.g., Si, GaAs, InP)	High electron mobility, tunable bandgap, high surface-to-volume ratio	Transistors, solar cells, sensors, optoelectronics
Biological Nanowire	Protein-based (e.g., pili, flagella)	Biocompatible, self-assembling, conductive	Biomedical devices, biosensors, biocompatible electronics
Oxide Nanowire	Inorganic oxide material (e.g., ZnO, TiO_2)	High surface-to-volume ratio, piezoelectric properties, catalytic activity	Sensors, photocatalysts, energy storage devices
Hybrid Nanowire	Combination of different materials (e.g., semiconductor and metal)	Unique properties from the combination of materials	Optoelectronics, energy storage devices, sensors
Metal Nanowire	Pure metallic material (e.g., Ag, Au, Cu)	High conductivity, plasmonic properties	Electronics, sensors, optical devices

are looking into novel approaches to make use of the special qualities of biological nanowires for a range of uses in nano- and biotechnology. According to their compositions, characteristics, and uses, different varieties of nanowires are compared in Table 4.1 as well.

4.2 Properties of Nanowires

Nanowires have special electrical, optical, and mechanical properties because of their small size, high surface-to-volume ratio, and quantum confinement effects [71–75]. They are very promising for a variety of applications in electronics, photonics, energy conversion, as well as biological sensing and imaging due to these exceptional qualities. Researchers in many domains have focused heavily on the special characteristics of nanowires, and work is still being done to create novel techniques for creating and modifying these tiny structures as well as to investigate their potential uses in a variety of sectors [76–80].

4.2.1 Electrical Properties of Nanowire

Nanowires have unique electrical characteristics because of their small size, high surface area-to-volume ratio, and quantum confinement effects. Some of the important electrical properties of nanowires are:

Conductivity: Due to the tiny cross-sectional area of nanowires, which lowers the number of imperfections and impurities in the crystal lattice, they can display excellent electrical conductivity. They are advantageous for producing high-performance electrical interconnects and devices because of this feature.

Resistance: The diameter, length, and substance of a nanowire can all have an impact on its resistance. Due to the limitation of electron mobility within the wire, quantum confinement effects can result in a reduction in resistance for very thin nanowires. In contrast, a nanowire's resistance lowers as its diameter rises as a result of decreased electron scattering [81–83].

Carrier Mobility: High carrier mobility, which measures how quickly electrons or holes may travel through a material, can be found in nanowires. For the development of high-speed electronic components like transistors, this feature is crucial.

Bandgap: A nanowire's bandgap may be adjusted by altering its composition or diameter. This characteristic may be used to make electrical and photonic devices based on nanowires, such as solar cells and light-emitting diodes [84–87].

Schottky Barrier: An energy barrier called the Schottky barrier develops at the junction of a metal and a semiconductor. The Schottky barrier height in nanowires is influenced by the size and geometry of the wire, which affects how well nanowire-based electrical devices function.

Overall, nanowires are attractive prospects for a variety of applications in electronics, photonics, and energy conversion due to their distinctive electrical characteristics [88–91].

4.2.2 Mechanical Properties

Nanowires are tiny, have a high surface area-to-volume ratio, and have specific mechanical characteristics that set them apart from their bulk counterparts. Among nanowires' crucial mechanical characteristics are:

Strength: Due to the lack of flaws and the tiny size of their crystal lattice, nanowires can display tremendous strength. They help produce strong and light materials since they can tolerate severe stresses and strains [92–94].

Elasticity: Because of their small size and high aspect ratio, nanowires can exhibit a high degree of flexibility. They can be turned into flexible and elastic electrical devices and sensors because of this feature.

Plasticity: Due to their tiny size, nanowires are susceptible to plastic deformation under low pressures, which can result in the production of dislocations and flaws. For the creation of materials with high ductility and toughness, this feature may be helpful [95–99].

Fracture Toughness: The diameter, length, and substance of a nanowire can all have an impact on how difficult it is to fracture. In contrast to those formed of brittle materials, nanowires with smaller diameters and shorter lengths often have better fracture toughness.

Fatigue: Due to repeated stress cycles, nanowires may display fatigue behavior, which over time may result in mechanical failure. The fatigue resistance of nanowires can be improved, as some applications based on them may be constrained by this attribute.

In general, nanowires are attractive prospects for a variety of applications in materials science, nanotechnology, and engineering due to their distinctive mechanical characteristics. Nanowires must be carefully designed and used since flaws and surface effects can also have an impact on how they behave mechanically.

4.2.3 Optical Properties of Nanowire

Nanowires display unique optical properties due to their small size, high surface area-to-volume ratio, and quantum confinement effects. Among nanowires' significant optical characteristics are:

Absorption and Emission: Nanowires' configurable bandgap and quantum confinement characteristics allow them to absorb and emit light over a broad spectrum of wavelengths. They can be used to produce light-emitting diodes, lasers, and photodetectors thanks to this feature.
Plasmonics: Localized surface plasmon resonances, which produce stronger electromagnetic fields close to the nanowire's surface, can be supported by nanowires. Numerous applications in sensing, imaging, and energy conversion can make use of this characteristic [100–103].
Waveguiding: As light waveguides, nanowires may contain and direct light inside their structure. For making nanophotonic devices like nanolasers and nanosensors, this characteristic can be helpful.

4.2.4 Nonlinear Optical Properties

Because of their tiny size and strong optical intensity, nanowires can display nonlinear optical characteristics such as fourth-wave mixing and second-harmonic creation. Due to this characteristic, they may be used to create nonlinear optical devices like switches and frequency converters [104–106].

4.2.5 Photovoltaic Properties

Due to their extensive surface area and effective charge transfer, nanowires can demonstrate great photovoltaic efficiency. They can be used to build high-efficiency solar cells thanks to this characteristic.

Overall, the unique optical properties of nanowires make them promising candidates for a wide range of applications in photonics, energy conversion, sensing, and imaging. However, the optical properties of nanowires can also be affected by defects, surface states, and other environmental factors, which require careful consideration in their design and application.

4.3 Nanowire-FET

A nanowire serves as the channel connecting the source and drain electrodes in a Nanowire-FET (field-effect transistor). The nanowire is often formed of silicon or another semiconductor material, and its diameter is only a few nanometers or less. A gate electrode controls the flow of electrons across the nanowire, which is the foundation of a Nanowire-FET's functionality. The nanowire and the gate electrode are separated from one another by a thin insulating layer. When a voltage is applied to the gate electrode, the resulting electric field creates an electron channel in the nanowire, which subsequently in turn regulates the current flow across the source and drain terminals.

In comparison to traditional FETs, Nanowire-FETs have several potential benefits. For instance, they can be significantly smaller than traditional FETs, enabling higher device densities and better use of available space. They are appropriate for employing low-power devices since they can function at lower power levels. Furthermore, Nanowire-FETs may perform faster electrically in terms of quicker switching times and improved control over electron transport. However, there are further difficulties involved in creating NWFETs with high yield and repeatability, operating them, and integrating them with other components in intricate circuits. For instance, nanowires may be more prone to flaws and changes in their electrical characteristics due to their tiny size, which can make them challenging to produce. Additionally, because of their high surface area to volume ratio, nanowires may be more susceptible to surface effects, which might compromise their electrical efficiency [107–110].

4.4 Proposed Work (GaAs Nanowire-FET)

Simulation variables of GaAs Nanowire FET are given in Table 4.2. A GaAs nanowire-based field effect transistor (FET) is a particular kind of transistor in which the channel through which current travels is a nanowire constructed of gallium arsenide (GaAs). The source and drain electrodes are attached to each end of the nanowire, and a gate electrode is positioned close by. The FET is a three-terminal device. A GaAs Nanowire-FET works by using the electric field produced by the gate electrode to regulate the current passing across the nanowire. The characteristics of the nanowire and the interface between the nanowire and the gate dielectric are altered by creating an electric field by applying a voltage to the gate electrodes. As a result, the conductivity of the nanowire channel and, consequently, the current passing through it, are altered. GaAs is a common material for high-speed electronic devices because it is an III-V compound semiconductor with exceptional electron mobility. GaAs-based Nanowire-FETs provide several benefits over conventional planar FETs, including greater transconductance, reduced capacitance, and enhanced gate control. They are perfect for use in high-frequency applications like telecommunications and wireless communication because of these benefits. GaAs Nanowire-FETs are still a hot topic for research and development, and in the upcoming years, major advancements in device performance and dependability are anticipated. GaAs Nanowire-FET cross-section and schematic are shown in Figure 4.3.

At $V_{gs} = 0.3$ V, the electric field generated by the gate voltage is relatively weak. As a result, the current flow through the nanowire is also relatively weak. The graph will show a relatively flat electric field profile, indicating a

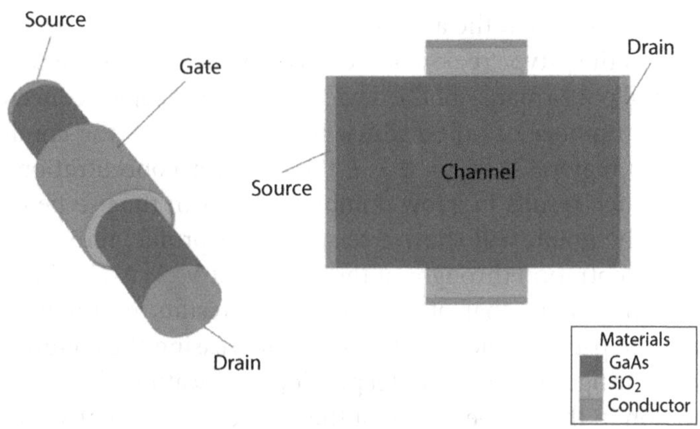

Figure 4.3 Cross-sectional view and 3-D view of GaAs nanowire-FET.

Table 4.2 Simulated Variables of GaAs NW-FET.

Variables	Values
L_g(Gate-Length)	50nm
L_s(Source-Length)	50nm
L_d(Drain-Length)	50nm
Work-Function(ϕ)	4.8eV
Dielectric Constant	3.9
Gate Doping	10^{20} cm^{-3}
Source Doping	10^{19} cm^{-3}

weak electric field throughout the channel region. At V_{gs} = 0.6 V, the electric field generated by the gate voltage is stronger than at V_{gs} = 0.3 V. This results in a stronger current flow through the nanowire. The Figure 4.4 will show a steeper slope in the electric field profile, indicating a stronger electric field in the channel region. At V_{gs} = 0.9 V, the electric field generated by the gate voltage is even stronger than at V_{gs} = 0.6 V. This results in an even stronger current flow through the nanowire. The graph will show an even steeper slope in the electric field profile, indicating a very strong electric field in the channel region. In general, as the gate-source voltage (V_{gs}) increases, the electric field in the channel region also increases, resulting in a stronger current flow through the nanowire. The electric field graph of a GaAs-based Nanowire-FET at different V_{gs} values reflects this behavior [111–113].

In a GaAs-based FET, the acceptor concentration determines the number of holes in the p-type region of the device, which affects the conductivity and overall performance of the device. A higher acceptor concentration increases the number of holes, which in turn increases the conductivity of the p-type region. At V_{gs} = 0.3 V, the acceptor concentration is relatively low, which results in a low number of holes in the p-type region of the device. The graph will show a relatively flat profile, indicating a low acceptor concentration throughout the device. At V_{gs} = 0.6 V, the acceptor concentration is higher than at V_{gs} = 0.3 V. This results in a higher number of holes in the p-type region of the device, increasing the conductivity of the region. The figure 4.4 show a steeper slope, indicating a higher acceptor concentration in the p-type region of the device. At V_{gs} = 0.9 V, the acceptor concentration is even higher than at V_{gs} = 0.6 V. This results in an even

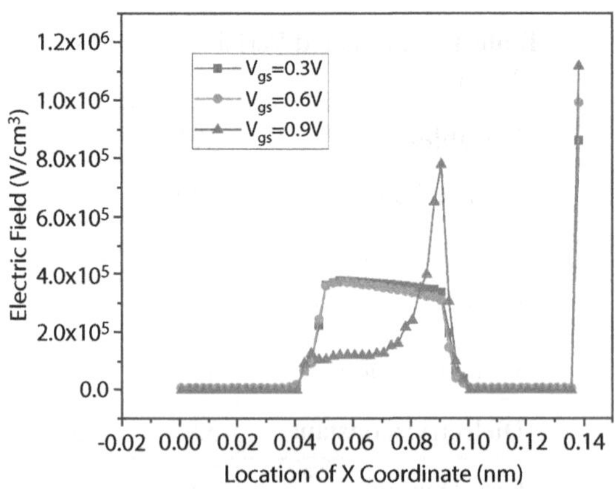

Figure 4.4 Electric representation of GaAs nanowire-FET at different V_{gs} = 0.3 V, 0.6 V, 0.9 V.

higher number of holes in the p-type region of the device, further increasing the conductivity of the region. The graph will show an even steeper slope, indicating a very high acceptor concentration in the p-type region of the device. V_{gs} increases, and the acceptor concentration in the p-type region of the device also increases, leading to a higher number of holes and increased conductivity. Figure 4.5 shows the acceptor concentrations graph at different $V_{gs.}$ The acceptor concentration graph of a GaAs-based Nanowire-FET at different V_{gs} values reflects this behavior.

In a GaAs-based FET, the gate voltage controls the flow of current between the source and drain terminals by creating an electric field that attracts or repels electrons from the source to the channel, allowing them to flow from the source to the drain. The strength of this electric field depends on the V_{gs}. Now let's consider the conduction current density graph of a GaAs-based Nanowire-FET at different V_{gs} values. At V_{gs} = 0.3 V, the electric field generated by the gate voltage is relatively weak, resulting in a weak flow of electrons through the channel. The graph will show a relatively flat profile, indicating a low conduction current density throughout the channel region. At V_{gs} = 0.6 V, the electric field generated by the gate voltage is stronger than at V_{gs} = 0.3 V. This results in a stronger flow of electrons through the channel. The graph will show a steeper slope, indicating a higher conduction current density in the channel region. At V_{gs} = 0.9 V, the electric field generated by the gate voltage is even stronger than at V_{gs} = 0.6 V. This results in an even stronger flow of electrons

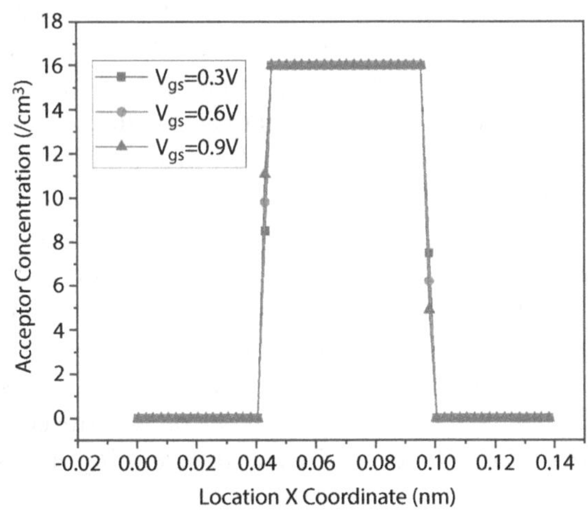

Figure 4.5 Acceptor concentration representation of GaAs nanowire-FET at different V_{gs} = 0.3 V, 0.6 V, 0.9 V.

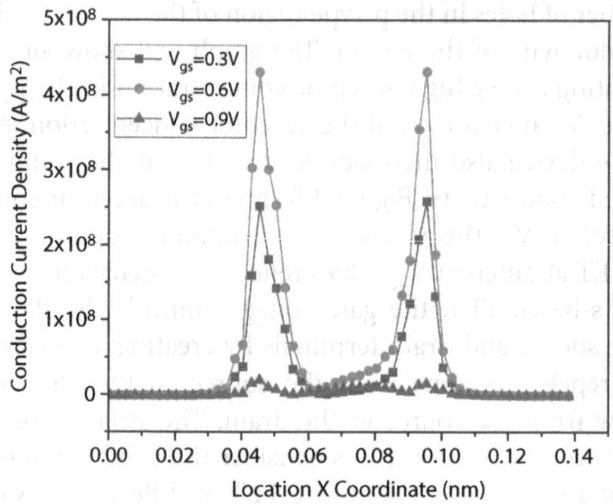

Figure 4.6 Conduction current density graph of GaAs nanowire-FET at different V_{gs} = 0.3 V, 0.6 V, 0.9 V.

through the channel. Figure 4.6 show an even steeper slope, indicating a very high conduction current density in the channel region. In general, as the V_{gs} increases, the electric field in the channel region increases, leading to a stronger flow of electrons and increased conduction current density. The conduction current density graph of a GaAs-based Nanowire-FET at different V_{gs} values reflects this behavior.

Now, when V_{gs} = 0.3 V, the gate voltage is relatively low, resulting in a weak electric field and a low electric potential in the channel region. The graph will show a relatively flat profile, indicating a low electric potential throughout the channel region. At V_{gs} = 0.6 V, the gate voltage is higher than at V_{gs} = 0.3 V, resulting in a stronger electric field and higher electric potential in the channel region. The graph will show a steeper slope, indicating a higher electric potential in the channel region. At V_{gs} = 0.9 V, the gate voltage is even higher than at V_{gs} = 0.6 V, resulting in an even stronger electric field and an even higher electric potential in the channel region. Figure 4.7 will show an even steeper slope, indicating a very high electric potential in the channel region. In general, as the gate-source voltage (V_{gs}) increases, the electric potential in the channel region increases, reflecting the stronger electric field created by the gate voltage. The electric potential graph of a GaAs-based Nanowire-FET at different V_{gs} values reflects this behavior.

When V_{gs} = 0.3 V, the energy band diagram of the GaAs FET shows a small depletion region between the channel and the source. The depletion

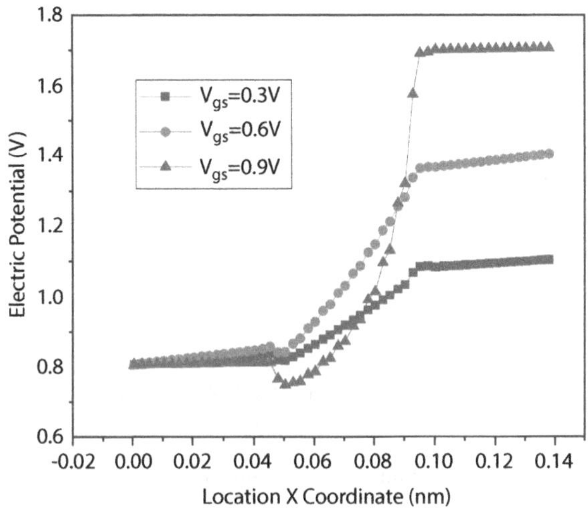

Figure 4.7 Electric potential graph of GaAs nanowire-FET at different V_{gs} = 0.3 V, 0.6 V, 0.9 V.

region occurs due to the potential difference between the gate and the channel. As a result, the gate controls the flow of electrons through the channel. When the gate voltage is small, only a small number of electrons can pass through the channel, resulting in low current flow. At V_{gs} = 0.6 V, the energy band diagram of the GaAs FET shows a larger depletion region between the source and the channel. The gate voltage is higher than before, resulting in a larger potential difference and a larger depletion region. This causes the channel to become narrower, reducing the number of electrons that can flow through it. As a result, the current flow through the channel decreases. At V_{gs} = 0.9 V, the energy band diagram of the GaAs FET shows an even larger depletion region between the source and the channel. The gate voltage is high enough to create a fully depleted channel. Figure 4.8 shows the energy band diagram of GaAs FET in off state and Figure 4.9 shows the on state. This means that there are no free electrons in the channel, and no current can flow through it. The device is effectively switched off at this voltage. Furthermore, as the gate voltage increases, the potential difference between the gate and the channel increases, creating a larger depletion region and reducing the number of free electrons in the channel. This reduces the current flow through the device, effectively turning it off. When the gate voltage is increased to a sufficiently high value, the depletion region becomes large enough to fully deplete the channel, turning the device off completely. Table 4.3 presents the parameters extraction of GaAs nanowire-FET.

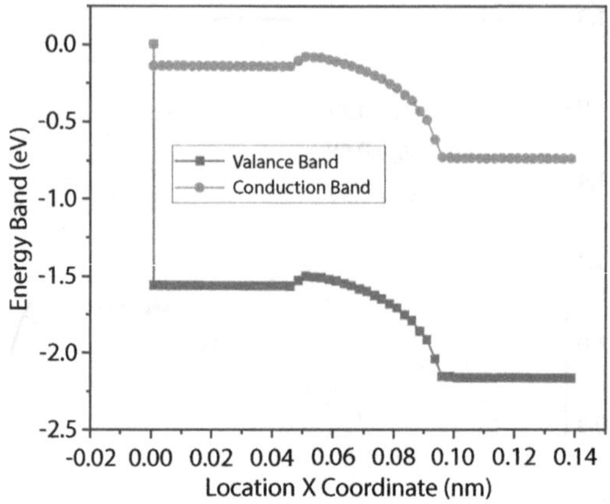

Figure 4.8 The energy band diagram of GaAs nanowire-FET in OFF state.

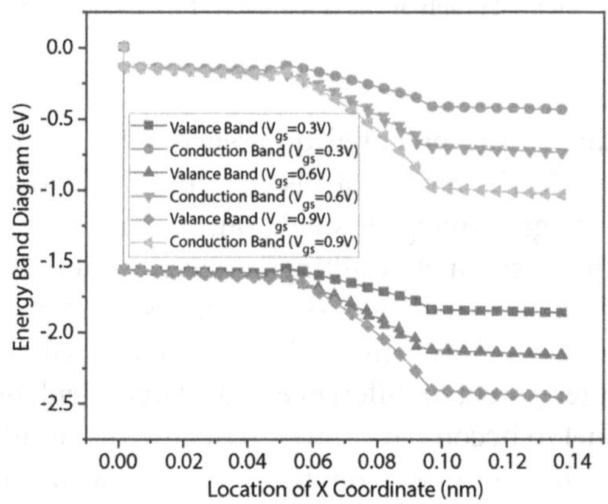

Figure 4.9 The energy band diagram of GaAs nanowire-FET in ON state.

Table 4.3 Parameters Extraction of GaAs Nanowire-FET.

Parameters	Unit	Si-Nanowire-FET [Ref.10]	Ge-Nanowire-FET [Ref.10]	GaAs Nanowire-FET (Proposed Work)
Subthreshold Swing	mV/dec	74.50	64.38	49.09
I_{ON}	A/μm	2.25×10^{-4}	1.73×10^{-5}	1.04×10^{-5}
I_{OFF}	A/μm	5.71×10^{-12}	9.40×10^{-13}	1.16×10^{-12}
I_{ON}/I_{OFF}	-	3.94×10^{7}	1.84×10^{7}	8.9×10^{6}
Transconductance	μS	9.8	9.8	9.8

4.5 Conclusion

Due to its high carrier mobility, low power consumption, and scalability, the GaAs Nanowire-FET is a promising candidate for future electronic devices. Since the nanowire has a singular one-dimensional structure, it has better electrostatic control, which results in better gate control, higher on/off ratios, and lower subthreshold swing. However, further research is still needed to optimize the fabrication process and improve device performance. With continued efforts and advancements in the field, the GaAs Nanowire-FET holds great potential for next-generation electronics. The simulation process will be done on the TCAD tool.

References

1. Chin CY, Bo-Yi L, Meng CH (2008) Scaling Study of Nanowire and Multi-Gate MOSFETs. *IEEE* 978-1-4244-2186-2
2. S.W. Chung, J.-Y. Yu, and J. R. Heath, "Silicon nanowire devices," *Appl. Phys. Lett.*, vol. 76, no. 15, pp. 2068–2070, Apr. 2000.
3. J.P. Colinge and S.H. Dhong , "Prospective for nanowire transistors", *IEEE Custom Integrated Circuits Conference (CICC)*, Page(s): 1 – 8, 2013.
4. Xiangchen Chen and Cher Ming Tan, "Modeling and analysis of gateall-around silicon Nanowire-FET", in *Microelectron. Reliab.*, vol. 54, pp.1103-1108, July 2013.
5. Singh Mandeep, Chaudhary Tarun, Raj Balwinder "Design and Simulation Analysis of Silicon Germanium Nanowire-FET for Low Power Applications" *Silicon* 15:1491–1497, (2023) https://doi.org/10.1007/s12633-022-02101-y
6. Anil K. Bansal and Abhisek Dixit, "Advances in logic device scaling," *IETE Tech. Rev.*, April 2015.
7. M. M. Frank, "High-k/Metal gate innovations enabling continued CMOS scaling", *ESSDERC*, pp. 50-58, 2011.
8. A.Bhardwaj, P.Kumar, B.Raj, and S.Anand, "Design and Performance Optimization of Doping-Less Vertical Nanowire TFET Using Gate Stack Technique," *J. Electron. Mater. (JEMS)*, Springer, Vol. 41, No. 7, pp. 4005–4013, 2022.
9. J.Singh, and B.Raj, "Tunnel Current Model of Asymmetric MIM Structure Levying Various Image Forces to Analyze the Characteristics of Filamentary Memristor", *Appl. Phys. A*, Springer, Vol. 125, No. 3, pp. 203.1 to 203.11, 2019.
10. C.Goyal, J. S.Ubhi, and B.Raj, "Low Leakage Zero Ground Noise Nanoscale Full Adder Using Source Biasing Technique", *J. Nanoelectron. Optoelectron., American Scientific Publishers*, Vol. 14, pp. 360–370, 2019.
11. G.Singh, R. K.Sarin, and B.Raj, "A Novel Robust Exclusive-OR Function Implementation in QCA Nanotechnology with Energy Dissipation Analysis", *J. Comput. Electron.*, Springer, Vol. 15,No. 2, pp. 455–465, 2016.

12. T.Wadhera, D.Kakkar, G.Wadhwa, and B.Raj, "Recent Advances and Progress in Development of the Field Effect Transistor Biosensor: A Review", *J. Electron. Mater.*, Springer, Vol. 48, No. 12, pp. 7635–7646, 2019.
13. G.Wadhwa, P.Kamboj, and B.Raj, "Design Optimisation of Junctionless TFET Biosensor for High Sensitivity", *Adv. Nat. Sci.: Nanosci. Nanotechnol.*, Vol. 10, No. 7, p. 045001, 2019.
14. P.Bansal, and B.Raj, "Memristor Modeling and Analysis for Linear Dopant Drift Kinetics", *J. Nanoeng. Nanomanuf.*, American Scientific Publishers, Vol. 6, pp. 1–7, 2016.
15. A.Singh, M.Khosla, and B.Raj, "Circuit Compatible Model for Electrostatic Doped Schottky Barrier CNTFET," *J. Electron. Mater.*, Springer, Vol. 45, No. 12, pp. 4825–4835, 2016.
16. Ashima, D.Vaithiyanathan, and B.Raj, "Performance Analysis of Charge Plasma induced Graded Channel Si Nanotube", *J. Eng. Res. (JER)*, EMSME Special Issue, Vol.9, pp. 146–154, 2021.
17. S.Tomar, V. K.Magraiya, and B.Raj, "Scaling of Access and Data Transistor for High Performance DRAM Cell Design", *Quantum Matter*, Vol. 2, pp. 412–416, 2013.
18. N.Jain, and B.Raj, "Parasitic Capacitance and Resistance Model Development and Optimization of Raised Source/Drain SOI FinFET Structure for Analog Circuit Applications", *J. Nanoelectron. Optoelectron.*, ASP, USA, Vol. 13, pp. 531–539, 2018.
19. S.Singh, S. K.Vishvakarma, and B.Raj, "Analytical Modeling of Split-Gate Junction-Less Transistor for a Biosensor Application", *Sens. Bio-Sens.*, Elsevier, Vol. 18, pp. 31–36, 2018.
20. M.Gopal, and B.Raj, "Low Power 8T SRAM Cell Design for High Stability Video Applications", *ITSI Transaction on Electrical and Electronics Engineering*, Vol. 1, No. 5, pp. 91–97, 2013.
21. B.Raj, J.Mitra, D. K.Bihani, V.Rangharajan, A. K. Saxena, and S. Dasgupta, "Analysis of Noise Margin, Power and Process Variation for 32 nm FinFET Based 6T SRAM Cell", *J. Comput. (JCP)*, Academy Publisher, FINLAND, Vol. 5, No. 6, PP.1-6, 2010.
22. D.Sharma, R.Mehra, and B.Raj, "Comparative Analysis of Photovoltaic Technologies for High Efficiency Solar Cell Design", *Superlattices Microstruct.*, Elsevier, Vol. 153, pp. 106861, 2021.
23. P.Kaur, A. S.Buttar, and B.Raj, "A Comprehensive Analysis of Nanoscale Transistor Based Biosensor: A Review", *Indian J. Pure Appl. Phys.*, Vol. 59, pp. 304–318, 2021.
24. D.Yadav, B.Raj, and B.Raj, "Design and Simulation of Low Power Microcontroller for IoT Applications", *J. Sens. Lett.*, ASP USA,Vol. 18, pp. 401–409, 2020.
25. S.Singh, and B.Raj, "A 2-D Analytical Surface Potential and Drain Current Modeling of Double-Gate Vertical t-Shaped Tunnel FET", *J. Comput. Electron.*, Springer, Vol. 19, pp. 1154–1163, 2020.

26. J.Singh, and B.Raj, "An Accurate and Generic Window function for Nonlinear Memristor Model" *J. Comput. Electron.*, Springer, Vol. 18, No. 2, pp. 640–647, 2019.
27. M.Kaur, N.Gupta, S.Kumar, B.Raj, and A. K.Singh, "Comparative RF and Crosstalk Analysis of Carbon Based Nano Interconnects", *IET Circuits, Devices Syst.*, Vol. 15, No. 6, pp. 493–503, 2021.
28. N.Kandasamy, F.Ahmad, D.Ajitha, B.Raj, and N.Telagam, "Quantum Dot Cellular Automata Based Scan Flip Flop and Boundary Scan Register", *IETE J. Res.*, Vol. 69, P P. 535-548, 2023.
29. S. K.Sharma, B.Raj, and M.Khosla, "Enhanced Photosensitivity of Highly Spectrum Selective Cylindrical Gate In1-xGaxAs Nanowire MOSFET Photodetector", *Mod. Phys. Lett. B*, Vol. 33, No. 12, p. 1950144, 2019.
30. J.Singh, and B.Raj, "Design and Investigation of 7T2M NVSARM with Enhanced Stability and Temperature Impact on Store/Restore Energy", *IEEE Trans. Very Large Scale Integr. Syst.*, Vol. 27, No. 6, pp. 1322–1328, 2019.
31. A. K.Bhardwaj, S.Gupta, B.Raj, and A.Singh,"Impact of Double Gate Geometry on the Performance of Carbon Nanotube Field Effect Transistor Structures for Low Power Digital Design", *J. Comput. Theor. Nanosci.*, ASP USA, Vol. 16, pp. 1813–1820, 2019.
32. N.Jain, and B.Raj, "Thermal Stability Analysis and Performance Exploration of Asymmetrical Dual-k Underlap Spacer (ADKUS) SOI FinFET for Security and Privacy Applications", *Indian J. Pure Appl. Phys. (IJPAP)*, Vol. 57, pp. 352–360, 2019.
33. A.Singh, M.Khosla, and B.Raj, "Design and Analysis of Dynamically Configurable Electrostatic Doped Carbon Nanotube Tunnel FET", *Microelectronics J.*, Elsevier, Vol. 85, pp. 17–24, 2019.
34. N.Jain, and B.Raj, "Dual-k Spacer Region Variation at the Drain Side of Asymmetric SOI FinFET Structure: Performance Analysis towards the Analog/RF Design Applications", *J. Nanoelectron. Optoelectron.*, American Scientific Publishers, Vol. 14, pp. 349–359, 2019.
35. J.Singh, S.Sharma, B.Raj, and M.Khosla, "Analysis of Barrier Layer Thickness on Performance of In1-xGaxAs Based Gate Stack Cylindrical Gate Nanowire MOSFET", *JNO*, ASP USA, Vol. 13, pp. 1473–1477, 2018.
36. N.Jain, and B.Raj, "Analysis and Performance Exploration of High-k SOI FinFETs over the Conventional Low-k SOI FinFET toward Analog/RF Design", *J. Semicond. (JoS)*, IOP Science,Vol. 39, No. 12, pp. 124002-1-7, 2018.
37. C.Goyal, J. S.Ubhi, and B.Raj, "A Reliable Leakage Reduction Technique for Approximate Full Adder with Reduced Ground Bounce Noise", *J. Math. Probl. Eng.*, Hindawi, Vol. 2018, Article ID 3501041, PP.1-16, 2018.
38. Anuradha, J. Singh, B.Raj, and M.Khosla, "Design and Performance Analysis of Nano-Scale Memristor-Based Nonvolatile SRAM", *J. Sens. Lett.*, American Scientific Publishers USA, Vol. 16, pp. 798–805, 2018.

39. G.Wadhwa, and B.Raj, "Parametric Variation Analysis of Charge-Plasma-Based Dielectric Modulated JLTFET for Biosensor Application", *IEEE Sens. J.*, Vol. 18, No. 15, PP. 6070 – 6077, 2018.
40. J.Singh, and B.Raj, "Comparative Analysis of Memristor Models for Memories Design", *JoS*, IoP, Vol. 39, No. 7, PP. 074006-1-12, 2018.
41. D.Yadav, S. S.Chouhan, S. K.Vishvakarma, and B.Raj, "Application Specific Microcontroller Design for IoT Based WSN", *Sens. Lett.*, ASP, Vol. 16, pp. 374–385, 2018.
42. G.Singh, R. K.Sarin, and B.Raj, "Fault-Tolerant Design and Analysis of Quantum-Dot Cellular Automata Based Circuits", *IEEE/IET Circuits, Devices Syst.*, Vol. 12, pp. 638–664, 2018.
43. J.Singh, and B.Raj, "Modeling of Mean Barrier Height Levying Various Image Forces of Metal Insulator Metal Structure to Enhance the Performance of Conductive Filament Based Memristor Model", *IEEE Nanotechnol.*, Vol. 17, No. 2, pp. 268–275, 2018.
44. A.Jain, S.Sharma, and B.Raj, "Analysis of Triple Metal Surrounding Gate (TM-SG) III-V Nanowire MOSFET for Photosensing Application", *Opto-Electron. J.*, Elsevier, Vol. 26, No. 2, P P. 141–148, 2018.
45. A.Jain, S.Sharma, and B.Raj, "Design and Analysis of High Sensitivity Photosensor Using Cylindrical Surrounding Gate MOSFET for Low Power Sensor Applications", *Eng. Sci. Technol. Int. J.*, Elsevier's, Vol. 19, No. 4, pp. 1864–1870, 2016.
46. A.Singh, M.Khosla, and B.Raj, "Analysis of Electrostatic Doped Schottky Barrier Carbon Nanotube FET for Low Power Applications," *J. Mater. Sci.: Mater. Electron.*, Springer, Vol. 28, P P. 1762–1768, 2017.
47. G.Saiphani Kumar, A.Singh, B.Raj, "Design and Analysis of Gate All around CNTFET Based SRAM Cell Design", *J. Comput. Electron.*, Springer, Vol. 17, No.1, P P. 138–145, 2018.
48. G. P.Singh, B. S.Sohi, and B.Raj, "Material Properties Analysis of Graphene Base Transistor (GBT) for VLSI Analog Circuits", *Indian J. Pure Appl. Phys. (IJPAP)*, Vol. 55, pp. 896–902, 2017.
49. S.Kumar and B.Raj, "Estimation of Stability and Performance Metric for Inward Access Transistor Based 6T SRAM Cell Design Using n-type/p-type DMDG-GDOV TFET", *IEEE VLSI Circuits Syst. Lett.*, Vol. 3, No. 2, PP. 25–39, 2017.
50. S.Sharma, A.Kumar, M.Pattanaik, and B.Raj, "Forward Body Biased Multimode Multi-Treshold CMOS Technique for Ground Bounce Noise Reduction in Static CMOS Adders", *Int. J. Inf. Electron. Eng.*, Vol. 3, No. 3, PP. 567–572, 2013.
51. H. Singh, P. Kumar, and B. Raj, "Performance Analysis of Majority Gate SET Based 1-bit Full Adder", *International J. Comput. Commun. Eng. (IJCCE)*, IACSIT Press Singapore, ISSN: 2010-3743, Vol. 2, No. 4, PP.1-8, 2013.
52. AK. Bhardwaj, S. Gupta, and B. Raj, "Investigation of Parameters for Schottky Barrier (SB) Height for Schottky Barrier Based Carbon Nanotube Field Effect

Transistor Device", *J. Nanoelectron. Optoelectron.*, ASP, Vol. 15, P P. 783–791, 2020.

53. P. Bansal, and B. Raj, "Memristor: A Versatile Nonlinear Model for Dopant Drift and Boundary Issues", *JCTN*, American Scientific Publishers, Vol. 14, No. 5, P P. 2319–2325, 2017.
54. N. Jain, and B. Raj, "An Analog and Digital Design Perspective Comprehensive Approach on Fin-FET (Fin-Field Effect Transistor) Technology - A Review", *Reviews in Advanced Sciences and Engineering (RASE)*, ASP, Vol. 5, PP. 1–14, 2016.
55. S.Sharma, B.Raj, and M.Khosla, "Subthreshold Performance of In1-xGaxAs Based Dual Metal with Gate Stack Cylindrical/Surrounding Gate Nanowire MOSFET for Low Power Analog Applications", *J. Nanoelectron. Optoelectron.*, American Scientific Publishers, USA, Vol. 12, P P. 171– 176, 2017.
56. B.Raj, A. K.Saxena, and S.Dasgupta, "Analytical Modeling for the Estimation of Leakage Current and Subthreshold Swing Factor of Nanoscale Double Gate FinFET Device" *Microelectron. Int.*, UK, Vol. 26, PP. 53–63, 2009.
57. S. S.Soniya, G.Wadhwa, and B.Raj, "An Analytical Modeling for Dual Source Vertical Tunnel Field Effect Transistor", *Int. J. Recent Technol. Eng, (IJRTE)*, Vol. 8, No. 2, 2019.
58. S.Singh, and B.Raj, "Design and Analysis of Hetrojunction Vertical T-Shaped Tunnel Field Effect Transistor", *J. Electron. Mater.*, Springer, Vol. 48, No. 10, PP. 6253–6260, 2019.
59. C.Goyal, J. S.Ubhi, and B.Raj, "A Low Leakage CNTFET Based Inexact Full Adder for Low Power Image Processing Applications", *Int. J. Circuit Theory Appl.*, Wiley, Vol. 47, No. 9, PP. 1446–1458, 2019.
60. B.Raj, A. K.Saxena, and S.Dasgupta, "A Compact Drain Current and Threshold Voltage Quantum Mechanical Analytical Modeling for FinFETs", *J. Nanoelectron. Optoelectron. (JNO)*, USA, Vol. 3, No. 2, PP. 163–170, 2008.
61. G.Wadhwa, and B.Raj, "An Analytical Modeling of Charge Plasma Based Tunnel Field Effect Transistor with Impacts of Gate Underlap Region", *Superlattices Microstruct.*, Elsevier, Vol. 142, PP. 106512-22, 2020.
62. S.Singh, and B.Raj, "Modeling and Simulation Analysis of SiGe Heterojunction Double GateVertical t-Shaped Tunnel FET", *Superlattices Microstruct., Elsevier*, Vol. 142, P P. 106496-104, 2020.
63. A.Singh, D. K.Saini, D.Agarwal, S.Aggarwal, M.Khosla, and B.Raj, "Modeling and Simulation of Carbon Nanotube Field Effect Transistor and Its Circuit Application", *J. Semicond. (JoS)*, IOP Science, Vol. 37, PP. 074001–074006, 2016.
64. N.Jain, and B.Raj, "Device and Circuit Co-Design Perspective Comprehensive Approach on FinFET Technology - A Review", *J. Electron Devices*, Vol. 23, No. 1, PP. 1890–1901, 2016.
65. S.Kumar and B.Raj, "Analysis of ION and Ambipolar Current for Dual-Material Gate-Drain Overlapped DG-TFET", *J. Nanoelectron. Optoelectron.*, American Scientific Publishers, USA, Vol. 11, P P. 323– 333, 2016.

66. N.Anjum, T.Bali, and B.Raj, "Design and Simulation of Handwritten Multiscript Character Recognition", *Int. J. Adv. Res. Comput. Commun. Eng.*, Vol. 2, No. 7, PP. 2544–2549, 2013.
67. S.Sharma, B.Raj, and M.Khosla, "A Gaussian Approach for Analytical Subthreshold Current Model of Cylindrical Nanowire FET with Quantum Mechanical Effects", *Microelectronics J.*, Elsevier, Vol. 53, PP. 65–72, 2016.
68. K.Singh, and B.Raj, "Performance and Analysis of Temperature Dependent Multi-Walled Carbon Nanotubes as Global Interconnects at Different Technology Nodes", *J. Comput. Electron.*, Springer, Vol. 14, No. 2, PP. 469–476, 2015.
69. S.Kumar and B.Raj, "Compact Channel Potential Analytical Modeling of DG-TFET Based on Evanescent–Mode Approach", *J. Comput. Electron.*, Springer, Vol. 14, No. 2, P P. 820–827, 2015.
70. K.Singh, and B.Raj, "Temperature Dependent Modeling and Performance Evaluation of Multi-Walled CNT and Single-Walled CNT as Global Interconnects", *J. Electron. Mater.*, Springer, Vol. 44, No. 12, PP. 4825–4835, 2015.
71. V. K. Sharma, M. Pattanaik, and B. Raj, "INDEP Approach for Leakage Reduction in Nanoscale CMOS Circuits", *Int. J. Electron.*, Taylor & Francis, Vol. 102, No. 2, P P. 200–215, 2014.
72. K.Singh, and B.Raj, "Influence of Temperature on MWCNT Bundle, SWCNT Bundle and Copper Interconnects for Nanoscaled Technology Nodes", *J. Mater. Sci.: Mater. Electron.*, Springer, Vol. 26, No. 8, PP. 6134–6142, 2015.
73. N.Anjum, T.Bali, and B.Raj, "Design and Simulation of Handwritten Gurumukhi and Devanagri Numerical Recognition", *Int. J. Comput. Appl.*, Published by Foundation of Computer Science, New York, USA,Vol. 73, No. 12, PP. 16–21, 2013.
74. S. K.Verma, S.Singh, G.Wadhwa and B.Raj, "Detection of Biomolecules Using Charge-Plasma Based Gate Underlap Dielectric Modulated Dopingless TFET", *Trans. Electr. Electron. Mater. (TEEM)*, Springer, Vol. 21, PP. 528–535, 2020.
75. N.Jain, and B.Raj, "Impact of Underlap Spacer Region Variation on Electrostatic and Analog/RF Performance of Symmetrical High-k SOI FinFET at 20 nm Channel Length", *J. Semicond. (JoS)*, IOP Science, Vol. 38, No. 12, PP. 122002-10, 2017.
76. S.Singh, and B.Raj, "Analytical Modeling and Simulation Analysis of T-Shaped III-V Heterojunction Vertical T-FET", *Superlattices Microstruct.*, Elsevier, Vol. 147, PP. 106717-25, 2020.
77. G.Singh, R. K.Sarin, and B.Raj, "Design and Analysis of Area Efficient QCA Based Reversible Logic Gates", *J. Microprocess. Microsyst.*, Elsevier, Vol. 52, PP. 59–68, 2017.
78. A.Singh, M.Khosla, and B.Raj, "Compact Model for Ballistic Single Wall CNTFET under Quantum Capacitance Limit", *J. Semicond. (JoS)*, IOP Science,Vol. 37, P P. 104001–104008, 2016.

79. S.Khandelwal, V.Gupta, B.Raj, and R. D.Gupta, "Process Variability Aware Low Leakage Reliable Nano Scale DG-FinFET SRAM Cell Design Technique", *J. Nanoelectron. Optoelectron.*, Vol. 10, No. 6, P P. 810–817, 2015.
80. V. K. Sharma, M. Pattanaik, and B. Raj, "ONOFIC Approach: Low Power High Speed Nanoscale VLSI Circuits Design", *Int. J. Electron.*, Taylor & Francis, Vol. 101, No. 1, PP. 61–73, 2014.
81. S. Khandelwal, B. Raj, and R. D. Gupta, "FinFET Based 6T SRAM Cell Design: Analysis of Performance Metric, Process Variation and Temperature Effect", *J. Comput. Theor. Nanosci.*, ASP, USA, Vol. 12, PP. 2500–2506, 2015.
82. S. Singh, Y. Shekhar, R. Jagdeep, S. Anurag, and B. Raj, "Impact of HfO2 in Graded Channel Dual Insulator Double Gate MOSFET", *J. Comput. Theor. Nanosci.*, American Scientific Publishers, Vol. 12, No. 6, PP. 950–953, 2015.
83. V. K.Sharma, M.Pattanaik, and B.Raj, "PVT Variations Aware Low Leakage INDEP Approach for Nanoscale CMOS Circuits", *Microelectron. Reliab.*, Elsevier, Vol. 54, PP. 90–99, 2014.
84. B. Raj, A. K. Saxena and S. Dasgupta, "Quantum Mechanical Analytical Modeling of Nanoscale DG FinFET: Evaluation of Potential, Threshold Voltage and Source/Drain Resistance", *Elsevier's J. Mater. Sci. Semicond. Process.*, Elsevier, Vol. 16, No. 4, P P. 1131–1137, 2013.
85. M. Gopal, S. S. D. Prasad, and B. Raj, "8T SRAM Cell Design for Dynamic and Leakage Power Reduction", *Int.l J. Comput. Appl.*, Published by Foundation of Computer Science, New York, USA,Vol. 71, No. 9, PP. 43–48, 2013.
86. M. Pattanaik, B. Raj, S. Sharma, and A. Kumar, "Diode Based Trimode Multi-Threshold CMOS Technique for Ground Bounce Noise Reduction in Static CMOS Adders", *Adv. Mat. Res., Trans Tech Publications*, Switzerland, Vol. 548, P P. 885–889, 2012.
87. B.Raj, A. K.Saxena, and S.Dasgupta, "Nanoscale FinFET Based SRAM Cell Design: Analysis of Performance Metric, Process Variation, Underlapped FinFET and Temperature Effect", *IEEE Circuits Syst. Mag.*, Vol. 11, No. 2, PP. 38–50, 2011.
88. V. K. Sharma, M. Pattanaik, and B.Raj, "Leakage Current ONOFIC Approach for Deep Submicron VLSI Circuit Design", International Journal of Electrical, Computer, Electronics and Communication Engineering, World Academy of Sciences, Engineering and Technology, Vol. 7, No. 4, PP. 239–244, 2013.
89. T.Chawla, M.Khosla, and B.Raj, "Design and Simulation of Triple Metal Double-Gate Germanium on Insulator Vertical Tunnel Field Effect Transistor", *Microelectronics J.*, Elsevier, Vol. 114, PP. 105125-32, 2021.
90. P. Kaur, S. S. Gill, and B. Raj, "Comparative Analysis of OFETs Materials and Devices for Sensor Applications", *J. Silicon*, Springer, Vol. 14, PP. 4463–4471, 2022.
91. S. K.Sharma, P.Kumar, B.Raj, and Balwinder Raj, "In1-xGaxAs Double Metal Gate-Stacking Cylindrical Nanowire MOSFET for Highly Sensitive Photo Detector", *J. Silicon*, Springer, Vol. 14, P P. 3535–3541,2022.

92. B.Raj, A. K.Saxena, and S.Dasgupta, "Analytical Modeling of Quasi Planar Nanoscale Double Gate FinFET with Source/Drain Resistance and Field Dependent Carrier Mobility: A Quantum Mechanical Study", *J. Comput. (JCP)*, Academy Publisher, FINLAND, Vol. 4, No. 9, PP.1-8, 2009.
93. S.Bhushan, S.Khandelwal, and B.Raj, "Analyzing Different Mode FinFET Based Memory Cell at Different Power Supply for Leakage Reduction", Seventh International Conference on Bio-Inspired Computing: Theories and Application, (BIC-TA 2012), organized IIITM Gwalor, PP. 89–100, 14-16 Dec 2012.
94. J.Singh, and B.Raj, "Temperature Dependent Analytical Modeling and Simulations of Nanoscale Memristor", *Eng. Sci. Technol. Int. J.*, Elsevier's, Vol. 21, PP. 862–868, 2018.
95. S.Singh, S.Bala, B.Raj, and Balwinder Raj, "Improved Sensitivity of Dielectric Modulated Junctionless Transistor for Nanoscale Biosensor Design", *Sens. Lett.*, ASP, Vol. 18, PP. 328–333, 2020.
96. V.Kumar, S. K.Vishvakarma, and B.Raj, "Design and Performance Analysis of ASIC for IoT Applications", *Sens. Lett.*, ASP, Vol. 18, P P. 31–38, 2020.
97. A.Jaiswal, R. K.Sarin, B.Raj, and S.Sukhija, "A Novel Circular Slotted Microstrip-fed Patch Antenna with Three Triangle Shape DEffected Ground Structure for Multiband Applications", *Adv. Electromagn. (AEM)*, Vol. 7, No. 3, PP. 56–63, 2018.
98. G.Wadhwa, and BRaj, "Label Free Detection of Biomolecules Using Charge-Plasma-Based Gate Underlap Dielectric Modulated Junctionless TFET", *J. Electron. Mater. (JEMS)*, Springer, Vol. 47, No. 8, PP. 4683– 4693, 2018.
99. G.Singh, R. K.Sarin, and B.Raj, "Design and Performance Analysis of a New Efcient Coplanar Quantum-Dot Cellular Automata Adder", *Indian J. Pure Appl. Phys. (IJPAP)*, Vol. 55, P P. 97–103, 2017.
100. A.Singh, M.Khosla, and B.Raj, "Design and Analysis of Electrostatic Doped Schottky Barrier CNTFET Based Low Power SRAM", *Int. J. Electron. Commun.*, (AEÜ), Elsevier,Vol. 80, P P. 67–72, 2017.
101. P.Kaur, V.Pandey, and B.Raj, "Comparative Study of Efcient Design, Control and Monitoring of Solar Power Using IoT", *Sens. Lett.*, ASP, Vol. 18, P P. 419–426, 2020.
102. G.Wadhwa, P.Kamboj, J.Singh, B.Raj, "Design and Investigation of Junctionless DGTFET for Biological Molecule Recognition", *Trans. Electr. Electron. Mater.*, Springer, Vol. 22, PP. 282–289, 2020.
103. T.Chawla, M.Khosla, and B.Raj, "Optimization of Double-Gate Dual Material GeOI-Vertical TFET for VLSI Circuit Design", *IEEE VLSI Circuits Syst. Lett.*, Vol. 6, No. 2, PP. 13–25, 2020.
104. S.Singh, M.Khosla, G.Wadhwa, and B.Raj, "Design and Analysis of Double-Gate Junctionless Vertical TFET for Gas Sensing Applications", *Appl. Phys. A*, Springer,Vol. 127, No. 16, PP.883-890, 2021.

105. I.Singh, B.Raj, M.Khosla, and B. K.Kaushik, "Potential MRAM Technologies for Low Power SoCs", SPIN World Scientific Publisher, SCIE; Vol. 10, No. 04, PP. 2050027-35, 2020.
106. S.Singh, and B.Raj, "Parametric Variation Analysis on Hetero-Junction Vertical t-Shape TFET for Suppressing Ambipolar Conduction", *Indian J. Pure Appl. Phys.*, Vol. 58, P P. 478–485, 2020.
107. S. S.Soniya, G.Wadhwa, and B.Raj, "Design and Analysis of Dual Source Vertical Tunnel Field Effect Transistor for High Performance", *Trans. Electr. El ectron. Mater.*, Springer, Vol. 21, P P. 74–82, 2019.
108. M.Kaur, N.Gupta, S.Kumar, B.Raj, and A. K.Singh, "RF Performance Analysis of Intercalated Graphene Nanoribbon Based Global Level Interconnects", *J. Comput. Electron.*, Springer, Vol. 19, PP.1002–1013, 2020.
109. G.Wadhwa, B.Raj, "Design and Performance Analysis of Junctionless TFET Biosensor for High Sensitivity" *IEEE Nanotechnol.*, Vol.18, PP. 567–574, 2019.
110. J.Singh, and B.Raj, "Enhanced Nonlinear Memristor Model Encapsulating Stochastic Dopant Drift", *JNO*, ASP, Vol.14, PP. 958–963, 2019.
111. N. Singh, A. Agarwal, L. K. Bera, T. Y. Liow, R. Yang, S. C. Rustagi, C. H. Tung, R. Kumar, G. Q. Lo, N. Balasubramanian, and D.-L. Kwong, "High-performance fully depleted silicon nanowire (diameter 5 nm) gate-all-around CMOS devices," *IEEE Electron Device Lett.*, vol. 27, no. 5, pp. 383–385, May 2006.
112. J. Guo, J.Wang, E. Polizzi, S. Datta, andM. Lundstrom, "Electrostatics of nanowire transistors," *IEEE Trans. Nanotechnol.*, vol. 2, no. 4, pp. 329-334, Dec. 2003.
113. Dunwei Wang, Qian Wang, Ali Javey, Ryan Tu, Hongjie Dai, Hyoungsub Kim, Paul C. McIntyre, Tejasrishnamohan and Krishna C. Saraswat, "Germanium nanowire field-effect transistors with SiO2 and high-k HfO2 gate dielectrics", *Appl. Phys. Lett.*, Volume 83, Number 12, September 2003.

5

Graphene Nanoribbon for Future VLSI Applications: A Review

Himanshu Sharma

Department of ECE, Chandigarh University, Mohali, India

Abstract

The Moore law for scaling of technological nodes has led to the major advancement in chip design and functionality. With the downscaling of technological nodes, the efficiency of gate becomes insignificant compared to the interconnect performance. The reliability issue play a major role at nano-scaled technological nodes since future interconnects requires higher current density with compact cross-sectional dimensions. Formerly, copper was employed as interconnect due to high current density and less resistivity but as technology downscales below 45nm its resistance increases due to the reduction of MFP. The scaling of interconnect dimensions effects the delay time as well as power dissipation. The electronics circuit operational speed is affected by the signal delay at the output of wire. The power dissipation is also a key factor in VLSI ICs. In today's world, everyone demands higher power standby and speed in electronic circuits. The motivation for minimizing delay in signal and power dissipation at global lengths for various technological nodes varies according to applications. The overall performance of electronic circuit depends on the product of power dissipation and delay. Both are independent parameters but to estimate actual performance of an interconnect, their product must be taken into account for elite integrated circuits. This chapter has proposed MLGNR as suitable and alternative material in on-chip interconnects for future technological nodes compared to copper interconnects.

***Keywords*:** Nanoribbon, graphene material, interconnect, MOSFET, VLSI, scaling, repeaters, lumped model

Email: himanshu_744@yahoo.com

Balwinder Raj, Suman Lata Tripathi, Tarun Chaudhary, K. Srinivasa Rao and Mandeep Singh (eds.) Integrated Devices for Artificial Intelligence and VLSI, (101–124)

5.1 Introduction

Human society and civilization have been strongly transformed into an era of communications and digital computing. Integrated electronic circuits have been developed, measured by the number of transistors and connecting wires on a single chip. The phenomenon growth has been achieved with downscaling or scaling the size of metal oxide semiconductor (MOS) transistors. The rapid shrinking has been detected in the transistors from the previous four decades. The MOS transistor being smaller in size, consume less power and operate faster as a result brought improvement in the efficiency, power consumption and consistency of electronics circuit. With the downscaling of technology nodes the efficiency of gate becomes less pronounced compared to the interconnect performance due to insignificant parasitic effects. The very large scale integration (VLSI) is the methodology used for creating integrated circuits (ICs), where several transistors are integrated into a single chip. The performance of the VLSI circuits can be improved by the interconnects used to connect different devices of electrical circuit [1, 2]. The interconnects can be characterized as a bunch of wires that are used to connect the transistors in order to execute the work related to the devices in the VLSI design chip.

The study of on-chip interconnects is a challenging area in the field of VLSI integrated circuits.The reliability for adopting the paramount material as an interconnect, such as copper and aluminum, is challenging [3–5]. Downscaling the dimensions of the interconnects affects the parasitic parameters. The performance of the interconnects in an IC primarily depends on the current carrying capability and its parasitic parameters (capacitance (C), inductance (L), resistance (R)) [3, 6–10]. The continuous scaling of technology nodes introduces several new ideas in VLSI-ICs, which are incorporated to connect millions of active devices in the ICs [11–18]. The structure of an interconnect is shown in Figure 5.1, which is driven by a complementary MOS (CMOS) inverter and connected to a capacitive load. To evaluate the performance, an interconnect has to be

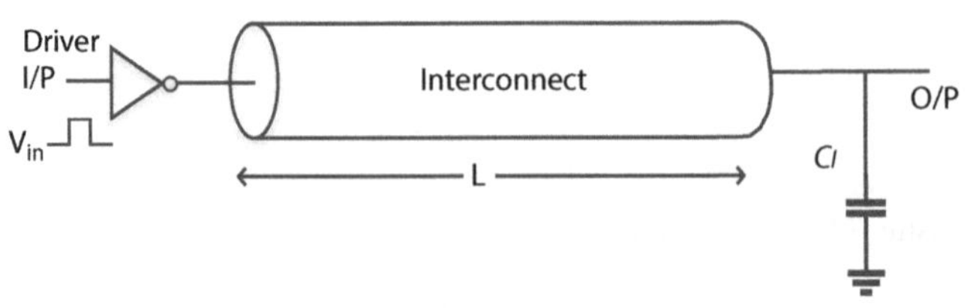

Figure 5.1 Structure of an interconnect.

replaced by an R-L-C electrical circuit model. The parasitic of the interconnects rely on the cross-sectional dimensions and properties of materials used as the VLSI interconnect [19–22].

The ICs are fabricated depending on the length as well as cross-sectional dimensions of interconnects, which are classified into three types i.e., global, intermediate, and local interconnects as shown in Figure 5.2.

The cross-sectional structure shows several active devices connected through the VLSIinterconnect layers [15].

Local interconnects are the first level interconnects. These are obtained at the length between few micrometers to tens of micrometers [15–25]. Such interconnects connect the gate, source, and drain in an MOS technology. Local interconnects have higher resistance in comparison to global interconnects because they are utilized for small length and cross- sectional dimensions.

Intermediate interconnects are advantageous over local interconnects because of its lower resistance. Such interconnects are also termed as

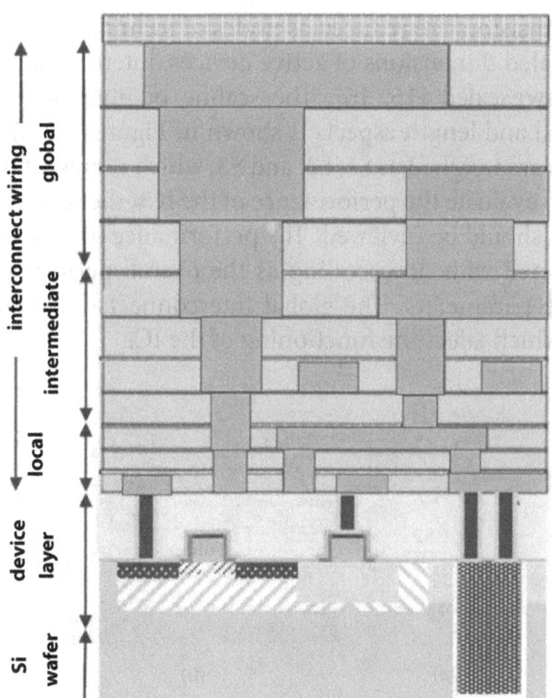

Figure 5.2 Cross-sectional structure of MOS showing interconnect levels [15].

semi-global interconnects. They cover both small as well as long length having range between 20μm to 400μm [26, 27]. These are used to connect devices within a block.

Global interconnects are generally lengthy interconnects having length greater than 500μm. They often cover a large distance between devices [27, 28]. The global interconnects are used to distribute power and connect the various functional blocks including ground, and clock [25]. To understand the performance of interconnects, power dissipation and signal delay are the key parameters of global interconnects.

5.1.1 Significance of Nano-Scale Reign

Efficient packaging of CMOS transistors and the interconnections between them are significant challenges in the manufacturing of VLSI ICs. With the growing demand for packaging, obtaining a high performance is crucial. Scaling process in VLSI circuits minimizes the feature size of the device to a sub-micrometer level depending on the technology parameters while the geometric ratios are maintained. Scaling is categorized in two parts, i.e., device scaling and interconnect scaling. In device scaling, dimensions of the CMOS transistors are reduced as nodes of technology are downscaled. With downscaled dimensions of active devices, interconnects dimensions should be downscaled [15, 16]. The scaling of interconnects from the cross-sectional and length aspects is shown in Figures 5.3(a) and (b). The interconnects are weighed as S1, S2, and S3, which represent different scaling factors. To evaluate the performance of the ICs, the scaling effect on the interconnects should be reviewed. The performance of local interconnects does not affected with downscaling as the overall performance depends on the device parameters. The global interconnects have large parasitic parameters, which affect the functioning of the IC.

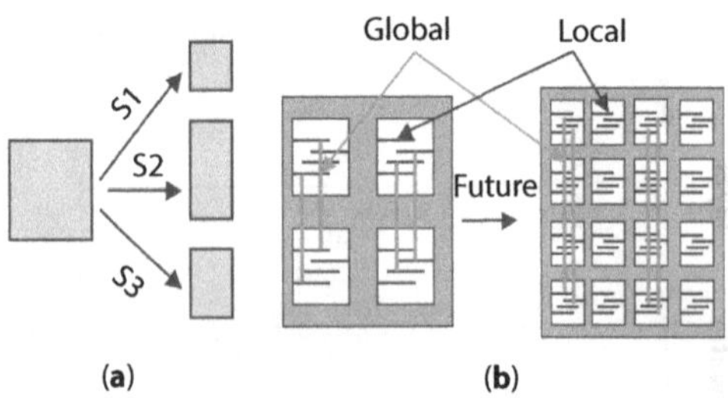

Figure 5.3 (a) Wire cross-section of interconnects (b) IC corresponding to scaling of wire length [15, 16].

In global interconnects, if scaling is performed by a factor 1/S, then the delay time increases by $S^2.S^2$, where S_C is the scaling factor of the chip size [3, 4, 29].

5.1.2 Importance of Repeaters

As discussed in Section 5.2, the advancement in VLSI technology led to the miniaturization of chip size. The parasitic parameters of local interconnects are very small than a driver,i.e., transistor parasitic. The parasitic parameters of interconnects raises linearly with increasing interconnect lengths, thus, affecting the performance parameters of interconnects with respect to the power, delay, and PDP (Power Delay Product). Delay in interconnects becomes larger than that of the drivers at global levels. In order to decreasethe latency, repeaters are suggested between the lengthy interconnects [3, 30–33]. Figure 5.4 depicts an interconnect that is broken up into smaller subsections called repeaters, driven by CMOS inverter [34]. The time constant (RC) is the signal delay of interconnect. The downscaling of technological nodes decreases the conductivity of integrated circuits, which further increases the delay in signals. Thus, to enhance the performance of interconnects at global lengths, optimum number of repeaters with optimum size should be deployed at regular intervals [3, 30–32, 34–43].

5.1.3 Interconnect Models

The connection between the active devices in an IC through interconnect wires behaves as a transmission line. A circuit model exists for every transmission line, which has parasitic parameters. The actual performance of circuit model is evaluated through its parasitic parameters [44]. The circuit

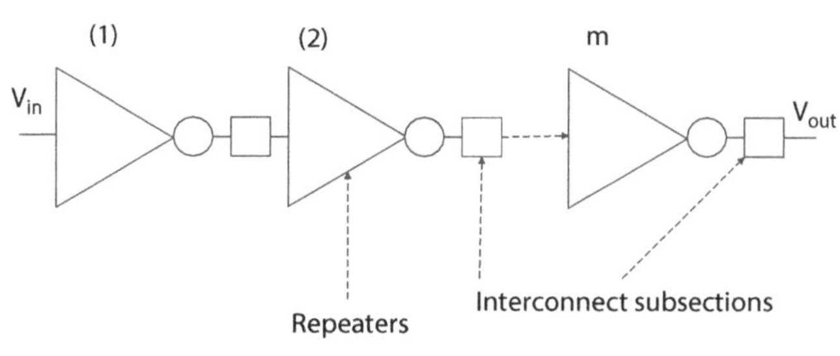

Figure 5.4 Lumped interconnect divided into subsections with optimum number of repeaters (m).

models can be simple or complex but depend on the arrangementof the structure as shown in Figure 5.4.

5.1.4 Lumped Model

Models can be simple or complex but depend on the arrangement of the Structure as shown in figure 5.4 . It is an interconnect model that consists of parasitic parameters (resistance, inductance, andcapacitance). Figure 5.5 displays the various lumped model configurations. Further, these models are categorized into L-shaped, T-shaped, and Π-shaped depending on the probability that they are connected or laid out [45]. In recent studies, RLC circuit models were preferred over RC models because of their better performance. Figure 5.5(a) shows three different lumped circuit models, i.e., L, T, and Π shaped.

5.1.5 Distributed Model

The distributed model comprised of sub-sections of lumped models, as shown in Figure 5.6.

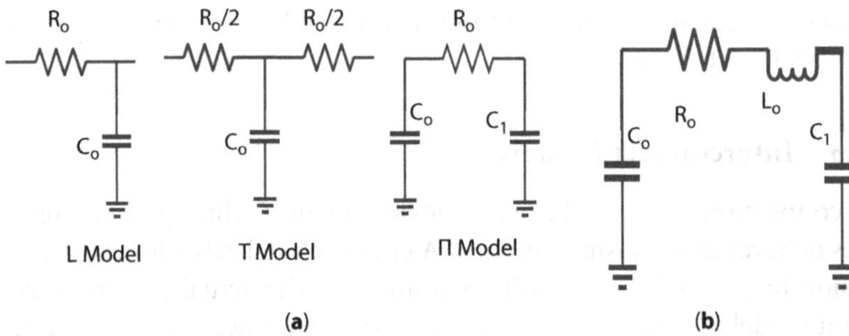

Figure 5.5 (a) Different lumped RC interconnect circuit models (b) RLC lumped model.

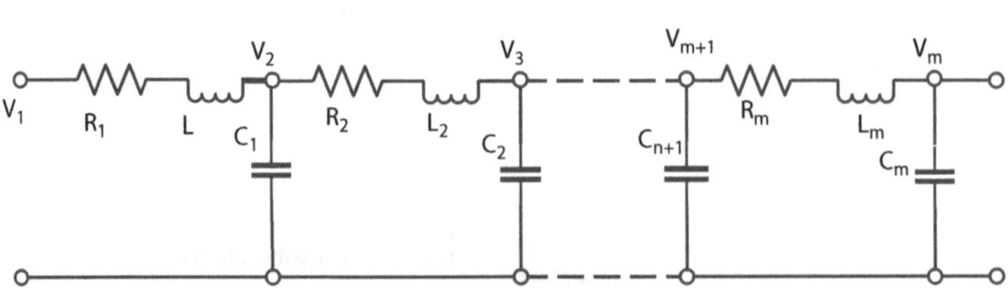

Figure 5.6 Distributed model (RLC with m segments).

Repeaters are inserted to drive subsections of long distributed models. With longer interconnects, this architecture improves drive capacity while decreasing total delay. The optimum size with optimum number of repeaters are considered for the simulations at differentnodes of technology [46–53].

5.1.6 Aluminum and Copper as Interconnects

Aluminum material was used in previous studies as an interconnect in silicon ICs becauseof the following properties:

- Dry etching
- Ease of deposition
- Low resistivity
- Good Conductivity
- Does not contaminate Si
- Excellent adhesion to dielectrics

It makes excellent ohmic contact with silicon as well. Despite the numerous advantages, ithas a few disadvantages:

- Electro-migration
- Hillocks (shorts between levels)
- Higher scattering
- Short mean free path

With the downscaling of technological nodes [24, 54], the interconnect current density increases. At high current densities, electro-migration occurs (it is the transport of material performed by the progressive development of the particles in a conduit because of the energy exchange between directing electrons and directing metal molecules). Further, the resistivity of aluminum is 2.7μΩcm, which is very high in comparison with 1.7μΩcm of the copper resistivity. Hence, aluminum as interconnect at micro-scaled technological nodes cannot produce higher current density owing to its high resistivity and electromigration [6, 18, 55–58].

At such a stage, a better interconnect was required. Therefore, copper was considered as an interconnect owing to its higher conductivity and is further contrary to electro-migration than aluminum [59]. The comparison of various copper properties with similar types of conducting elements as depicted in Table 5.1.

Table 5.1 Properties comparison of various conducting materials [24].

Properties/ Materials	**Copper (Cu)**	**Aluminum (Al)**	**Gold (Au)**	**Silver (Ag)**	**Tungsten (W)**
Resistivity (μΩ-cm)	1.7	2.7	2.35	1.6	5.65
Melting Point (K)	1357	933	1264	1062	3387
Corrosion in Air	Poor	Good	Excellent	Poor	Good
Adhesion to SiO2	Poor	Good	Poor	Poor	Poor

It is concluded from Table 5.1 that copper was appropriate as VLSI interconnect owing to its high melting point and less resistivity among other conducting materials at micro-scaled technological nodes. Although, technologies downscales to less than 45nm, copper resistance along the cross-sectional dimensions increases sharply due to the small MFP (Mean Free Path) of electrons and stuffy effects such as the boundary scattering of the surface and grain, resulting in an increasing electro-migration and decreasing current density issues [18, 46, 60]. The difficulty in increasing resistance of copper as an interconnect is challenging at the global level interconnects. Hence, an alternative and potential fabrication material is required to improve the performance of the VLSI circuits [61–68]. Carbon nanomaterials are emerging as a suitable interconnect material at global lengths as an alternative to copper interconnects [60, 69–71].

5.1.7 Graphene Nanoribbon as Interconnects

Recently, the nanomaterials made from carbon are proposed as on-chip interconnects in VLSIused for developing ICs for nano-scaled technological nodes [16, 24, 72, 73]. Earlier, copper was employed as interconnect for global, intermediate, and local levels. With the developmentof technological nodes, it becomes essential to rise the current density and reducing the dimensions of cross-sectional area of the interconnects. As discussed, stuffy effects like the boundary scattering of surface and grain in the copper [60, 64, 74] results in increasing electro-migration, & resistivity; and reduces the current-density. Therefore, a suitable interconnect material is required to be used as interconnect to reduce the issues corresponding with copper.

Further, the nano-materials made from carbon exhibits a significant effect similar to that of thein VLSI interconnects owing to its thermal, electrical as well as mechanical aspects [60, 75–77]. Nanomaterials like, Carbon Nano Tube (CNT) and Graphene Nano Ribbon (GNR), are resulted to be better alternatives other than using copper, for the nano-scaled nodes oftechnology. Since last many years, CNTs are tested well and different parameters of modelinghas also been presented in the studies [28, 52, 78–82]. "Multi-Wall Carbon Nanotubes" (MWCNTs) and "Single-Wall Carbon Nanotubes" (SWCNTs) are the two sets of CNTs, withwell researched electrical capabilities [28, 79]. CNT interconnects depict a smaller delay in signal than that of the copper used in applications of VLSI [52]. In addition to this, the effect of cross-talk and the tedious processes of development are the main issues reported inMWCNTs and SWCNTs [83, 84].

Table 5.2 lists the properties of GNRs, MWCNT, SWCNT, and copper corresponding to interconnect applications.

Table 5.2 shows that the GNR and CNTs are better than copper with respect to the MFP, maximum current density, and thermal conductivity. Graphene sheets are also known to be the unfolded type of Carbon Nanotube. The Graphene nanoribbons are acquired by sorting the graphene made from a thin carbon atom sheet and is used for structuring CNTs [60, 85].It also shows same types of characteristics as that of a CNT, although its process of fabrication is much easier than that of the CNTs [85–87]. Graphene is considered as an emerging and alternative material for replacing copper in VLSI interconnect [86–92].

Table 5.2 Comparison of GNR, SWCNT, MWCNT, and Copper [85].

Properties	GNR	SWCNT	MWCNT	Copper
Mean free path(nm) @ room temp.	>1000	1000	$2.5X10^4$	40
Max current density(A/cm^2)	$>10^8$	1010	1010	$\sim10^7$
Thermal conductivity($\times10^3$ W/mK)	3-5	1.75-5.8	3	0.385
Melting point(K)	3800 (graphite)			1357

5.1.8 Classification of GNRs

GNRs are classified into various types with respect to its chirality, lattice structures, and conductivity.

5.1.9 Fundamental Physics

As per its chirality, the Graphene Nanoribbons are classified into two types: (1) zz-GNRs (zigzag-edged); (2) ac-GNRs (Armchair-edged). The arm-chaired Graphene Nanoribbon can be semi-conducting or metallic according to (*N*) number of hexagonal rings present across its width [69], that is it shows metallic features with $N = 3p - 1$ and semi-conducting with $N = 3p$ or $3p + 1$,where p is an integer showing specular features of the edges of GNR (Figure 5.7(a)). When compared with the arm-chaired, the zig-zag edge GNR are observed to be independent of *N* and metallic in nature (Figure 5.7(b)).

5.1.10 According to Structure and Conductivity

GNRs can be classified according to the number of layers [93–95]:

(a) Single-layer and
(b) Multi-layer

Single-layer graphene nanoribbons (SLGNRs) consist of a single sheet of graphene. Figure 5.8(a) shows a single layer GNR, which is located at *d* distance from the level of ground. The SLGNR structure helps to construct MLGNR interconnect.

The MLGNR structure is derived by stacking layers of GNR in a parallel combination, as shown in Figure 5.8(b). As per the compact physical circuit model of GNRs as interconnects, MLGNR has lower resistance than the SLGNR [96].

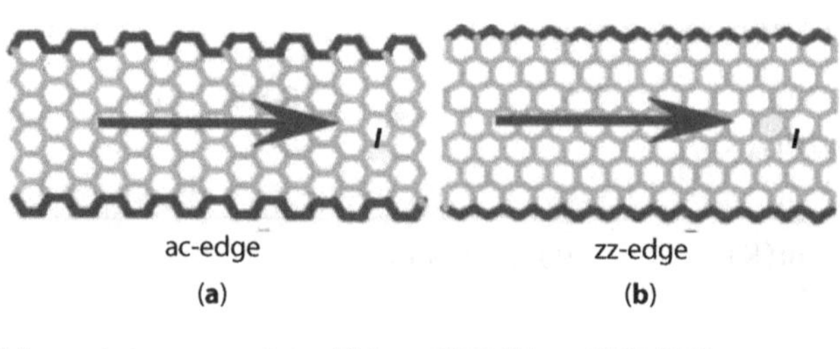

Figure 5.7 Schematic interpretation of (a) ac-GNR (b) zz-GNR [85].

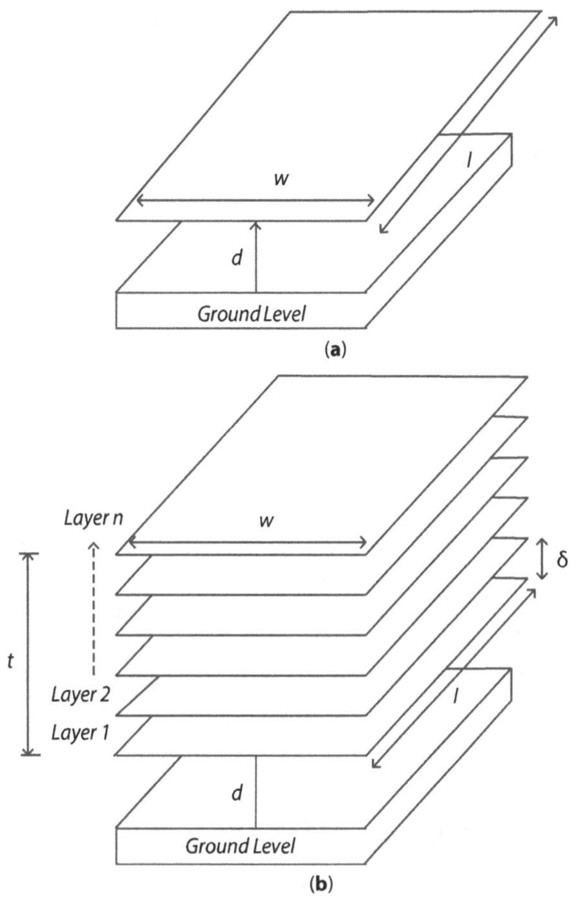

Figure 5.8 Interconnect structures (a) SLGNR and (b) MLGNR.

Therefore, MLGNR could be employed as an outstanding and suitable material for therealization of On-Chip VLSI interconnects.

5.1.11 GNR Field Effect Transistor (GNRFET)

GNRFETs can be created by placing a GNR channel with metallic contacts at the proper sourceand drain, as shown in Figure 5.9. A few notable similarities they have with metallic CNTs include their high carrier mobility for ballistic transport, compatibility with high k dielectrics,high carrier velocities for quick switching, and good thermal conductivity [97]. In 2D graphene, zero bandgap semimetal characteristics are observed. But by arranging the graphene into a few-nm-wide GNR regime, a bandgap can be

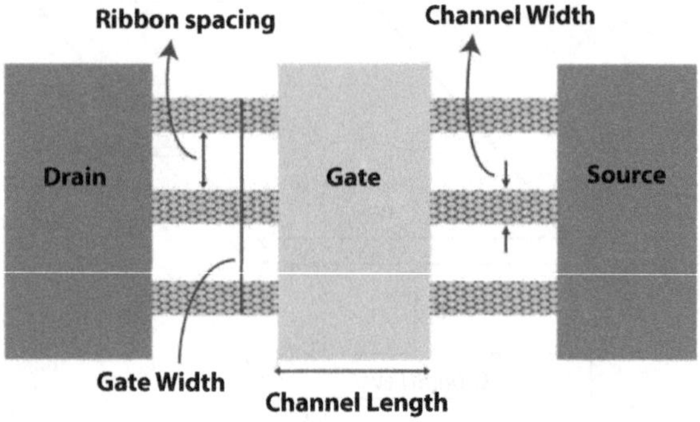

Figure 5.9 GNR based FET [99].

produced. Hydrogen atoms passivate dangling bondsat the ribbon's edges, which affects whether the bonds are armchair or zigzag. The performance of GNRFETs can be hampered by defects and unpredictability in thin nm-wide geometries. Variations could result from the inability to control the GNR width, oxide thickness, and line edge roughness (LER) during production [98]. The first theoretically and physically described graphene-based transistor structure was the Schottky barrier type (SB-GNRFET), and as new topologies have been studied, a growing number of GNRFET configurations have evolved.

5.1.12 Model Development of GNRFET

To estimate the findings of the various futuristic models were developed to estimate and analyze the optimized parametric values that required altering to understand the different effectsof processes without evaluating one by one. Figure 5.10 presents extra peak electric field (EPF).

The UIUC model for the University of Illinois at Urbana-Champaign is used for the modelingof a graphene nanoribbon field-effect transistor (GNRFET) using HSPICE. The model was proposed by Morteza Gholipour *et al.* and Ying-Yu Chen *et al.*, respectively [100, 101]. The model includes model definitions, HSPICE netlists, global parameters for the models, suggested design parameters, SB-GNRFET and MOS-GNRFET libraries for single-gate,double-gate, and subband, as well as subband global parameters and the MATLAB file that generates the subband information used in the models. It also contains the output file for the constructed example. There are also references available for MOS and SB-GNRFET.

The SPICE compatible compact model for the Schottky barrier (SB-GNRFET), an analytical model focused on physics for its IV properties,

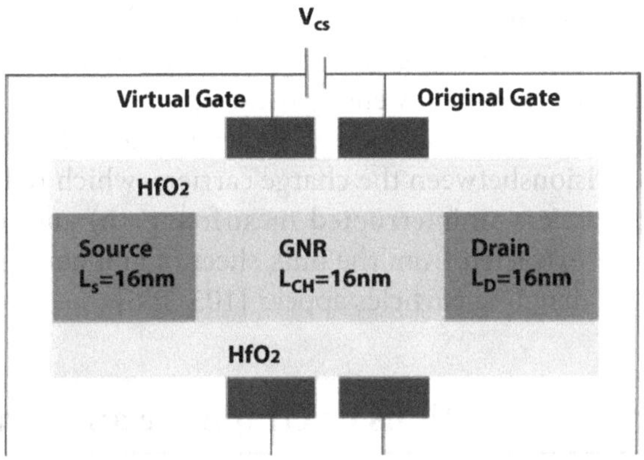

Figure 5.10 GNRFET with extra peak electric field (EPF) [100].

is displayed in the proposed model file. This model allows for the approximate tunneling effect of the Schottky barrier. Numerous characteristics, including LER, oxide thickness, channel width, and channel length, are taken into account by the analytical model. The carrier current and carrier charge density are roughly calculated to construct a compact model of the SB-GNRFET, producing better results while upholding the compact nature.

The model validates graphene-based circuit level simulations while also taking into account various design parameters, process changes, and LER impacts.

5.1.13 Pros and Cons of GNRFET

According to reports, graphene has the highest electron mobility (15,000 cm2V1s1), which is essentially temperature-independent from 10 K to 100 K. Its decreased effective mass is whatcauses the improved carrier mobility. However, acoustic phonon dispersion greatly lowers carrier mobility in graphene. Surface optical phonons of the surrounding materials or chargedimpurities at graphene interfaces (known as Coulomb's scattering) can cause scattering when graphene is sandwiched between a substrate and a dielectric. Furthermore, carrier mobility is drastically decreased at bandgaps greater than 0.5 eV. The mobilities in a uniformly doped channel GNRFET are higher than in an inversion mode channel as a result of the decrease [102–104].

The energy band gap shrinks with increasing width, requiring less energy for charge carriers tocross it and enter the CB. Even when turned off, it can conduct electrons due to its low energybandgap.

While widening the channel increases current and reduces delay, it also causes more collisionsbetween the charge carriers, which reduces ballistic transport (the carriers' uninterrupted meanfree path) and lowers speed. When GNRs are extracted from the bulk sheet of graphene, LERbecomes problematic and dangling particles appear [105, 106].

5.2 Future Applications of Graphene and Graphene-Based FETs

Oscillators, radiometers, mixers, and frequency multiplication devices might all be made withgraphene because of its ambipolar features, which allow both electrons and holes to participate in conduction. The ideal "square law" behavior of graphene FETs at I (VG) 1 VG 2, which is the ambipolar point, would suppress odd order harmonic. In the field of medicine, graphene-based bioelectric sensors and bioimaging systems are being studied for DNA sequencing. By utilizing its electrical properties, artificial implants could interface to neurological systems andbe used as spinal surgical equipment [107]. Graphene-based FETs, like doped FETs, could offer LNA with a greater dynamic range in comparison to conventional devices like heterojunction bipolar transistors and metal-semiconductor field-effect transistors [108]. The ability of graphene to conduct energy could lead to faster battery charging, which could be utilized in photovoltaics for defensive purposes. Paints and polymers that can store solar energy and power homes could be made with graphene. Graphene-based CPUs could be employed incomputer chips because of the bandgap characteristics of the material. Given that it has a 100 times lower contact resistance and a better electron mobility than conventional materials, graphene could be employed as transparent electrodes for organic thin-film transistors. As a transparent and theoretically flexible electrode component, graphene could be employed in touchscreens [109]. Graphene will also find applications in future interconnects made of GNR and metal.

References

[1] N. Srivastava, H. Li, F. Kreupl, and K. Banerjee, "On the applicability of single walled carbon nanotubes as VLSI interconnects," *IEEE Trans. Nanotechnol.*, vol. 8, pp. 542–559, 2009.

[2] N. Srivastava and K. Banerjee, "Interconnect challenges for nanoscale electronic circuits," *JOM*, vol. 56, pp. 30–31, 2004.

[3] J. Meindl, "Beyond Moore's law: The interconnect era," *Comput. Sci. Eng.*, vol. 5, no. 1, pp. 20–24, 2003.

[4] H. B. Bakoglu and J. D. Meindl, "Optimal interconnection circuits for VLSI *IEEE Trans. Electron Devices*, vol. 32, pp. 903-909, 1985.

[5] H. B. Bakoglu, *Circuits, Interconnections, and Packaging for VLSI*, Addison Wesley Pub(Sd), 1990.

[6] L. W. Schaper and D. Amey, "Improved electrical performance required for future MOS packaging," *IEEE Trans. Components, Hybrids Manuf. Technol.*, vol. 6, pp. 283-289, 1983.

[7] T. Sakurai, "Closed-form expressions for interconnection delay, coupling and crosstalk in modern electronics," *IEEE Trans. Electron Devices* vol. 40, pp. 118-124, 1993.

[8] A. B. Kahng and S. Muddu, "Efcient gate delay modeling for large inter connect loads," In *Multi-Chip Module Conference Proceedings, IEEE*, pp 202-207, 1996.

[9] F. L. Traversa and F. Bonani, "Improved harmonic balance implementation of Floquet analysis for nonlinear circuit simulation," *AEU-Int. J. Electron. Commun.*, vol. 66, pp. 357-363, 2012.

[10] M. B. Lin, *Introduction to VLSI Systems: A Logic, Circuit, and System Perspective*: CRCPress, 2011.

[11] A. Deutsch, G. V. Kopcsay, P. J. Restle, H. H. Smith, G. Katopis, W. D. Becke *et al.*, "When are transmission-line effects important for on-chip interconnects?," *IEEE Trans. Microw. Theory Tech.*, vol. 45 pp. 1836-1846, 1997.

[12] M. K. Rai and S. Sarkar, "Influence of tube diameter on carbon nanotube interconnect delay and power output," *Phys. Status Solidi A*, vol. 208 pp. 735-739, 2011.

[13] S. K. Vishwakarma, B. Raj, A. K. Saxena, R. Singh, C. R. Panda, and S. Dasgupta, "Evaluation of threshold voltage for 30 nm Symmetric Double Gate (SDG) MOSFET and it's variation with process parameters," *J. Comput. Theor. Nanosci.*, vol. 5, pp. 619-629, 2008.

[14] J. S. Roychowdhury, A. R. Newton, and D. O. Pederson, "Algorithms for the transient simulation of lossy interconnect," *Trans. Comput.-Aided Des. Integr. Circuits Syst.*, vol. 13, pp. 96-104, 1994.

[15] International Technology Roadmap for Semiconductors, 2011 Edition *http://public.itrs.net.*

[16] International Technology Roadmap for Semiconductors, 2013 Edition, *http://public.itrs.net.*

[17] G. Kaushal, B. Murgan, M. Pattanaik, C. N. Raghuram, and S. S. Rathod, "Soft error- resilient RHBD16T SRAM cell in 32nm technology," In *Innovative Applications of Nanowires for Circuit Design*, pp. 171-188, 2021.
[18] K. C. Saraswat and F. Mohammadi, "Effect of scaling of interconnections on the time delay of VLSI circuits," *IEEE Trans. Electron Devices*, vol. 29, pp. 645-650, 1982.
[19] Y. S. Duksh, B. K. Kaushik, S. Sarkar, and R. Singh, "Analysis of propagation delay and power with variation in driver size and number of shells in multi walled carbon nanotube interconnects," *J. Eng. Des. Technol.*, vol. 11, pp. 19-33, 2013.
[20] J. D Meindl, "Interconnect opportunities for gigascale integration," *IEEE Micro*, vol. 23, pp. 28-35, 2003.
[21] Man D. Mann, A. Javey, J. Kong, Q. Wang, and H. Dai, "Ballistic transport in metallic nanotubes with reliable Pd ohmic contacts," *Nano Lett.*, vol. 3, pp. 1541-1544, 2003.
[22] J. D. Meindl, J. A. Davis, P. Zarkesh-Ha, C. S. Patel, K. P. Martin and P. A. Kohl, "Interconnect opportunities for gigascale integration," *IBM J. Res. Dev.*, vol. 46, pp. 245-263, 2002.
[23] B. Q. Wei, R. Vajtai and P. M. Ajayan, "Reliability and current carrying capacity of carbon nanotubes," *Appl. Phys. Lett.*, vol. 79, pp. 1172-1174, 2001.
[24] H. Li, C. Xu, and K. Banerjee, "Carbon nanomaterials: The ideal interconnect technology for nexts-generation ICs," *IEEE Des. Test Comput.*, vol. 27, pp. 20-31. 2010.
[25] A. Singh, M. Khosla, and B. Raj, "Compact model for ballistic single wall CNTFET under quantum capacitance limit," *J. Semicond.*, vol. 37, pp. 104001-104008, 2016.
[26] N. Srivastava and K. Banerjee, "Performance analysis of carbon nanotube interconnects for VLSI applications," In *International Conference on Computer-Aided Design, IEEE*, pp. 383-390, 2005.
[27] H. Li, C. Xu, N. Srivastava, and K. Banerjee, "Carbon nanomaterials for next-generation interconnects and passives: Physics, status, and prospects," *IEEE Trans. Electron Devices*, vol. 56, pp. 1799-1821, 2009.
[28] H. Li, W. Y. Yin, K. Banerjee, and J. F. Mao, "Circuit modeling and performance analysis of multi-walled carbon nanotube interconnects," *IEEE Trans. Electron Devices*, vol. 55, pp. 1328-1337, 2008.
[29] B. Wang, T. Q. Nguyen, A. T. Do, J. Zhou, M. Je, and T. T-H. Kim, "Design of an ultra-low voltage 9T SRAM with equalized bitline leakage and CAM-Assisted energy efficiency improvement," *IEEE Trans. Circuits Syst. I: Regul. Papers*, vol. 62, pp. 441-448, 2014.
[30] V. Adler and E. G. Friedman, "Repeater design to reduce delay and power in resistive interconnect," *IEEE Trans. Circuits Syst. II: Analog Digital Signal Process.*, vol. 45, pp. 607–616, 1998.
[31] Anjana Bhardwaj, Pradeep Kumar, B. Raj, Sunny Anand, "Design and Performance Optimization of doping-less Vertical Nanowire TFET using

Gate Stack Technique", *J. Electron. Mater. (JEMS)* Springer, Vol. 41, Issuse 7, PP. 4005-4013, 2022.

[32] Jeetendra Singh, BRaj, "Tunnel Current Model of Asymmetric MIM Structure Levying Various Image Forces to Analyze the Characteristics of Filamentary Memristor", *Appl. Phys. A*, Springer, Vol. 125, No.3, pp. 203.1 to 203.11 Feb 2019.

[33] Candy Goyal, Jagpal Singh Ubhi and B Raj, Low Leakage Zero Ground Noise Nanoscale Full Adder using Source Biasing Technique, "*J. Nanoelectron. Optoelectron.*", American Scientific Publishers, Vol. 14, PP. 360–370, March 2019.

[34] Gurmohan Singh, R. K. Sarin and B Raj, "A Novel Robust Exclusive-OR Function Implementation in QCA Nanotechnology with Energy Dissipation Analysis", *J. Comput. Electron.*, Springer, Volume 15, Issue 2, PP. 455-465, June 2016.

[35] Tanu Wadhera, Deepti Kakkar, Girish wadhwa, B Raj, " Recent Advances and Progress in Development of the Field Effect Transistor Biosensor: A Review" *J. Electron. Mater.*, Springer, Volume 48, Issue 12, pp 7635–7646, December 2019.

[36] Girish Wadhwa, Priyanka Kamboj, Balwinder Raj" Design optimisation of junctionless TFET biosensor for high sensitivity" *Adv. Nat. Sci.: Nanosci. Nanotechnol.*, Vol. 10, PP. 045001 (7), (2019).

[37] Priya Bansal, BRaj, "Memristor Modeling and Analysis For Linear Dopant Drift Kinetics," *J. Nanoeng. Nanomanuf.*, American Scientifc Publishers, Vol. 6, pp. 1–7, 2016.

[38] Amandeep Singh, Mamta Khosla, BRaj, "Circuit Compatible Model for Electrostatic Doped Schottky Barrier CNTFET," *J. Electron. Mater.*, Springer, Vol. 45, Issue 12, pp 4825-4835, 2016.

[39] Ashima, D Vaithiyanathan, B Raj," Performance Analysis of Charge Plasma induced Graded Channel Si nanotube", *J. Eng. Res. (JER)*, EMSME Special Issue, Vol.9, pp. 146-154 Aug 2021.

[40] Abhishek Singh Tomar, Vijay Kumar Magraiya, and B Raj, Scaling of Access and Data Transistor for High Performance DRAM Cell Design, *Quantum Matter*, Vol. 2, PP. 412-416 Oct. 2013.

[41] Neeraj Jain, BRaj, "Parasitic Capacitance and Resistance Model Development and Optimization of Raised Source/Drain SOI FinFET Structure for Analog Circuit Applications, *J. Nanoelectron. Optoelectron.*, ASP, USA, Vol. 13, PP. 531–539, Apl 2018.

[42] Shradhya Singh, S. K. Vishvakarma and B Raj, "Analytical Modeling of Split-Gate Junction-Less Transistor for a Biosensor Application", *Sens. Bio-Sens.*, Elsevier, Vol. 18, Pages 31-36, April 2018.

[43] Maisagalla Gopal, and Balwinder Raj, Low Power 8T SRAM Cell Design for High Stability Video Applications, *ITSI Trans. Electr. Electron. Eng..*, Vol. 1(5), pp. 91-97, 2013.

[44] Balwinder Raj, Jatin Mitra, Deepak Kumar Bihani, Rangharajan V, A.K. Saxena and S. Dasgupta, "Analysis of Noise Margin, Power and Process Variation for 32 nm FinFET Based 6T SRAM Cell" *J. Comput. (JCP)*, Academy Publisher, FINLAND, vol. 5, issue 6, PP.1-8, 2010.

[45] Divya Sharma, Rajesh Mehra and B Raj, "Comparative Analysis of Photovoltaic Technologies for High Efficiency Solar Cell Design", *Superlattices Microstruct.*, Elsevier, Volume 153, P P. 106861, May 2021.

[46] Pawandeep Kaur, Avtar Singh Buttar, B Raj, "A comprehensive Analysis of Nanoscale Transistor based Biosensor: A Review, *Indian Journal of Pure and Applied Physics*, Vol. 59, pp. 304-318, April 2021.

[47] Divya Yadav, Balwant Raj and B Raj, "Design and Simulation of Low Power Microcontroller for IoT Applications", *J. Sens. Lett.*, ASP, Vol. 18, PP. 401–409, May 2020.

[48] Shailendra Singh, B Raj,"A 2-D Analytical Surface Potential and Drain current Modeling of Double-Gate Vertical t-shaped Tunnel FET", *J. Comput. Electron.*, Springer, Vol. 19, PP.1154–1163, Apl 2020.

[49] Jeetendra Singh, B Raj, "An Accurate and Generic Window function for Nonlinear Memristor Model" *J. Comput. Electron.*, Springer, Volume 18 Issue 2, PP. 640-647 June 2019.

[50] Manjit Kaur; Neena Gupta; Sanjeev Kumar; BRaj; Arun Kumar Singh, "Comparative RF and Crosstalk Analysis of Carbon Based Nano Interconnects" *IET Circuits, Devices Syst.*, Vol.15(6), PP. 493-503, Feb 2021.

[51] Nehru Kandasamy, Firdous Ahmad, D. Ajitha, B Raj, and Nagarjuna Telagam "Quantum Dot Cellular Automata based Scan Flip Flop and Boundary Scan Register, *IETE J. Res.*, Volume 66, PP. 535-548, 2020.

[52] Sharma, S. K.., Raj, B., Khosla, M., "Enhanced Photosensivity of Highly Spectrum Selective Cylindrical Gate In1-xGaxAs Nanowire MOSFET Photodetector, *Mod. Phys. Lett.-B*, Vol. 33, No. 12, PP. 1950144 (2019).

[53] Jeetendra Singh, BRaj, "Design and Investigation of 7T2M NVSARM with Enhanced Stability and Temperature Impact on Store/Restore Energy", *IEEE Trans. Very Large Scale Integr. Syst.*, Vol. 27, Issure 6, PP. 1322 - 1328 June 2019.

[54] Anil Kumar Bhardwaj, Sumeet Gupta, BRaj, Amandeep Singh, "Impact of Double Gate Geometry on the Performance of Carbon Nanotube Field Effect Transistor Structures for Low Power Digital Design", *J. Comput. Theor. Nanosci.*, ASP, Vol. 16, PP. 1813–1820, 2019.

[55] Neeraj Jain, BRaj, "Termal Stability Analysis and Performance Exploration of Asymmetrical Dual-k underlap Spacer (ADKUS) SOI FinFET for Security and Privacy Applications", *Indian J. Pure Appl. Phys. (IJPAP)*, Vol. 57, pp. 352-360, May 2019.

[56] Amandeep Singh, Mamta Khosla, BRaj, "Design and Analysis of Dynamically Confgurable Electrostatic Doped Carbon Nanotube Tunnel FET", *Microelectronics J.*, Elsevier, Volume 85, Pages 17-24, March 2019.

[57] Neeraj Jain and Balwinder Raj, "Dual-k Spacer Region Variation at the Drain Side of Asymmetric SOI FinFET Structure: Performance Analysis towards the Analog/RF Design Applications", communicated in "*J. Nanoelectron. Optoelectron.*", American Scientifc Publishers, Vol. 14, PP. 349–359, March 2019.

[58] Jeetendra Singh, Sanjeev Sharma, B Raj, Mamta Khosla Analysis of barrier layer thickness on performance of In1-xGaxAs based Gate Stack Cylindrical Gate Nanowire MOSFET, *JNO*, ASP, Vol. 13, PP 1473-1477, Oct 2018.

[59] Neeraj Jain, B Raj, "Analysis and Performance Exploration of High-k SOI FinFETs Over the Conventional Low-k SOI FinFET toward Analog/RF Design", *J. Semicond. (JoS)*, IOP Science, Vol.39, Issue 12, PP-124002-1-7, Dec 2018.

[60] Candy Goyal, Jagpal Singh Ubhi and BRaj, A reliable leakage reduction technique for approximate full adder with reduced ground bounce noise, *J. Math. Probl. Eng.*, Hindawi, Volume 2018, Article ID 3501041, 16 pages, 15 Oct 2018.

[61] Anuradha, Jeetendra Singh, B Raj and Mamta Khosla, "Design and Performance Analysis of Nano-scale Memristor-based Nonvolatile SRAM" *J. Sens. Lett.*, American Scientific Publishers, Vol. 16, PP 798– 805, Oct 2018.

[62] Girish Wadhwa, BRaj, "Parametric Variation Analysis of Charge-Plasma-based Dielectric Modulated JLTFET for Biosensor Application" *IEEE Sens. J.*, VOL. 18, NO. 15, AUGUST 1, PP. 6070 - 6077, 2018.

[63] Jeetendra Singh, BRaj, "Comparative Analysis of Memristor Models for Memories Design", *JoS*, IoP, Vol. 39, No. 7, P P. 074006-1-12, July 2018.

[64] Divya Yadav, Shailesh Singh Chouhan, Santosh Kumar Vishvakarma and BRaj, "Application Specific Microcontroller Design for IoT based WSN", *Sens. Lett.*, ASP, Vol. 16, P P. 374–385, May 2018..

[65] Gurmohan Singh, R. K. Sarin and BRaj, "Fault-Tolerant Design and Analysis of Quantum-Dot Cellular Automata Based Circuits", *IEEE/IET Circuits, Devices & Systems*, Vol. 12, PP. 638 - 64, 2018.

[66] Jeetendra Singh, BRaj, "Modeling of Mean Barrier Height Levying Various Image Forces of Metal Insulator Metal Structure to Enhance the Performance of Conductive Filament Based Memristor Model", *IEEE Nanotechnol.*, Vol. 17, No. 2, P P. 268-267, March 2018.

[67] Aakash Jain, Sanjeev Sharma, BRaj, "Analysis of Triple Metal Surrounding Gate (TM-SG) III-V Nanowire MOSFET for Photosensing Application", *Opto-Electron. J.*, Elsevier, Volume 26, Issue 2, Pages 141-148, May 2018.

[68] Aakash Jain, Sanjeev Sharma, B Raj, "Design and Analysis of High Sensitivity Photosensor Using Cylindrical Surrounding Gate MOSFET For Low Power Sensor Applications", *Eng. Sci. Technol. Int. J.*, Elsevier's, Volume 19, Issue 4, Pages 1864–1870, December 2016.

[69] Amandeep Singh, Mamta Khosla, B Raj, "Analysis of Electrostatic Doped Schottky Barrier Carbon Nanotube FET for Low Power Applications," *J. Mater. Sci.: Mater. Electron.*, Springer, Vol. 28, PP. 1762–1768, 2017.

[70] G. Saiphani Kumar, Amandeep Singh, BRaj, "Design and Analysis of Gate All Around CNTFET based SRAM cell Design", *J. Comput. Electron.*, Springer, Volume 17, Issue 1, pp 138–145, March 2018.

[71] Gurinder pal Singh, B. S. Sohi, Balwinder Raj, "Material Properties Analysis of Graphene Base Transistor (GBT) for VLSI Analog Circuits, *Indian J. Pure Appl. Phys. (IJPAP)*, Vol. 55, pp. 896-902, Dec-2017.

[72] Amandeep Singh, Mamta Khosla, B Raj, "Comparative Analysis of Carbon Nanotube Field Effect Transistor and Nanowire Transistor for Low Power Circuit Design," *J. Nanoelectron. Optoelectron.*, American Scientific Publishers, USA, Vol. 11, PP. 388-393, June 2016.

[73] Sunil Kumar and B Raj, "Estimation of Stability and Performance metric for Inward Access Transistor based 6T SRAM Cell Design using n-type/p-type DMDG-GDOV TFET" *IEEE VLSI Circuits Syst. Lett.*, Volume3, Issue2, PP 25-39, June2017

[74] Shashikant Sharma, Anjan Kumar, Manisha Pattanaik and B Raj, "Forward Body Biased Multimode Multi-Threshold CMOS Technique for Ground Bounce Noise Reduction in Static CMOS Adders" *Int. J. Inf. Electron. Eng.*, pp. 567-572, Vol. 3, issue 3, 2013.

[75] Hamendra Singh, Pankaj Kumar, Balwinder Raj, "Performance Analysis of Majority Gate SET Based 1-bit Full Adder", *International J. Comput. and Communication Engineering* (IJCCE), IACSIT Press Singapore, ISSN: 2010-3743, Volume 2, No. 4, PP.1-8, 2013.

[76] Anil Kumar Bhardwaj, Sumeet Gupta and B Raj, "Investigation of Parameters for Schottky Barrier (SB) Height for Schottky Barrier Based Carbon Nanotube Field Effect Transistor Device", *J. Nanoelectron. Optoelectron.*, ASP, Vol. 15, PP. 783–791, July 2020.

[77] Priya Bansal, BRaj, "Memristor: A Versatile Nonlinear Model For Dopant Drift And Boundary Issues," *JCTN*, American Scientifc Publishers, Vol. 14 (5), P P. 2319-2325, May 2017.

[78] Neeraj Jain, BRaj, "A n Analog and Digital Design Perspective Comprehensive Approach on Fin-FET (Fin-Field Effect transistor) Technology - A Review", *Rev. Adv. Mater. Sci. (RASE)*, ASP, Vol. 5, PP. 1-14, 2016.

[79] Sanjeev Sharma, B Raj, Mamta Khosla, "Subthreshold Performance of In1-xGaxAs based Dual Metal with Gate Stack Cylindrical/Surrounding Gate Nanowire MOSFET for Low Power Analog Applications, *J. Nanoelectron. Optoelectron.*, American Scientifc Publishers, USA, Vol.12, P P. 171–176, 2017.

[80] Balwinder Raj, A. K. Saxena and S. Dasgupta, "Analytical Modeling for the Estimation of Leakage Current and Subthreshold Swing Factor of Nanoscale Double Gate FinFET Device" *Microelectron. Int.*, UK, vol. 26, pp. 53-63, 2009.

[81] A. Mafucci, G. Miano, and F. Villone, "A new circuit model for carbon nano-tube interconnects with diameter-dependent parameters," *IEEE Trans. Nanotechnol.*, vol. 8, pp. 345-354, 2008.

[82] F. Ferranti, G. Antonini, T. Dhaene, L. Knockaert, and A. Orlandi, "Compact and accurate models of large single-wall carbon-nanotube interconnects," *IEEE Trans. Electromagn. Compat.*, vol. 53, pp. 1025-1033, 2011.

[83] S. N. Pu, W. Y. Yin, J. F. Mao, and Q. H. Liu, "Crosstalk prediction of single- and double- walled carbon-nanotube (SWCNT/DWCNT) bundle interconnects," *IEEE Trans. Electron Devices,* vol. 56, pp. 560-568, 2009

[84] F. Liang, G. Wang, and H. Lin, "Modeling of crosstalk effects in multiwall carbon nanotube interconnects," *IEEE Trans. Electromagn. Compat.*, vol. 54, pp. 133-139, 2011.

[85] C. Xu, H. Li, and K. Banerjee, "Modelling, analysis and design of graphene nano-ribbon interconnects," *IEEE Trans. Electron Devices*, vol. 56, pp. 1567-1568, 2009.

[86] S. Rakheja, V. Kumar, and A. Naeemi, "Evaluation of the potential performance of graphene nanoribbons as on-chip interconnects," In *Proceedings of the IEEE,* vol. 101, pp. 1740–1765, 2013.

[87] W. Choi, I. Lahiri, R. Seelaboyina, and Y. S. Kang, "Synthesis of graphene and its applications: A Review," *Crit. Rev. Solid State Mater. Sci.*, vol. 35, pp. 52–71, 2010.

[88] X. Wang, Y. Ouyang, L. Jiao, H. Wang, L. Xie, J. Wu, J. Guo, and H. Dai, "Graphene nanoribbons with smooth edges behave as quantum wire," *Nature Nanotechnology*, vol. 6, pp. 563–567, 2011.

[89] M. Politou, I. Asselberghs, B. Soree, C. S. Lee, S. Sayan, D. Lin, P. Pashaei, C. Huyghebaert, P. Raghavan, I. Radu, *et al.*, "Single and multilayer graphene wires as alternative interconnects," *Microelectronic Engineering*, vol. 156, pp. 131–135, 2016.

[90] A. Mafucci and G. Miano, "Transmission line model of graphene nanoribbon interconnects," *Nanosci. Nanotechnol. Lett.*, vol. 5, pp. 1207–1216, 2013.

[91] A. Mafucci and G. Miano, "Electrical properties of graphene for interconnect applications," *Applied Sciences*, vol. 4, pp. 305–317, 2014.

[92] S. H. Nasiri, M. K. Moravvej-Farshi, and R. Faez, "Stability analysis in graphene nanoribbon interconnects," *IEEE Electron Device Lett.*, vol. 31, pp. 1458–1460, 2010.

[93] X. Chen, D. Akinwande, K. J. Lee, G. F. Close, S. Yasuda, B. C. Paul, S. Fujita, J. Kong, and H. S. P. Wong, "Fully integrated graphene and carbon nanotube interconnects for gigahertz high-speed CMOS electronics," *IEEE Trans. Electron Devices*, vol. 57, pp. 3137-3143, 2010.

[94] A. Naeemi and J. D. Meindl, "Conductance modeling for graphene nanoribbon (GNR) interconnects," *IEEE Electron Device Lett.*, vol. 28, pp. 428-431, 2007.

[95] A. Hazra and S. Basu, "Graphene nanoribbon as potential on-chip interconnect material—A review," *C—Journal of Carbon Research*, vol. 4, p. 49-1-49-27, 2018.

[96] A. Naeemi and J. D. Meindl, "Compact physics-based circuit models for graphene nanoribbon interconnects," *IEEE Trans. Electron Devices*, vol. 56, pp. 1822- 1833, 2009.

[97] K. Singh and B. Raj, "Performance and analysis of temperature dependent multi-walled carbon nanotubes as global interconnects at different technology nodes," *J. Comput. Electron.*, vol. 14, pp. 469-476, 2015.

[98] E. Pop, "The role of electrical and thermal contact resistance for Joule breakdown of single-wall carbon nanotubes," *Nanotechnology*, vol. 19, pp. 2952021-2952025, 2008.

[99] P. J. Wessely, F. Wessely, E. Birinci, U. Schwalke, and B. Riedinger, "Transfer-free fabrication of graphene transistors," *J. Vac. Sci. Technol. B: Nanotechnol. Microelectron.: Mater. Process. Measure. Phenom.*, vol. 30, 2012.

[100]M. Gholipur, Y. Y. Chen, A. Sangai, and D. Chen, "Highly accurate SPICE-compatible modeling for single-and double-gate GNRFETs with studies on technology scaling," *Design, Automation & Test in Europe Conference & Exhibition (DATE)*, pp. 1-6, 2014.

[101]C. H. O. Chen, S. Park, T. Krishna, S. Subramanian, A. P. Chandrakasan, and L.S. Peh, "SMART: A single-cycle reconfigurable NoC for SoC applications." In *2013 Design, Automation & Test in Europe Conference & Exhibition (DATE)*, pp. 338-343, 2013.

[102]A. V. Voitsekhovskii, S. N. Nesmelov, S. M. Dzyadukh, D. I. Gorn, S. A. Dvoretsky, N. N. Mikhailov, ... And M. V. Yakushev, "II-VI Semiconductor-Based Unipolar Barrier Structures for Infrared Photodetector Arrays," In *Handbook of II-VI Semiconductor-Based Sensors and Radiation Detectors*, vol. *2*, pp. 135-154, 2023.

[103]J. Lee, T. J. Ha, H. Li, K. N. Parrish, M. Holt, A. Dodabalapur, ... and D. Akinwande, "25 GHz embedded-gate graphene transistors with high-K dielectrics on extremely flexible plastic sheets," *ACS nano*, vol. *7*, pp 7744-7750, 2013.

[104]K. N. Parrish, and D. Akinwande, D. (2011). Erratum:"Impact of contact resistance on the transconductance and linearity of graphene transistors" (2011)]. *Appl. Phys. Lett.*, vol *99*, pp. 183505, 2013.

[105]S. Fregonese, C. Maneux, and T. Zimmer, 'A versatile compact model for ballistic 1D transistor: Applications to GNRFET and CNTFET," In *2009 International Semiconductor Device Research Symposium*, pp. 1-2, 2009.

[106]M. R. Moslemi, M. H. Sheikhi, K. Saghaf, and M. K. Moravvej-Farshi, "Electronic properties of a dual-gated GNR-FET under uniaxial tensile strain," *Microelectron. Reliab.*, vol. *52*, pp. 2579-2584, 2012.

[107]K. Tamersit and F. Djefal, "Double-gate graphene nanoribbon field-effect transistor for DNA and gas sensing applications: simulation study and sensitivity analysis," *IEEE Sens. J.*, vol. 16, pp. 4180-4191, 2016.

[108]J. S. Moon, D. K. Gaskill, P. Campbell, and P. Asbeck, "Graphene-on-SiC and graphene-on-Si transistors and RF applications," In *2011 IEEE MTT-S International Microwave Symposium*, pp. 1-4, 2011.

[109]C. Sire, F. Ardiaca, S. Lepilliet, J. W. T. Seo, M. C. Hersam, G. Dambrine, G., ... & V. Derycke, "Flexible gigahertz transistors derived from solution-based single-layer graphene," *Nano Lett.*, vol. 12, pp. 1184-1188, 2012.

6

Ferroelectric Random Access Memory (FeRAM)

B. Vimala Reddy, Tarun Chaudhary, Mandeep Singh and Balwinder Raj*

Department of ECE, NIT Jalandhar, Punjab, India

Abstract

We examine the recent development of ferroelectric memory, along with ferroelectric random access memory (FRAM). Sophisticated non-volatile memories including resistive random access memory (ReRAM), phase-change random access memory (PRAM), and magnetoresistive random access memory (MRAM) have all been expanded, but FRAM is the first of these to be commercially available. Currently, a few Mb of extremely dependable FRAM is accessible. FRAM has been used in sophisticated smartcards, RFID tags, and other electronic devices because of its superior electric features, which include fast read/write speeds around 50 ns, low power consumption, and high switching durability around 10^{13}. We also discuss recently created materials, manufacturing techniques, and circuit technologies that are anticipated to solve the scalability issue with FRAM.

***Keywords*:** FeRAM, ferroelectric, SRAM, polarization, PZT

6.1 Introduction

The polarization state of the ferroelectric material is used to store information in ferroelectric memory, which includes ferroelectric random access memory (FRAM or FeRAM). Ferroelectric materials have bistable polarization states that can be easily switched between even when there isn't an extrinsic electric field present. This field is typically applied at a rate of around 100 kV/cm. Such as FRAM offers superior electric characteristics

**Corresponding author*: rajb@nitj.ac.in

Balwinder Raj, Suman Lata Tripathi, Tarun Chaudhary, K. Srinivasa Rao and Mandeep Singh (eds.) Integrated Devices for Artificial Intelligence and VLSI, (125–156) © 2024 Scrivener Publishing LLC

to standard FLASH memory, such as fast read/write speeds around 50 ns, low power consumption, and high switching durability around 10^{13}; it has been used in RFID tags, authentication systems, and sophisticated smartcards.

Pulvari proposed a ferroelectric material (barium titanate) based method of information storage in 1951 [1]. It is intriguing that he suggested operating this ferroelectric memory at a temperature that is not many degrees less than the Tc (Curie temperature) in order to attain minimum power consumption and fast write/read speeds. 4 years after this suggestion, in 1955, Ross put forth a semiconductor-based ferroelectric memory [2]. Numerous electronic companies throughout the world began challenging the advancement of FRAM in the late 1980s, with the intention of creating "a dream memory" that is non-volatile and has excellent read/write speeds. Some new businesses in the US started working on developing functional Si- based capacitor-type FRAMs in the late 1980s [3, 4]. In the late 1990s, a number of US and Japanese businesses produced FRAM in large quantities with memory densities ranging from 16 to 256 Kb [5, 6, 8]. FRAMs with capacities up to 8 Mb and several varieties of integrated FRAMs have recently entered the market. This paper seeks to provide an overview of FRAM's development over time.

6.1.1 Basic Characteristics of Ferroelectric Capacitors

To better understand the parts that follow, the basic characteristics of ferroelectric capacitors are briefly discussed in this section. The unprompted polarization of ferroelectric materials can be reversed by using outside electric field. A polarization-electric field (P-E) hysteresis curve is produced when a ferroelectric capacitor is exposed to such an electric field (Figure 6.1). The value of the electric field as a result of which the ferroelectric capacitor's macroscopic polarization vanishes is known as a coercive field (Ec). Remaining polarization without an electric field is referred to as residual polarization (Pr), where "±" denotes the polarization's direction. When reading data from a FRAM, the voltage produced by polarization-reversal current is measured.

Spontaneous polarization in a ferroelectric crystal results from the relative displacement of cations and anions. Both of these places in a unit cell are not charge-centered, therefore an electric dipole is created on its own. Figure 6.3 depicts schematically the connection between energy potential and ion displacement. By using an external electric field, the Ti^{4+} or Zr^{4+} ions may be shifted between the two stable locations, each of

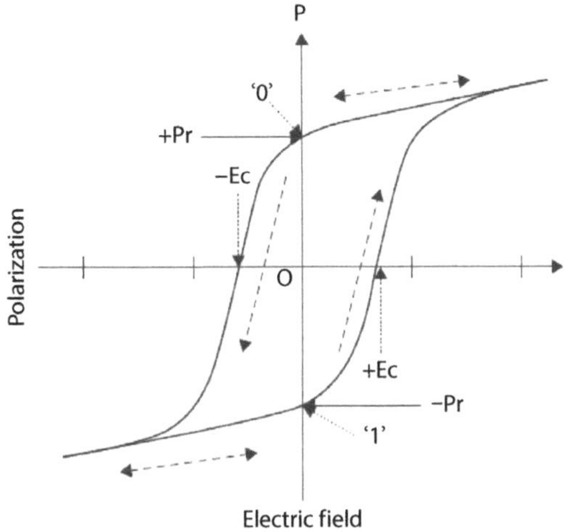

Figure 6.1 Polarization-electric field hysteresis curve [7].

which they occupy. Figure 6.2 shows Schematic PZT crystal structure. A bistable condition combines into a single steady state when the temperature rises. The Curie temperature is referred to as this crucial temperature (Tc). PZT's crystal structure is tetragonal below Tc, here the lengths of all the a-axis, b-axis, and c-axis are approximately identical. However, beyond Tc, the structure changes to cubic (phase change). Specifically, Figure 6.3 illustrates the energy potential for the *2nd* order phase transition, here free energy varies continuously with temperature.

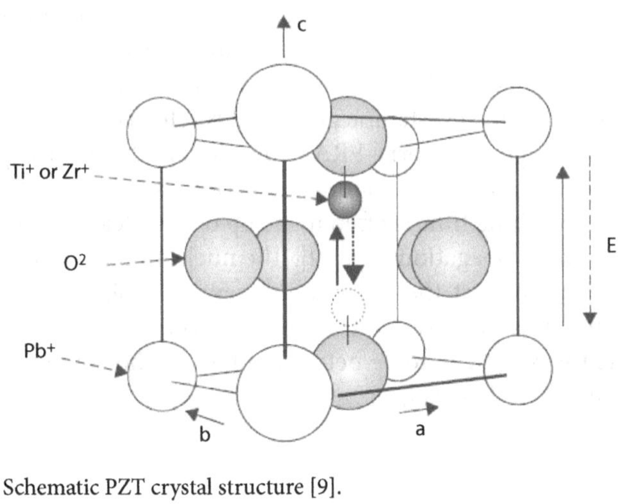

Figure 6.2 Schematic PZT crystal structure [9].

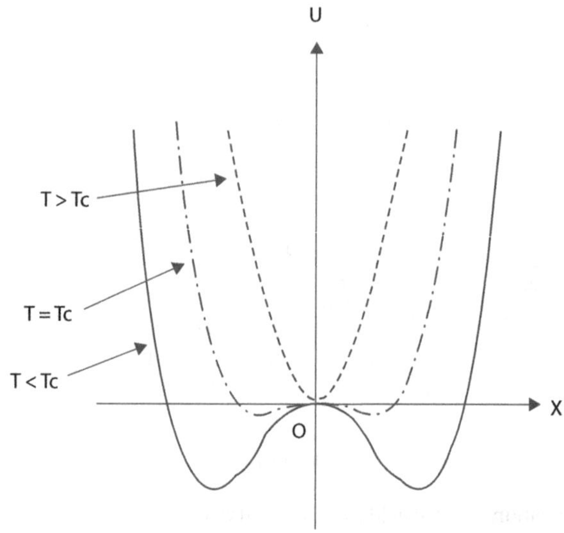

Figure 6.3 The relationship between ion displacement (X) and energy potential (U) for a 2nd order phase transition [10].

There is a first-order phase transition as well, in which the free energy varies abruptly with temperature [9]. The dependency of temperature on polarization is expressed as

$$\left(\sqrt{1-T_c}\right)(0) = P(T) \rightarrow 1 \tag{6.1}$$

Where P (0) is polarization at zero Kelvin and T is temperature [59–63]. The ions are located on the same side of the double-well when seen microscopically, in a small area of hundreds of nm (Figure 6.3). It is often smaller than a ferroelectric polycrystalline grain and is referred to as a domain.

One of the things that worry FRAM developers the most is polarizing switching speed. Ishibashi has conducted theoretical research [10]. As seen in Figure 6.4, switching time is often made up of a nucleation time, domain growth time, and propagation time [9]. The nucleation time is the amount of time required to start the polarization reversal point; relatively small nucleation domains evaporate during this period. The switching time of a PZT capacitor is noticed to be between a few ns and some tens of ns in experimental studies [11, 12].

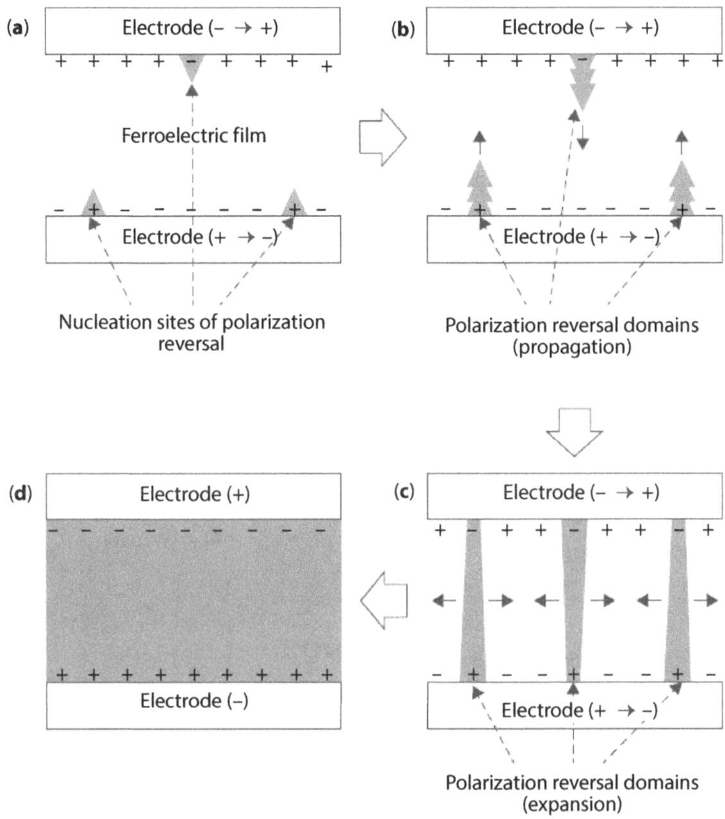

Figure 6.4 Switching model of polarization [11].

6.1.2 FRAM Fabrication Process

Ferroelectric capacitor and traditional CMOS manufacturing techniques were used to create commercial Si-base FRAM. A ferroelectric capacitor fabrication procedure is typically implanted in the middle of a metallization procedure (back end of line: BEOL) and a transistor fabrication procedure (front end of line: FEOL). Because logic transistors may be used in FRAM memory cells "as is", the FRAM method is easier than the Flash memory technique, which requires a separate high voltage transistor manufacturing step. This is beneficial for making FRAM integrated logic devices in particular. For the same reason, noble metals like Ir, Pt, and IrO2 are typically used as capacitor electrodes (and also these metals have high work functions). Perovskite-base electrodes that can withstand high temperatures have also been suggested [19].

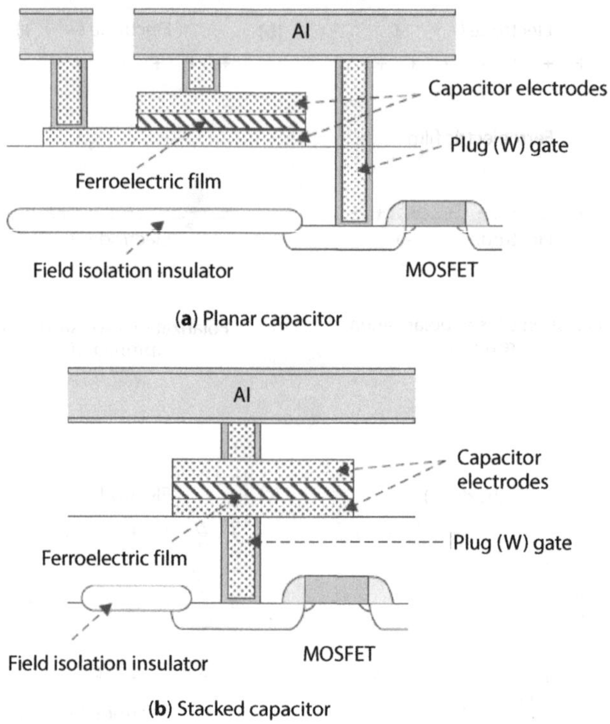

Figure 6.5 Capacitor fabrication structure [26].

Ferroelectric capacitor structures come in two varieties as shown in Figure 6.5. After planarizing a dielectric layer over a MOSFET, a ferroelectric capacitor is positioned above field insulator layers [8, 14]. After the dielectric interlayer is deposited over the capacitor, electrical connections are created between the MOSFETs and the capacitor electrodes. Despite the unit memory cell area being greater, this procedure is easier than the stacked capacitor approach [20–23].

The electrode in the bottom of a ferroelectric capacitor is positioned straight onto the contact metal (W) linking a MOSFET when using stacked capacitors. An extremely fine electrode structure in the bottom and/or materials must be employed in the high temperature ferroelectric procedure to stop the contact metal from oxidizing. For a bottom electrode, oxygen barrier metals like TiN, Ir, $IrO2$, and/or their compounds are often utilized. Although the manufacturing technique and materials used are a little more complex compared to a traditional planar CMOS approach, the stacked capacitor offers the benefit of a compact memory cell area.

Physical vapor deposition (PVD), which mostly uses MOCVD (metal organic chemical vapor deposition) [24, 25], sputter deposition [8], or CSV (chemical solution deposition), which primarily uses spin coating [5], were used to generate ferroelectric films. PVD is often used to deposit electrodes. Because of their cheap source costs and high throughput, PVD and CSD have lower deposition costs than MOCVD, but the MOCVD process offers the benefit of perfect source material controllability.

Due to the non-reactivity of noble metals, which are utilized for both the top and bottom electrodes of the ferroelectric capacitor, and the extremely low vapor pressure of ferroelectric byproducts, traditional RIE (reactive ion etching) cannot be used. Therefore, argon-mixed etching gases are used to physically enhance RIE.

6.2 Structure of Ferroelectric Memory Cells in Capacitor-Type FRAM Devices

6.2.1 A Capacitor-Type FRAM with a Memory Cell Resembling DRAM

A ferroelectric capacitor and a memory circuit resembling a DRAM circuit are included in this kind of FRAM [3, 4]. In comparison to FRAM, which uses the read/write operation voltage to apply a plate line, DRAM typically has a growth and cell structure with a voltage set at half of the provided voltage. There are two different types of memory circuits: The 1T1C, which consists of a single transistor, and the 2T2C, which consists of two transistors and two capacitors.

As a result of its two complementary memory units, 2T2C FRAM is extremely dependable despite having a larger memory area than 1T1C memory circuit, which is appropriate for high memory densities due to its tiny memory size.

Several electronics firms offered high density (≥1 Mb) FRAMs utilizing 1T1C stacked capacitors in the late 1990s and into the 2000s [15–18]. FRAMs of up to 8 Mb are currently available in the market. Figure 6.6 presents FRAM in the capacitor-type with a memory architecture similar to DRAM.

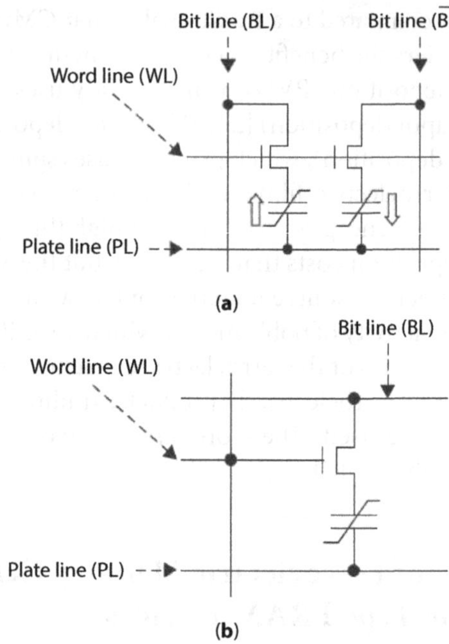

Figure 6.6 FRAM in the capacitor-type with a memory architecture similar to DRAM: (a) a 2T2C; and (b) a 1T1C [27].

6.3 Write/Read Operations in the FRAM Using a Capacitor-Type Memory Cell that Resembles a DRAM

It provides a brief explanation of the fundamental write/read process of a FRAM using a capacitor- type memory cell that resembles DRAM (Figure 6.7). The bit line is initially assigned to "0" or "1" in a write operation. A voltage pulse is then supplied to the plate line once the selection transistor has been turned on. When the PL is at the maximal level (Figure 6.7(a)), data "0" is stored on the capacitor; when the PL is at the minimal level (Figure 6.7(b)), data "1" is stored. This process enables the capacitor to receive the entire supply voltage in both directions while sufficiently altering the polarization. The BL is first charged at 0 V and then set to floating during a read operation. The BL voltage rises as a result of charge flowing from the cell capacitor to the BL after the selection transistor is turned on and the PL is driven to the high level [39–40].

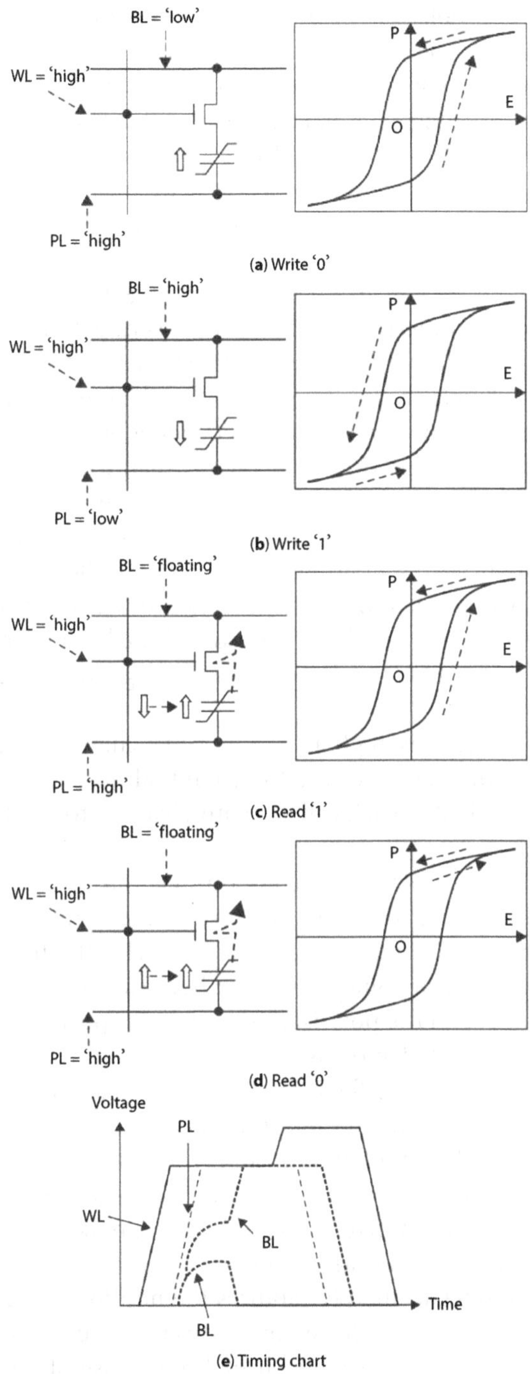

Figure 6.7 Read/write operation of FRAM and timing chart [28].

When reading "1" polarization reversal charges flow to the BL (Figure 6.7(c)), but when reading "0" linear charges flow to the BL (Figure 6.7(d)). Because the polarization reversal charge is greater than the linear charge, the rise in the BL voltage from detecting "1" is often more than that from detecting "0". As it requires polarization reversal to read data, the read operation described above for FRAM is frequently referred to as "a destructive read" [60–63].

When using a 2T2C-type amplifier, FRAM compares the levels of the capacitors to determine whether it is "0" or "1" however when using a 1T1C-type amplifier, FRAM compares the bit line voltage to the reference voltage to identify the state. The ferroelectric capacitor's state determines the amount of polarization reversal charge, which might decrease if the capacitor is damaged by imprint, depolarization, fatigue, as described further. As a result, the 1T1C FRAM's sensing margin is highly dependent on the reference voltage.

Consequently, 1T1C FRAM is less durable than 2T2C FRAM. The term "High-Z", which refers to a high impedance, is used to describe this reading procedure. In this design, the externally provided voltage Vdd is applied to every cell capacitor as the floating bit line reduces the applied voltage by [Cb/(Cb+Cs)], where Cb and Cs are the BL's and the capacitor's respective capacitances. Kawashima *et al.* have presented the bit line ground sensing (BGS) method to increase the detecting percentage in the higher densities FRAMs [28].

Since the plate line is high in the p-MOS charge transfer's sensing method and the BL state is closer to ground, almost the whole external voltage may be applied to each cell capacitor, leading to a noticeably larger sensing margin.

The design of the FRAM depends on accurate polarization switching behavior prediction. However, because of the complicated hysteretic behavior, there are no suitable simulation models. The hysteresis loop is dependent on the rate of voltage variation because the applied voltage has a substantial impact on the polarization switching speed.

In generally presented models, the coercive voltage (the voltage that corresponds to the coercive field- is considered as a fixed value), while it really fluctuates depending on the conditions. A circuit simulation model that replicates the behavior of polarization switching under any voltage settings has been suggested [29–35]. A collection of capacitor elements with leftover polarization and a (fixed) coercive voltage make up the parallel element concept. By providing each component a voltage dependent switching period, the model accurately captures the nonlinear characteristics of the capacitor across a wide voltage range. This model depicts the ferroelectric capacitor's impaired states, which makes it a useful tool for creating FRAM circuits that are resistant to ferroelectric deterioration [64–68].

6.4 Other Capacitor-Type FRAM

Two ferroelectric capacitors and six transistors make up the "non-volatile SRAM" (6T2C) that Dimmer and Eaton proposed [30]. When an external power supply is present, it functions as normal SRAM (non-destructive read), but when it is not, it stores data on the ferroelectric capacitors. As a result, this kind of FRAM has the capacity to read and write data at high speeds that are nearly as fast as those of traditional SRAM. Using a non-volatile SRAM having just a 6T4C structure, Masui *et al.* enhanced the "recall operation" of 4T2C FRAM [31]. They suggested using their 6T4C non-volatile SRAM having a 4 nanoseconds data speed in a highly efficient dynamically programmed gate array. Non-volatile logic has been proposed to minimize power consumption by distributing storage components over the logic-circuit plane. Distributed system components' standby signals can be controlled with the help of ferroelectric-based non-volatile storage components [32].

A "chain FRAM" that consists of the parallel connection of one ferroelectric capacitor and one transistor was presented as a way to minimize the size of memory cells. A block-selecting transistor and several memory cells linked in series make up a memory cell block. According to sources, a chain FRAM design may attain a unit cell coverage of $4F^2$ [33].

6.5 FRAM of FET Type

A ferroelectric insulating film is placed to the FET's gate in a single transistor type FRAM (Figure 6.8). Since the polarization pattern of the ferroelectric film affects the transistor threshold voltage data is recorded in the gate ferromagnetic film and taken out as transistors drain current. It is fortunate because this kind of FRAM supports non-destructive read operations and is totally scalable with transistor design. A thin layer of triglycine sulphate was placed among a liquid crystal transistor and a gold base electrode to form a "solid state variable resistor," as Moll and Tarui termed it in 1963 [35].

Wu developed a FET-type FRAM using Bi4Ti3O12 in 1974 using a metal-ferroelectric-Si surface gate arrangement. Metallic Electric-semiconductor refers to this gate structure [13]. When fabricating transistors, ferroelectric material and silicon undergo a chemical reaction that alters the transistor's properties in MFS structure [38–41].

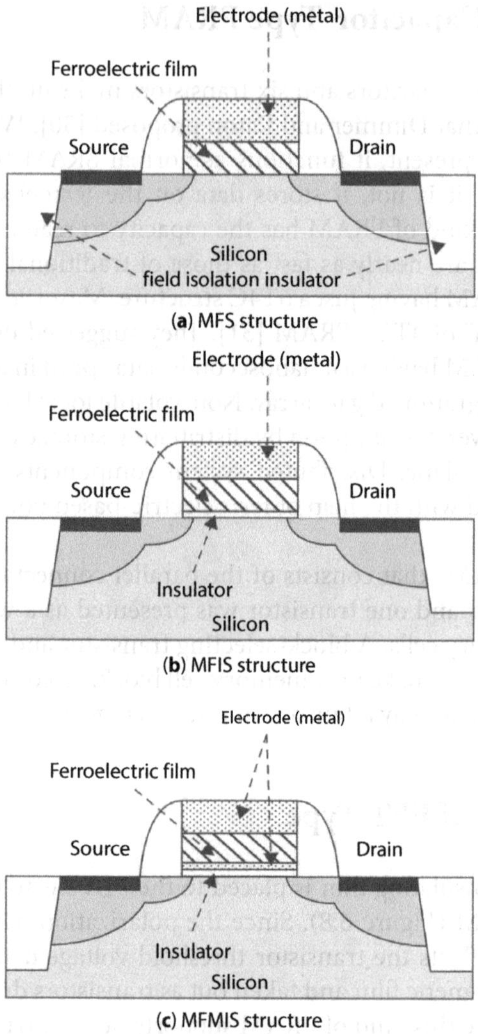

Figure 6.8 FRAM of FET-type [34].

MFMIS substrates and MFIS substrates were suggested as solutions to this deterioration [36–38]. Though several testing have been carried out, these FRAMs have a limited capacity for data retention. The use of an MFIS gate structure, according to Takahashi *et al.*'s (2005) research, resulted in a significant improvement in data store retention lasting up to 30 days [42]. In order to adapt FET-type FRAM to NAND FLASH circuits,

Sakai proposed the usage of MFIS-type ferromagnetic in 2008. This might result in a reduction in the wipe away voltage compared to conventional Flash memory [43].

6.6 Memory Utilizing a Ferroelectric Tunnel Junction

Memory utilizing a Ferroelectric tunnel junction is a sandwich of two metals with a several nanometer thick ferroelectric coating. The value of current all over a semiconducting sheet is severely impacted by the propagation direction. Non-destructive reading is anticipated to be possible with this memory. Esaki *et al.* first suggested tunnel connections using a ferroelectric barrier in 1971 [44]. After the FTJ (Pt/PbZr0.52Ti0.48O3 or BaTiO3/SrRuO3) was reported to operate at room temperature in 2002, in numerous reported tests, a high ON/OFF rate and low power usage were sought after [42]. Regarding this type of memory, the consistent and efficient control of nanometer-thick ferroelectric materials continues to be challenged [45, 46].

6.6.1 Previous Ferroelectric Memory Designs

Initial ferroelectric memory designs comprised not just the 2T2C cell mentioned, that is the type of dominant ferroelectric memory in use nowadays, but also prior less reliable cells like the cross-point matrix array, that experienced losing of data due to the half select impact. This also comprised multiple kinds of shadow RAMs that took on with the market's existing battery backup SRAM devices. Ferroelectric Shadow RAMs also became available in logic circuits as macros [69–73].

6.7 Cross Point Matrix Array

Previous research on ferroelectric memory employed a basic cross point matrix array structure, as seen in Figure 6.9 [47].

The half-select effect harmed these basic arrays. The voltage was applied to all of the cells across the chosen column and chosen row, although only chosen cell observed the voltage supplied to each of rows and columns. As a result of losing information caused by half select signals given to not chosen cells, the amount of read and write cycles was dramatically decreased compared to what might be predicted for just one cell of a ferroelectric array.

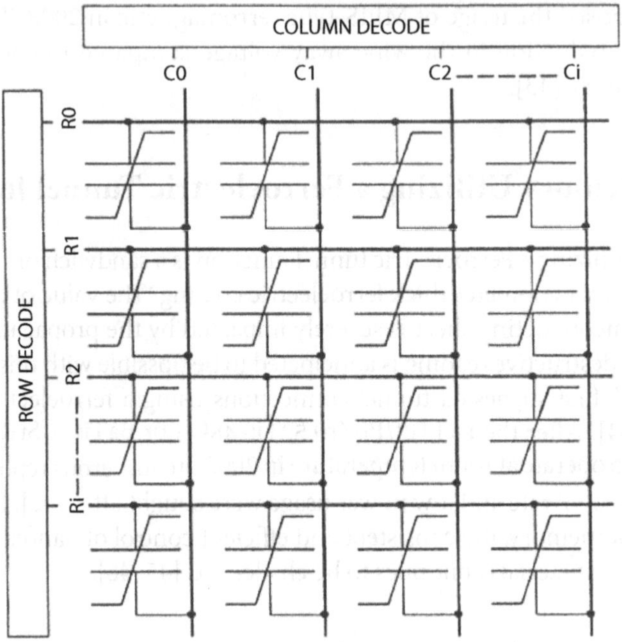

Figure 6.9 Basic cross point matrix array [47].

Previous ferroelectric memories were additionally affected with extreme endurance, impacts at 2windividual unit range, that were eventually mitigated by advancements in ferroelectric substance characteristics.

6.8 Ferroelectric Shadow RAMs

Ramtron demonstrated a shadow RAM that connected a transistor to a ferroelectric capacitor in series to separate it from the bit line [47]. This separation helps to keep disrupt pulses away from the not chosen ferroelectric capacitors. As illustrated in Figure 6.10 [47], the capacitors which were linked to the SRAM cell's storage node.

They presented the notion of a cycled a shared plate link for a ferroelectric memory cell with two capacitors in the same research. Because this is a shadow RAM, the ferroelectric standby capacitors are configured only on power up and power down, bypassing the fatigue constraints associated with SRAM cells [74–80].

By powering down the circuit, the transistors which are in series are switched on along with the shared plate kept at "0," causing the ferroelectric

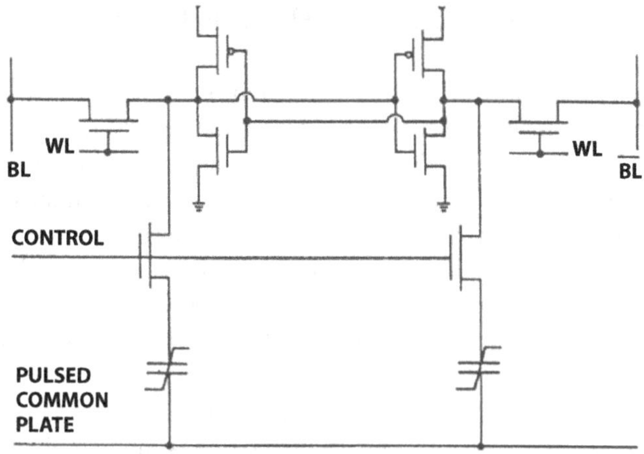

Figure 6.10 Shadow RAM using a cycled shared plate link and a separating transistor connected in series to a ferroelectric capacitor [47].

capacitor at the highest node to be configured high. The shared plate is subsequently driven to Vdd, causing the ferroelectric capacitor at the lowest node to be configured low while the second capacitor remains unaffected. The data of SRAM latch is now encoded into both of the capacitors. When both capacitors are powered up, the shared plate is pulsating from 0 to Vdd, with the bit lines equalized to ground. This switches one ferroelectric capacitor while not another, resulting in a bit line imbalance that can be detected and encoded into the Static RAM [48].

Whereas one of the main benefits of the shadow RAM idea is the fact that ferroelectric fatigue disadvantages are prevented because the capacitors are configured only on power up and power down, among its disadvantages the area cost of the ten devices employed in the cell (8 transistors and 2 capacitors).

To conserve space on some of such transistors, NEC proposed a shadow RAM macro in 2000 that deleted the series transistors from the ferroelectric capacitors [49, 50].

As a result, every ferroelectric capacitor is directly coupled to an Static RAM cell storage node. The plate connections were tuned to Vdd/2 to ensure only a slight voltage imbalance through the capacitors, also the ferroelectric capacitors' coercive voltage was set to larger than Vdd/2. The goal was to prevent polarization changes also thereby fatigue. Another feature of this cell was its quick operation and lack of demand for enhancing circuits [81–83].

6.9 2T2C Ferroelectric RAM Architecture

Krysalis, an early pioneer in this field, demonstrated the 2T2C FRAM design in practically its present configuration for the first time in 1988 [51]. Figure 6.11 [51] depicts an illustrated block diagram of this component.

A dual ended sensory method and a self referencing differentiated signal are used in the portion. The two accessing transistors are controlled by the shared word-line. Two capacitor plates are controlled by the same driving line. In fundamental functionality, this 2T2C capacitor is practically unaltered today.

Figure 6.12 depicts a design of the standard 2T2C data route as it is used nowadays.

The sensing amplifier is adjusted to the required condition before writing. After activating the word line, the plate line is made high. The capacitor of the sense amplifier's grounded side is polarized to the Q(0) state. When the drive line is grounded, the capacitor of the sense amplifier's high side is polarized to the Q(1) state.

For reading the cell, the bit lines have been precharged to ground and floated, the sensing amplifier is turned off, and the plate line is connected to a voltage high. Whenever the word line is triggered, just a single capacitor switches polarization, resulting in a differential charge flowing on the bit

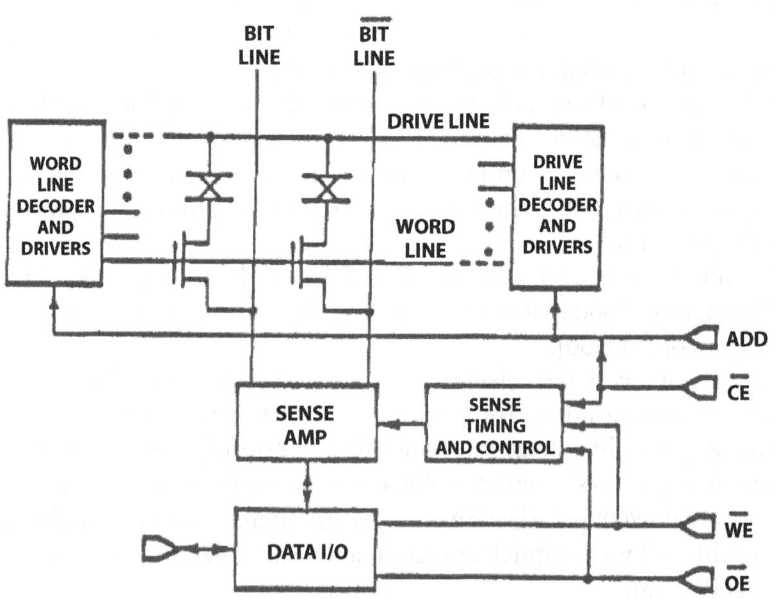

Figure 6.11 Previous 2T2C Ferroelectric RAM Architecture [51].

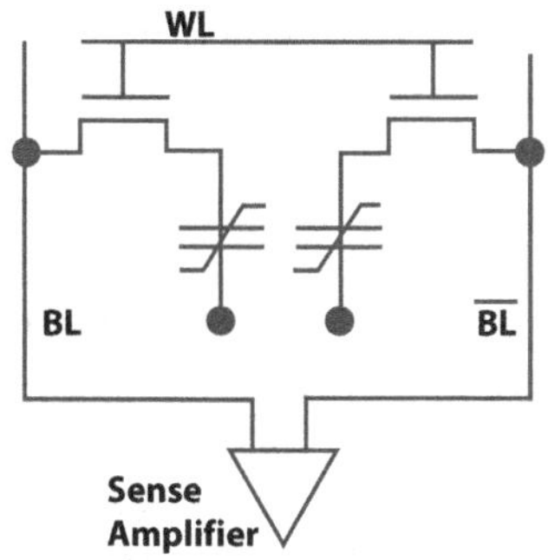

Figure 6.12 Circuit illustration of 2T2C Ferroelectric RAM cell architecture [52].

line and bit line bar. When the sensing amplifier is turned on, the higher bit line voltage is generated by the switched capacitor, while the other side is driven low. If the plate line becomes low, the switched capacitor is reset and the initial information will return to the cell. This restoration is performed as part of the writing procedure and is not visible to the user [84–88].

Krysalis demonstrated a more detailed architectural picture of the data path of the 2T2C ferroelectric memory in 1989 [53]. This 16K-bit component was constructed in n-well CMOS which has a chip enable access period of 200 ns. As illustrated in Figure 6.13, it had a capacitor plate line that ran parallel to the bit lines and linked all bits in a single column. The operation is substantially the same as in the previous section.

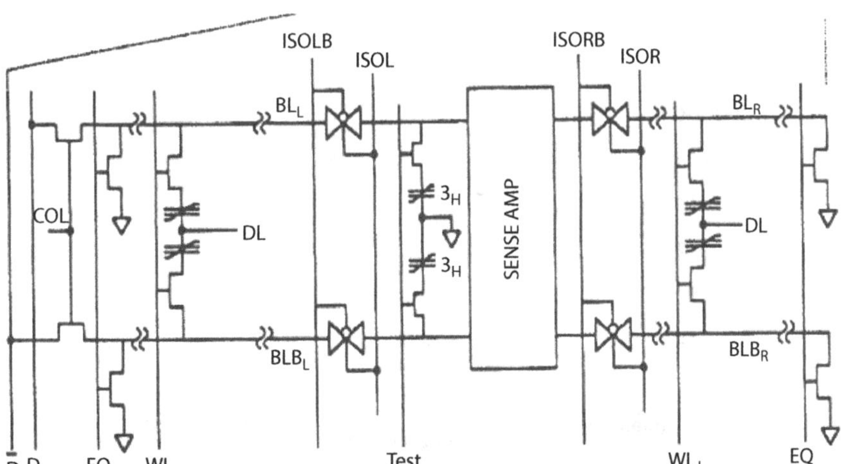

Figure 6.13 A 16K-bit ferroelectric RAM data path circuit diagram [53].

In the dual ended sensing technique, the two 1T1C cell components which were employed as a 2T2C cell to provide a differential signal between the bit line bar and the bit line which is self-referencing. Since the capacitor plate line was shared by the whole column, neglected column bits received the plate signal at the time of read and write operations. It had no effect on the capacitors within the "0" state, but it did affect the capacitors within the "1" state's residual polarization. This component included a voltage sensor circuit that detected when the 5V Vdd dipped below 3.6V.

NEC provided another instance of a FeRAM device of 2T2C in 1998. They demonstrated a synchronized FeRAM macro designed for use in the microcontroller to substitute the EEPROM previously utilized in this application. Figure 6.14 [54] depicts the corresponding synchronous timing diagrams associated with this macro.

These time graphs show how the timings of write and read operations of FeRAM are comparable since both include switching the ferroelectric capacitor to determine its polarization direction [54].

The benefits of this synchronized FRAM were its rapid write and read speeds, as well as the power savings from not having to incorporate the high voltage on chip voltage generators necessary on the EEPROM for write and erase onto the microcontroller chip [89–94].

Because the time for read and write operations is identical, a FeRAM macro that functions entirely synchronous with the on chip CPU core was developed. In this circuit, operation is therefore similar to that of an SRAM. An FeRAM macro of this type might be employed to substitute each the

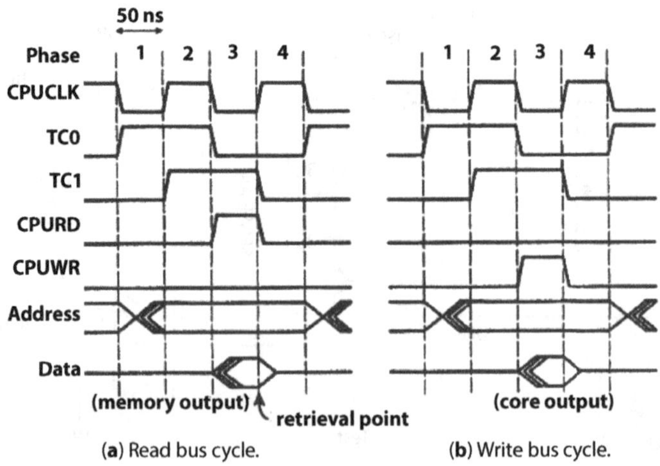

Figure 6.14 Timing diagrams of read and write operations in a synchronous FeRAM [54].

EEPROM and the Static RAM built inside this microcontroller. As such, it is the predecessor of today's embedded FeRAM-based microcontrollers.

6.9.1 Evaluation of FRAM Devices' Reliability

Depolarization, imprint, and fatigue are the three main types of electrical deterioration processes of ferroelectric capacitors. These degradations are seen as a deviation from P-E hysteresis (Figure 6.15). As Pr drops from the starting value (Figure 6.15(a)), depolarization is shown. As hysteresis moves away from its starting point, imprint is seen (Figure 6.15(b)). After storing the "1" state, P-E hysteresis, for instance, switches to the + direction. In Figure 6.15(c), fatigue is shown as the hysteresis decreases from its initial value. The main reason for these degradations has been the subject of several studies. It is thought that certain kinds of deterioration are induced by charged defects, such as oxygen vacancies or their associated flaws [9, 55, 56].

An appropriate testing method is essential for assessing the aforementioned deterioration of ferroelectric capacitors in highly dependable mass manufacturing of FRAM. With the use of a retention test in Figure 6.16(a), which comprises writing, saving, and reading stored data, the ferroelectric capacitor's ability to preserve the original recorded data is evaluated. If the ferroelectric capacitor is destroyed by depolarization, the data accessing procedure in this experiment could not be successful.

An imprint test, illustrated in Figure 6.16(b), examines whether a ferroelectric capacitor can overwrite the data stored as the opposite data by recording and storing 1 and then overwriting the recorded 1 as 0.

The steps in this test are to write data, save it first, edit the opposing data, and then get the modified data. In this test, it's possible that the reading procedure with the altered data will fail if the ferroelectric capacitor's hysteresis shifts Figure 6.16(b), since doing the reverse is now more challenging due to a rise in Ec. An examination of the ferroelectric capacitor's read/write performance is done after several 0–1 switches using a switching/fatigue retention test Figure 6.16(c). Commonly produced FRAMs have switching capabilities more than 10^{13}, or 10^{7} times that of Flash memory.

6.9.2 Comparative Analysis of FeRAM to Other Memory Technologies

Ramtron's FeRAM now falls into two separate groups: serial FeRAM and parallel FeRAM. Serial FeRAM is separated into two series: I2C 2-line

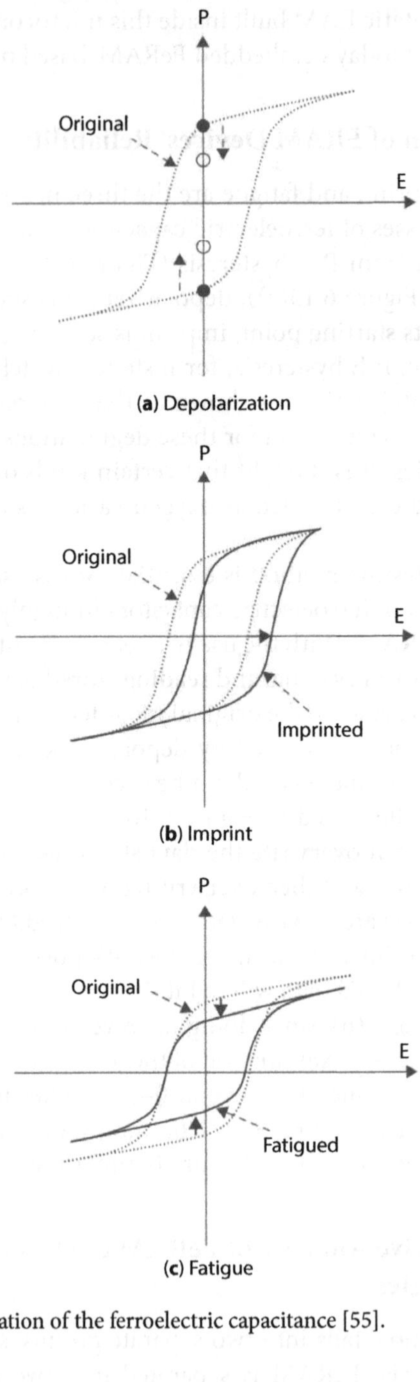

Figure 6.15 Deterioration of the ferroelectric capacitance [55].

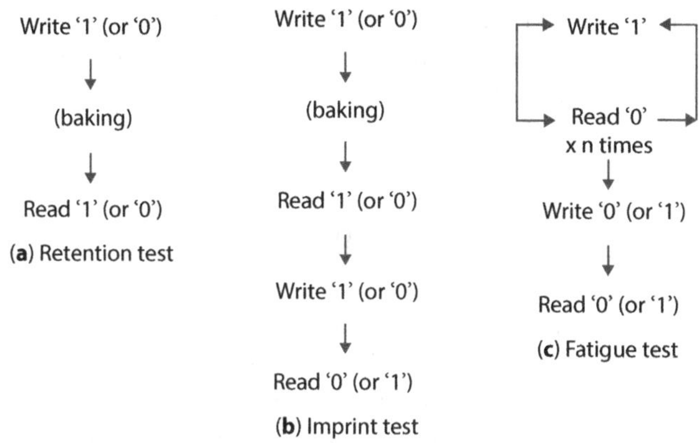

Figure 6.16 Testing procedures of FRAM [56].

FM24 series and SPI 3-line FM25 series. Serial FeRAM pins and timing are interchangeable with the standard 24xx and 25xx EEPROM pins and timing.

FeRAM devices offer the benefits of RAM and ROM, as well as high read and write speeds; they may also be utilized as non-volatile memory. Because of the limitations of ferroelectric particles, the number of accesses is restricted, after which FeRAM becomes volatile. The maximum access duration specified is 10 billion, although this does not indicate that FeRAM will be terminated if it exceeds this limit. FeRAM isn't non-volatile in this sense, but it may still be utilized as regular RAM [116–117].

6.10 FeRAM vs. EEPROM

As an alternative to EEPROM, FeRAM can be employed. FeRAM access speed is considerably quicker than EEPROM performance. When employing FeRAM, it has to be verified that there is no damage once 10 billion accesses are down to FeRAM in the system [95–98].

6.11 FeRAM vs. Static RAM

Static RAM outperforms FeRAM in terms of speed, cost, and accessibility; yet, FeRAM has certain benefits in terms of overall design.

Non-volatile FeRAM may store starting programs as well as configuration data. If all of the memory in the application have a maximum access time of 70ns, one piece of FeRAM can be employed to finish the system, simplifying the system layout.

6.12 FeRAM vs. Dynamic RAM

Dynamic RAM is appropriate for situations in where density and cost are more essential than the speed of access. Dynamic RAM, for example, is the greatest option for graphics display memory. The amount of pixels to be saved is high, and the recovery time is not critical. Use volatile Dynamic RAM memory if you don't need to store previous data at the following boot. When compared to FeRAM, the use and price of Dynamic RAM are appropriate. To summarize, Dynamic RAM can't be completely replaced with FeRAM.

6.13 FeRAM vs. Flash Memory

Flash memory is currently the most often utilized program memory since it is more accessible and less expensive to use. The program memory has to be non-volatile and rewriteable, although the utilization of FRAM is constrained by time to access [57].

6.13.1 Uses of FRAM Devices

Commercialized FRAMs may be classified into two groups based on their use for memory: standalone FRAM and embedded FRAM. The latter are geared for specialized customers, whilst the former only test memory and are designed for general users [99–103].

Storage capacities for commonly available standalone FRAM varies from a several Kb to several Mb. FRAMs are frequently used to replace traditional EEPROMs in order to save energy and enhance read/write performance. For instance, electric power meters employ stand-alone FRAMs for their memory since they need to rewrite data several times per second. Due to a lack of rewriting capacity (10^5), EEPROM cannot be used for this application. FRAMs are frequently used in place of battery-backed SRAMs, which has several advantages because it is not only more affordable but also better for the environment than battery-backed memory. FRAMs are also required for data monitoring, which is used to store technical information, in automation and automobiles because of their increased read/write and rewrite functionality [104–108].

FRAM-embedded LSIs come in a variety of forms, including sophisticated smart cards, RFID tags, and authentication LSIs. Greater communication distances between card/tag and reader/writer are possible in wireless applications due to decreased FRAM power consumption [109–115]. FRAM is also appropriate for smart card or RFID tag authentication since its fast read/write speeds allow for the adoption of powerful encryption methods. Compared to DRAM, EEPROM, and Flash memories, Gamma radiation is thought to be far less likely to damage FRAM. This is so that data is not stored in FRAM as electrostatic force, which gamma radiation has a significant negative influence on [58]. As a result, FRAM may be used for RFID tags linked to gamma-sterilized medical equipment [115–118].

6.14 Conclusion and Upcoming Trends

Although scaling is still one of its largest challenges, FRAM is the first high-end non-volatile memory to be commercially available, ahead of MRAM, PRAM, and ReRAM. As of 2021, depending on the manufacturer and application, ferroelectric RAM (FeRAM) technology is being employed at a variety of technology nodes ranging from 130 to 40 nm.

For example, Cypress Semiconductor manufactures FeRAM products using the 130 nm technology node, whereas Rohm Semiconductor manufactures FeRAM products using the 180 nm technology node. Moreover, Toshiba Memory and Fujitsu Semiconductor have been working on FeRAM products using the 40 nanometer technology node. Nonetheless, 40 nm FeRAM technology, in general, has significant advantages over previous nodes, including increased density, reduced power consumption, and quicker access times. Toshiba Memory's 40 nanometer FeRAM technology has the following specifications: 4 MB capacity (megabits), 1.8 V is the operating voltage, access time is 30 nanoseconds, 10^{12} cycles of read/write endurance, temperature range of operation is -40°C to +85°C. This progress is somewhat gloomy. FRAM has the ability to encounter its technical challenges and surpass these technologies, though, due to the recent major advances in ferroelectric developments taking place of FET-type FRAM components, as outlined above, and with just a moderate advancement in comparison to other enhanced semiconductor materials.

References

1. Pulvari, C.F. (1951), 'An electrostatically induced permanent memory', *J. Appl. Phys.*, 22: 1039–44.
2. Ross, I.M. (1955), 'Semiconductive translating device', *US Patent*, PN 2791760.
3. Evans, J.T. and Womack, R. (1988), 'An experimental 512-bit non-volatile memory with ferroelectric storage cell', *IEEE J. Solid-St. Circ.*, 23: 1171–5.
4. Eaton, S.S., Butler, D.B., Parris, M., Wilson, D. and McNeillie, H. (1988), 'A ferroelectric non-volatile memory', *IEEE Int. Solid-State Cir. Conf. Tech.*, 130–1.
5. Nakamura, T., Nakao, Y., Kamisawa, A. and Takasu, H. (1995), 'A single-transistor ferroelectric memory cell', *IEEE Int. Solid-St. Cir. Conf. Tech.*, 68–9.
6. Arita, K., Shimada, Y., Uemoto, Y., Hayashi, S., Azuma, M. *et al.* (1996), 'Ferroelectric non-volatile memory technology with bismuth layer-structured ferroelectric materials', *IEEE Int. Ferro.*, 1: 13– 16.
7. Izyumskaya, N. & Alivov, Yahya & Cho, Seong-Jun & Morkoç, H. & Lee, Howard & Kang, Y.-S. (2007). Processing, Structure, Properties, and Applications of PZT Thin Films. Critical Reviews in Solid State and Materials Sciences - *Crit. Rev. Solid State Mat. Sci.* 32. 111-202. 10.1080/10408430701707347.
8. Yamazaki, T., Inoue, K., Miyazawa, H., Nakamura, M., Sashida, N. *et al.* (1997), 'Advanced 0.5 μm FRAM device technology with full compatibility of half-micron CMOS logic device', *Int. El. Device Meet.*, 613–16
9. Ishibashi, Y. and Takagi, Y. (1971), 'Note on ferroelectric domain switching', *J. Phys. Soc. Jpn.*, 31: 506–10.
10. Scott, J.F. (2000), *Ferroelectric Memories, Advanced Microelectronics*, 3rd edition, K. Itoh and T. Sakurai, eds., Springer, pp. 1–22.
11. Larsen, P., Kampschoer, G., Ulenaers, M., Spierings, G. and Cuppens, R. (1991), 'Nanosecond switching of thin ferroelectric films', *Appl. Phys. Lett.*, 59: 611–13.
12. Hase, T. and Shiosaki, T. (1991), 'Preparation and switching kinetics of Pb(Zr,Ti)O3 thin films deposited by reactive sputtering', *Jpn. J. Appl. Phys.*, 30: 2159–62.
13. Zhu, W., Liu, Z. Q., Lu, W., Tse, M.S., Tan, H.S. *et al.* (1996), 'A systematic study on structural and dielectric properties of lead zirconate titanate/ (Pb,La)(Zr(1-x)Ti(x))O3 thin films deposited by metallo- organic decomposition technology', *J. Appl. Phys.*, 79: 4283–91.
14. Chai, F.K., Schrimpf, R.D., Brews, J.R., Birnie III, D.P., Galloway, K.F. *et al.* (1995), 'Effects of scaling thickness and niobium doping level on ferroelectric thin film capacitor memory operation', *Int. El. Device Meet.*, 123–26.
15. Kawai, H. (1969), 'The piezoelectricity of poly(vinylidene fluoride Jpn)', *J. Appl. Phys.*, 8: 975–6.

16. Bergman, Jr., J.G., McFee, J.H. and Crane, G.R. (1971), 'Pyroelectricity and optical second-harmonic generation in polyvinylidene fluoride films', *Appl. Phys. Lett.*, 18: 203–6.
17. Schroeder, R., Majewski, L.A., Voigt, M. and Grell, M. (2005), 'Memory performance and retention of an all-organic ferroelectric-like memory transistor', *IEEE Electr. Device L.*, 26: 69–71.
18. Fujisaki, S., Ishiwara, H. and Fujisaki Y. (2007), 'Low-voltage operation of ferroelectric poly (vinylidene fluoride-trifluoroethylene) copolymer capacitors and metal- ferroelectric-insulator- semiconductor diodes', *Appl. Phys. Lett.*, 90: 162902.
19. Ghonge, S.G., Goo, E., Ramesh, R., Sands, T. and Keramidas, V.G. (1993), 'Microstructure of epitaxial La0.5Sr0.5Co3/ ferroelectric Pb0.9La0.1 (Zr0.2Ti0.8)0.975O3 / La0.5Sr0.5CoO3 heterostructures on LaAIO3', *Appl. Phys. Lett.*, 63: 1628–30.
20. Shoji, K., Moniwa, M., Yamashim, H., Kim, T., Kaga, T. *et al.* (1996), 'A 7.03-μm^2 Vcc/2-plate non- volatile DRAM cell with a Pt/PZT/Pt/TiN capacitor patterned by one- mask dry etching', *IEEE Symp. VLSI Tech.*, 28–9.
21. Moise, T.S., Summerfelt, S.R., Xing, G., Colombo, L., Sakoda, T. *et al.* (1999), 'Electrical properties of submicron (÷0.13 μm^2) Ir/PZT/Ir capacitors formed on W plugs', *Int. Elect. Device Meet.*, 940–2.
22. Lee, S.Y., Kim, H.H., Jung, D.J., Song, Y.J., Jang, N.W. *et al.* (2001), 'Highly scalable sub-10F^2 lTlC COB cell for high density FRAM', *IEEE Symp. VLSI Tech.*, 111–12.
23. Horii, Y., Hikosaka, Y., Itoh, A., Matsuura, K., Kurasawa, M. *et al.* (2002), '4 Mbit embedded FRAM for high performance System on Chip (SoC) with large switching charge, reliable retention and high imprint resistance', *Int. El. Device Meet.*, 539–42.
24. Van Buskirk, J.P.C, Roeder, F., Baum, T.H., Bilodeau, S.M., Russell, M.W. *et al.* (1998), 'Common and unique aspects of perovskite thin film CVD process', *Integr. Ferroelectr.*, 21: 273–98.
25. Funakubo, H., Aratani, M., Oikawa, T., Tokita, K. and Saito, K. (2002), 'Ferroelectricity of one-axis- preferred-oriented polycrystalline Pb(Zr,Ti)O3 films prepared by pulsed- metallorganic chemical vapor deposition', *J. Appl. Phys.*, 92: 6768–72.
26. Ishiwara, Hiroshi. (2012). FeFET and ferroelectric random access memories. 10.1093/acprof:oso/9780199584123.003.0012.
27. TABUCHI, Y. & HASEGAWA, S. & TAMURA, T. & HOKO, H. & Kato, Kazuhiko & ARIMOTO, Y. & Ishiwara, Hiroshi. (2007). Multi-bit programming technique for an MFIS-FET with a Pt/(Bi, Nd)(4) Ti3O12/HfO2/Si substrate structure. *Integr. Ferroelectr.* 89. 171-179. 10.1080/10584580601077740.
28. Kawashima, S., Endo, T., Yamamoto, A., Nakabayashi, K., Nakazawa, M. *et al.* (2002), 'Bit line GND sensing technique for low-voltage operation FeRAM', *IEEE J. Solid-St. Circ.*, 37: 592–8.

29. Tamura, T., Arimoto, Y. and Ishiwara, H. (2002), 'A new circuit simulation model of ferroelectric capacitor', *Jpn. J. Appl. Phys.*, 41: 2654–7.
30. Dimmler, K. and Eaton Jr., S.S. (1987), 'Memory cell with volatile and non-volatile portions having ferroelectric capacitors', *US Patent*, PN 4809225.
31. Masui, S., Ninomiya, T., Oura, M., Yokozeki, W., Mukaida, K. *et al.* (2003), 'A ferroelectric memory- based secure dynamically programmable gate array', *IEEE J. Solid-St. Circ.*, 38: 715–25.
32. Kimura, H., Hanyu, T., Kameyama, M., Fujimori, Y., Nakamura, T. *et al.* (2004), 'Complementary ferroelectric-capacitor logic for low-power logic-in-memory VLSI', *IEEE J. Solid-St. Circ.*, 39: 919– 26.
33. Shiga, H., Takashima, D., Shiratake, S., Hoya, K., Miyakawa, T. *et al.* (2010), 'A 1.6 GBs DDR2 128 Mb chain FeRAM with scalable octal bit line and sensing schemes', *J. Solid-St. Circ.*, 45: 142–52.
34. Yongbiao, Zhai & Feng, Zihao & Zhou, Ye & Han, Su-Ting. (2021). Energy-Efficient Transistors: Suppressing The Subthreshold Swing below The Physics Limit. *Mater. Horiz.* 8. 10.1039/D0MH02029J.
35. Moll, T L. and Tarui, Y. (1963), 'A new solid state memory resistor', *IEEE T. Electron Dev.*, ED-10: 338–9.
36. Ito, K. and Tsuchiya, H. (1977), 'Memory modes of ferroelectric field effect transistors', *Solid-State Electron.*, 20: 529–37.
37. Baginsky, I.L. and Kostsova, E.G. (1993), 'Information writing mechanisms in thin-film mfis-structures', *Ferroelectrics*, 143: 239–50.
38. Hirai, T., Teramoto, K., Nishi, T., Goto, T., and Tarui, Y. (1994), 'Formation of metal/ ferroelectric/insulator/semiconductor structure with a CeO2 buffer layer', *Jpn. J. Appl. Phys.*, 33: 5219–22.
39. Nakamura, T., Nakao, Y., Kamisawa, A. and Takasu, H. (1995), 'A single-transistor ferroelectric memory cell', *IEEE Int. Solid-State Cir. Conf. Tech.*, 68–70.
40. Tokumitsu, E., Fujii, G. and Ishiwara, H. (1999), 'Non-volatile ferroelectric-gate field- effect transistors using SrBi2Ta2O9/Pt/SrTa2O6/SiON/Si structures', *Appl. Phy. Lett.*, 75: 575–7.
41. Ishiwara, H. (2003), 'Recent progress in FET-type ferroelectric memories', *Int. El. Device Meet.*, 263– 6.
42. Takahashi, K., Aizawa, K., Park, B.E. and Ishiwara, H. (2005), 'Thirty-day-long data retention in ferroelectric-gate field-effect transistors with HfO2 buffer layers', *Jpn. J. Appl. Phys.*, 44: 6218–20.
43. Sakai, S., Takahashi, M., Takeuchi, K., Lia, Q., Horiuchi, T. *et al.* (2008), 'Highly scalable Fe(ferroelectric)-NAND cell with MFIS(metal-ferroelectric-insulator- semiconductor) structure for sub-10 nm tera-bit capacity NAND Flash memories', *Non-Volatile Semiconductor Memory Workshop*, Abstract pp. 103–5.
44. Esaki, L., Leibowitz, R.B. and Stiles, P.J. (1971), 'Polar switch', *IBM Tech. Discl. Bull.*, 13: 2161.

45. Contreras, J.R., Schubert, J., Kohlstedt, H. and Waser, R. (2002), 'Memory device based on a ferroelectric tunnel junction', *60th Device Research Conference,* pp. 97–8.
46. Garcia, V., Fusil, S., Bouzehouane, K., Enouz-Vedrenne, S., Mathur, N.D. *et al.* (2009), 'Giant tunnel electroresistance for non-destructive readout of ferroelectric states', *Nature*, 460: 81–4.
47. Eaton, Butler, Parris, Wilson and McNeillie, "A Ferroelectric Nonvolatile Memory," 1988 IEEE International Solid-State Circuits Conference, 1988 ISSCC. Digest of Technical Papers, San Francisco, CA, USA, 1988, pp. 130-, doi: 10.1109/ISSCC.1988.663665.
48. US Patent 4873664, S.S. Eaton, Jr., "Self Restoring Ferroelectric Memory", October 1989.
49. T. Miwa, *et al.*, "NV- SRAM: A Nonvolatile SRAM with Backup Ferroelectric Capacitors", (NEC), IEEE CICC, May 2000, pp65.
50. T. Miwa, *et al.*, "NV- SRAM: A Nonvolatile SRAM with Backup Ferroelectric Capacitors", (NEC), *IEEE J. Solid State Cir.*, Vol. 36, No. 3, March 2001.
51. J. T. Evans and R. Womack, "An experimental 512-bit nonvolatile memory with ferroelectric storage cell," in *IEEE J. Solid State Cir.*, vol. 23, no. 5, pp. 1171-1175, Oct. 1988, doi: 10.1109/4.5940.
52. N. Tanabe, *et al.*, "A High Density I T I C Cell with Vcc/2 Reference Level for High Stable FeRAMs", (NEC), IEDM, December 1997.
53. R. Womack and D. Tolsch, "A 16 kb ferroelectric nonvolatile memory with a bit parallel architecture," *IEEE J. Solid State Cir. Conf.*, 1989 ISSCC. Digest of Technical Papers, New York, NY, USA, 1989, pp. 242-243, doi: 10.1109/ISSCC.1989.48273.
54. T. Miwa, *et al.*, "An Embedded FeRAM Macro Cell for a Smart Card Microcontroller", (NEC), *IEEE CICC,* May 1998.
55. Mihara, T., Watanabe, H., and Araujo, C.A. (1993), 'Evaluation of imprint properties in sol-gel ferroelectric Pb(ZrTi)O3 thin-film capacitors', *Jpn. J. Appl. Phys.*, 32: 4168–74.
56. https://www.apogeeweb.net/electron/ferroelectric-ram-technique.html
57. Tagantsev, A.K., Stolichnov, I., Setter, N. and Cross, J.S. (2004), 'Nature of nonlinear imprint in ferroelectric films and long-term prediction of polarization loss in ferroelectric memories', *J. Appl. Phys.*, 96: 6616–23.
58. Philpy, S.C., Kamp A.D. and Derbenwick, G.F. (2004), 'Non-volatile and SDRAM ferroelectric memories', *Aerosp. Conf. Proc.*, 4: 2294–9.
59. Anjana Bhardwaj, Pradeep Kumar, B. Raj, Sunny Anand, "Design and Performance Optimization of doping-less Vertical Nanowire TFET using Gate Stack Technique", *J. Electron.Mater. (JEMS)* Springer, Vol. 41, Issuse 7, PP. 4005-4013, 2022.
60. Jeetendra Singh, BRaj, "Tunnel Current Model of Asymmetric MIM Structure Levying Various Image Forces to Analyze the Characteristics of Filamentary Memristor", *Appl. Phys. A*, Springer, Vol. 125, No.3, pp. 203.1 to 203.11 Feb 2019.

61. Candy Goyal, Jagpal Singh Ubhi and B Raj, Low Leakage Zero Ground Noise Nanoscale Full Adder using Source Biasing Technique, "*J. Nanoelectron. Optoelectron.*", American Scientific Publishers, Vol. 14, PP. 360–370, March 2019.
62. Gurmohan Singh, R. K. Sarin and B Raj, "A Novel Robust Exclusive-OR Function Implementation in QCA Nanotechnology with Energy Dissipation Analysis", *J. Comput.l Electron.*, Springer, Volume 15, Issue 2, PP. 455-465, June 2016.
63. Tanu Wadhera, Deepti Kakkar, Girish wadhwa, B Raj, "Recent Advances and Progress in Development of the Field Effect Transistor Biosensor: A Review" *J. Electron. Mater.*, Springer, Volume 48, Issue 12, pp 7635–7646, December 2019.
64. Girish Wadhwa, Priyanka Kamboj, Balwinder Raj" Design optimisation of junctionless TFET biosensor for high sensitivity" *Adv. Nat. Sci.: Nanosci. Nanotechnol.*, Vol. 10, PP. 045001 (7), (2019).
65. Priya Bansal, BRaj, "MEMRISTOR MODELING AND ANALYSIS FOR LINEAR DOPANT DRIFT KINETICS," *J. Nanoeng. Nanomanuf.*, American Scientific Publishers, Vol. 6, pp. 1–7, 2016.
66. Amandeep Singh, Mamta Khosla, BRaj, "Circuit Compatible Model for Electrostatic Doped Schottky Barrier CNTFET," *J. Electron. Mater.*, Springer, Vol. 45, Issue 12, pp 4825-4835, 2016.
67. Ashima, D Vaithiyanathan, B Raj," Performance Analysis of Charge Plasma induced Graded Channel Si nanotube", *J. Eng. Res. (JER)*, EMSME Special Issue, Vol.9, pp. 146-154 Aug 2021.
68. Abhishek Singh Tomar, Vijay Kumar Magraiya, and B Raj, Scaling of Access and Data Transistor for High Performance DRAM Cell Design, *Quantum Matter*, Vol. 2, PP. 412-416 Oct. 2013.
69. Neeraj Jain, BRaj, "Parasitic Capacitance and Resistance Model Development and Optimization of Raised Source/Drain SOI FinFET Structure for Analog Circuit Applications, *J. Nanoelectron. Optoelectron.*, ASP, USA, Vol. 13, PP. 531–539, Apl 2018.
70. Shradhya Singh, S. K. Vishvakarma and B Raj, "Analytical Modeling of Split-Gate Junction-Less Transistor for a Biosensor Application", *Sens. Bio-Sens.*, Elsevier, Vol. 18, Pages 31-36, April 2018.
71. Maisagalla Gopal, and Balwinder Raj, Low Power 8T SRAM Cell Design for High Stability Video Applications, *ITSI Trans. Electr. Electron. Eng.* Vol. 1(5), pp. 91-97, 2013.
72. Balwinder Raj, Jatin Mitra, Deepak Kumar Bihani, Rangharajan V, A.K. Saxena and S. Dasgupta, "Analysis of Noise Margin, Power and Process Variation for 32 nm FinFET Based 6T SRAM Cell" *J. Comput. (JCP)*, Academy Publisher, FINLAND, vol. 5, issue 6, PP.1-8, 2010.
73. Divya Sharma, Rajesh Mehra and B Raj, "Comparative Analysis of Photovoltaic Technologies for High Efficiency Solar Cell Design", *Superlattices Microstruct.*, Elsevier, Volume 153, PP. 106861, May 2021.

74. Pawandeep Kaur, Avtar Singh Buttar, B Raj, "A comprehensive Analysis of Nanoscale Transistor based Biosensor: A Review, *Indian J. Pure Appl. Phys.*, Vol. 59, pp. 304-318, April 2021.
75. Divya Yadav, Balwant Raj and B Raj, "Design and Simulation of Low Power Microcontroller for IoT Applications", *J. Sens. Lett.*, ASP, Vol. 18, PP. 401–409, May 2020.
76. Shailendra Singh, B Raj,"A 2-D Analytical Surface Potential and Drain current Modeling of Double-Gate Vertical t-shaped Tunnel FET", *J. Comput. Electron.*, Springer, Vol. 19, PP.1154–1163, Apl 2020.
77. Jeetendra Singh, B Raj, "An Accurate and Generic Window function for Nonlinear Memristor Model" *J Comput. Electron.*, Springer, Volume 18 Issue 2, PP. 640-647 June 2019.
78. Manjit Kaur; Neena Gupta; Sanjeev Kumar; BRaj; Arun Kumar Singh, "Comparative RF and Crosstalk Analysis of Carbon Based Nano Interconnects" *IET Circuits, Devices Syst.*, Vol.15(6), PP. 493-503, Feb 2021.
79. Nehru Kandasamy, Firdous Ahmad, D. Ajitha, B Raj, and Nagarjuna Telagam "Quantum Dot Cellular Automata based Scan Flip Flop and Boundary Scan Register, *IETE J. Res.*, Volume 66, PP. 535-548, 2020.
80. Sharma, S. K.., Raj, B., Khosla, M., "Enhanced Photosensivity of Highly Spectrum Selective Cylindrical Gate In1-xGaxAs Nanowire MOSFET Photodetector, *Mod. Phys. Lett.* B, Vol. 33, No. 12, PP. 1950144 (2019).
81. Jeetendra Singh, BRaj, "Design and Investigation of 7T2M NVSARM with Enhanced Stability and Temperature Impact on Store/Restore Energy", *IEEE Trans. Very Large Scale Integr. Syst.*, Vol. 27, Issure 6, PP. 1322 - 1328 June 2019.
82. Anil Kumar Bhardwaj, Sumeet Gupta, BRaj, Amandeep Singh, "Impact of Double Gate Geometry on the Performance of Carbon Nanotube Field Effect Transistor Structures for Low Power Digital Design", *J. Comput. Theor. Nanosci.*, ASP, Vol. 16, PP. 1813–1820, 2019.
83. Neeraj Jain, BRaj, "Thermal Stability Analysis and Performance Exploration of Asymmetrical Dual-k underlap Spacer (ADKUS) SOI FinFET for Security and Privacy Applications", *Indian J. Pure Appl. Phys. (IJPAP)*, Vol. 57, pp. 352-360, May 2019.
84. Amandeep Singh, Mamta Khosla, BRaj, "Design and Analysis of Dynamically Configurable Electrostatic Doped Carbon Nanotube Tunnel FET", *Microelectronics J.*, Elesvier, Volume 85, Pages 17-24, March 2019.
85. Neeraj Jain and Balwinder Raj, "Dual-k Spacer Region Variation at the Drain Side of Asymmetric SOI FinFET Structure: Performance Analysis towards the Analog/RF Design Applications", communicated in "*J. Nanoelectron. Optoelectron.*", American Scientific Publishers, Vol. 14, PP. 349–359, March 2019.
86. Jeetendra Singh, Sanjeev Sharma, B Raj, Mamta Khosla Analysis of barrier layer thickness on performance of In1-xGaxAs based Gate Stack Cylindrical Gate Nanowire MOSFET, " *JNO, ASP*, Vol. 13, PP 1473-1477, Oct 2018.

87. Neeraj Jain, B Raj, "Analysis and Performance Exploration of High-k SOI FinFETs Over the Conventional Low-k SOI FinFET toward Analog/RF Design", *J. Semicond. (JoS)*, IOP Science, Vol. 39, Issue 12, PP- 124002-1-7, Dec 2018.
88. Candy Goyal, Jagpal Singh Ubhi and BRaj, A reliable leakage reduction technique for approximate full adder with reduced ground bounce noise, *Journal of Mathematical Problems in Engineering*, Hindawi, Volume 2018, Article ID 3501041, 16 pages, 15 Oct 2018.
89. Anuradha, Jeetendra Singh, B Raj and Mamta Khosla, "Design and Performance Analysis of Nano-scale Memristor-based Nonvolatile SRAM" *J. Sens. Lett.*", American Scientific Publishers, Vol. 16, PP 798– 805, Oct 2018.
90. Girish Wadhwa, BRaj, "Parametric Variation Analysis of Charge-Plasmabased Dielectric Modulated JLTFET for Biosensor Application" *IEEE Sens. J.*, VOL. 18, NO. 15, AUGUST 1, PP. 6070 – 6077, 2018.
91. Jeetendra Singh, BRaj, "Comparative Analysis of Memristor Models for Memories Design", *JoS*, IoP, Vol. 39, No. 7, PP. 074006-1-12, July 2018.
92. Divya Yadav, Shailesh Singh Chouhan, Santosh Kumar Vishvakarma and BRaj, "Application Specific Microcontroller Design for IoT based WSN", *Sens. Lett.*, ASP, Vol. 16, PP. 374–385, May 2018..
93. Gurmohan Singh, R. K. Sarin and BRaj, "Fault-Tolerant Design and Analysis of Quantum-Dot Cellular Automata Based Circuits", *IEEE/IET Circuits Devices Syst.*, Vol. 12, PP. 638 – 64, 2018.
94. Jeetendra Singh, BRaj, "Modeling of Mean Barrier Height Levying Various Image Forces of Metal Insulator Metal Structure to Enhance the Performance of Conductive Filament Based Memristor Model", *IEEE Nanotechnol.*, Vol. 17, No. 2, PP. 268-267, March 2018.
95. Aakash Jain, Sanjeev Sharma, BRaj, "Analysis of Triple Metal Surrounding Gate (TM-SG) III-V Nanowire MOSFET for Photosensing Application", *Opto-Electron. J.*, Elsevier, Volume 26, Issue 2, Pages 141-148, May 2018.
96. Aakash Jain, Sanjeev Sharma, B Raj, "Design and Analysis of High Sensitivity Photosensor Using Cylindrical Surrounding Gate MOSFET For Low Power Sensor Applications", *Engineering Science and Technology, an International Journal*, Elsevier's, Volume 19, Issue 4, Pages 1864–1870, December 2016.
97. Amandeep Singh, Mamta Khosla, B Raj, "Analysis of Electrostatic Doped Schottky Barrier Carbon Nanotube FET for Low Power Applications," *J. Mater. Sci.: Mater. Electron*, Springer, Vol.28, PP. 1762–1768, 2017.
98. G. Saiphani Kumar, Amandeep Singh, BRaj, "Design and Analysis of Gate All Around CNTFET based SRAM cell Design", *J. Comput. Electron.*, Springer, Volume 17, Issue 1, pp 138–145, March 2018.
99. Gurinder pal Singh, B. S. Sohi, Balwinder Raj, "Material Properties Analysis of Graphene Base Transistor (GBT) for VLSI Analog Circuits, *Indian J. Pure Appl. Phys. (IJPAP)*, Vol. 55, pp. 896-902, Dec-2017.
100. Amandeep Singh, Mamta Khosla, B Raj, "Comparative Analysis of Carbon Nanotube Field Effect Transistor and Nanowire Transistor for Low Power

Circuit Design," *J. Nanoelectron. Optoelectron.*, American Scientific Publishers, USA, Vol. 11, PP. 388-393, June 2016.

101. Sunil Kumar and B Raj, "Estimation of Stability and Performance metric for Inward Access Transistor based 6T SRAM Cell Design using n-type/p-type DMDG-GDOV TFET" *IEEE VLSI Circuits Syst. Lett.*, Volume3, Issue2, PP 25-39, June2017
102. Shashikant Sharma, Anjan Kumar, Manisha Pattanaik and B Raj, "Forward Body Biased Multimode Multi-Threshold CMOS Technique for Ground Bounce Noise Reduction in Static CMOS Adders" *Int. J. Inf. Electron. Eng.*, pp. 567-572, Vol. 3, issue 3, 2013.
103. Hamendra Singh, Pankaj Kumar, Balwinder Raj, "Performance Analysis of Majority Gate SET Based 1-bit Full Adder", *Int. J. Comput. Commun. Eng.* (IJCCE), IACSIT Press Singapore, ISSN: 2010-3743, Volume 2, No. 4, PP.1-8, 2013.
104. Anil Kumar Bhardwaj, Sumeet Gupta and B Raj, "Investigation of Parameters for Schottky Barrier (SB) Height for Schottky Barrier Based Carbon Nanotube Field Effect Transistor Device", *J. Nanoelectron. Optoelectron.*, ASP, Vol. 15, PP. 783–791, July 2020.
105. Priya Bansal, BRaj, "MEMRISTOR: A VERSATILE NONLINEAR MODEL FOR DOPANT DRIFT AND BOUNDARY ISSUES," *JCTN*, American Scientific Publishers, Vol. 14 (5), PP. 2319-2325, May 2017.
106. Neeraj Jain, BRaj, "An Analog and Digital Design Perspective Comprehensive Approach on Fin-FET (Fin-Field Effect transistor) Technology - A Review", *Reviews in Advanced Sciences and Engineering (RASE)*, ASP, Vol. 5, PP. 1-14, 2016.
107. Sanjeev Sharma, B Raj, Mamta Khosla, "Subthreshold Performance of In1-xGaxAs based Dual Metal with Gate Stack Cylindrical/Surrounding Gate Nanowire MOSFET for Low Power Analog Applications, *J. Nanoelectron. Optoelectron.*, American Scientific Publishers, USA, Vol.12, PP. 171–176, 2017.
108. Balwinder Raj, A. K. Saxena and S. Dasgupta, "Analytical Modeling for the Estimation of Leakage Current and Subthreshold Swing Factor of Nanoscale Double Gate FinFET Device" *Microelectronics Int.*, UK, vol. 26, pp. 53-63, 2009.
109. Soniya, Shailendra Singh, Girish Wadhwa and Balwinder Raj, "An Analytical Modeling for Dual Source Vertical Tunnel Field Effect Transistor" submitted to '*Int. J. Recent Technol. Eng. (IJRTE)*' at Volume-8 Issue-2, PP. 603-608, July 2019.
110. Shailendra Singh, B Raj, "Design and analysis of hetrojunction Vertical T-shaped Tunnel Field Effect Transistor", *J. Electron. Mater.*, Springer, Volume 48, Issue 10, pp 6253–6260, October 2019.
111. Candy Goyal, Jagpal Singh Ubhi and B Raj, "A Low Leakage CNTFET based Inexact Full Adder for Low Power Image Processing Applications",

Int. J. Circuit Theory Appl., Wiley, Volume 47, Issue 9, Pages 1446-1458, September 2019.

112. B Raj, A. K. Saxena and S. Dasgupta, "A Compact Drain Current and Threshold Voltage Quantum Mechanical Analytical Modeling for FinFETs" *J. Nanoelectron. Optoelectron. (JNO)*, USA, vol. 3, no. 2, pp. 163-170, 2008.
113. Girish Wadhwa, B Raj, "An Analytical Modeling of Charge Plasma based Tunnel Field Effect Transistor with Impacts of Gate underlap Region" *Superlattices Microstruct.*, Elsevier, Vol. 142, PP. 106512, June 2020.
114. Shailendra Singh, B Raj,"Modeling and Simulation analysis of SiGe hetrojunction Double GateVertical t-shaped Tunnel FET", *Superlattices Microstruct.*, Elsevier Volume 142, PP. 106496, June 2020.
115. Amandeep Singh, Dinesh Kumar Saini, Dinesh Agarwal, Sajal Aggarwal, Mamta Khosla, BRaj, "Modeling and Simulation of Carbon Nanotube Field Effect Transistor and its Circuit Application ," *J. Semicond. (JoS)*, IOP Science, Vol.37, PP. 074001-6, July 2016.
116. Neeraj Jain, Balwinder Raj, "Device and Circuit Co-Design Perspective Comprehensive Approach on FinFET Technology - A Review, *J. Electron Devices*, Vol. 23, Num. 1, pp 1890-1901, 2016.
117. Sunil Kumar and BRaj, "Analysis of I_{ON} and Ambipolar Current for Dual-Material Gate-drain Overlapped DG-TFET," *J. Nanoelectron. Optoelectron.*, American Scientific Publishers, USA, Vol.11, PP. 323-333, June 2016.
118. Naveed Anjum, Tarun Bali and B Raj, Design And Simulation Of Handwritten Multiscript Character Recognition, *Int. J. Adv. Res. Comput. Commun. Eng.*, Vol. 2 (7), PP. 2544-2549, July 2013.

7

Applications of AI/ML Algorithms in VLSI Design and Technology

Jaswinder Singh and Damanpreet Singh*

Department of Computer Science & Engineering, SLIET Longowal, Sangrur, (India)

Abstract

Very large-scale integration (VLSI) is the process of combining a large number of transistors on a single chip to form an integrated circuit. VLSI development is rapidly moving toward the efficient design of a single chip. However, VLSI design faces numerous challenges, including power optimization, design productivity, manufacturing variability and testability. Conventional methodologies are primarily manual, time-consuming and resource-intensive. Artificial Intelligence (AI) and Machine Learning (ML) techniques, on the other hand, provide automated learning approaches for dealing with large amounts of data in complex chip design. Furthermore, incorporating machine learning algorithms can improve system-on-chip performance. AI/ML is becoming more important in VLSI design due to its numerous applications in both logical and physical design. As a result, AI/ML has become an essential component in the VLSI industry. This chapter discusses popular ML algorithms, their mathematical foundations, performance metrics, and how AI and ML applications can transform VLSI design by allowing for high-speed, intelligent and efficient implementations.

***Keywords*:** Artificial intelligence, machine learning, supervised machine learning, unsupervised machine learning, VLSI design, classification, data clustering, performance measures

7.1 Introduction

The term Very Large Scale Integration (VLSI) refers to the process of creating an integrated circuit (IC) through integrating a large number of

**Corresponding author*: damanpreets@sliet.ac.in

Balwinder Raj, Suman Lata Tripathi, Tarun Chaudhary, K. Srinivasa Rao and Mandeep Singh (eds.) Integrated Devices for Artificial Intelligence and VLSI, (157–192) © 2024 Scrivener Publishing LLC

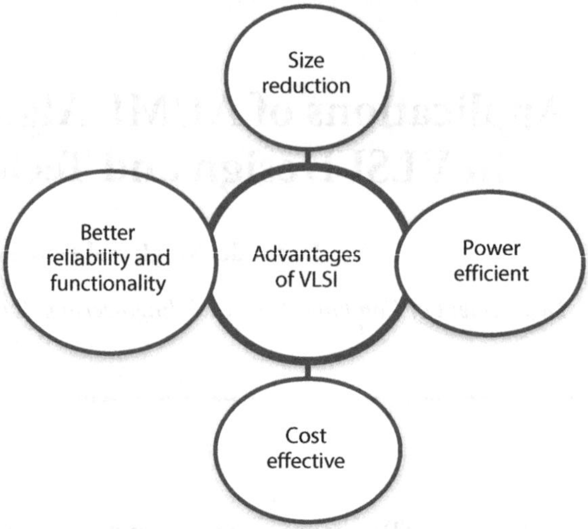

Figure 7.1 Advantages of VLSI design.

transistors onto a single chip [1, 2]. A majority of high-computing electronic circuits have been developed with the help of VLSI technology. Transistors, LEDs, and diodes are examples of integrated circuits that are made during the semiconductor fabrication process and are used in a variety of consumer electronics and electrical appliances.

The advancement of VLSI circuit design, such as microcontroller and microprocessors, has accelerated the growth of computer and communication applications during the past few decades [3]. Usually, components like primary memory, secondary memory and processing unit are needed for designing electronic circuits. The VLSI technology enables the creation of these components on a single chip. Due to the ongoing advancement of VLSI and its applications on diverse system architectures, the electronics industry has made impressive progress over the past several years. For instance, consumer electronics, telecommunication, and video or image processing applications are all developing rapidly. VLSI architecture is known for its small size, cheaper costs, reduced power consumption, excellent reliability, and high performance. In general, the benefits of using a circuit model based on VLSI shown in Figure 7.1. However, VLSI development is still advancing more quickly in the direction of minimizing both power and area [3]. Designing a low power VLSI circuit has a number of challenges, the primary one being how to maintain computing power while using less energy.

It has been found that there is a requirement for reducing the present power consumption of the hardware in order to incorporate Machine Learning (ML) algorithms to boost its efficiency in System-On-Chip (SoC). Hence, this makes Artificial Intelligence (AI) an integral part of the VLSI industry. The VLSI industry has started adopting the AI techniques in design automation [4]. AI is useful in several facets of VLSI logical and physical design, including manufacturing, miscorrelation, power analysis [5, 6], testing, mask synthesis, transition and connection latency, temperature reduction in SOC and the usage of CAD tools [7].

7.2 Artificial Intelligence and Machine Learning

AI has offered prominent solutions to several problems across a variety of fields. AI/ML is capable of recognizing trends and patterns in massive amounts of data. Patterns are also known as feature vectors, instances, observations, and data items [8]. Learning, reasoning, prediction, categorization, and perception are the goals of AI/ML. The electronic sector has been experiencing a burgeoning counterfeit market, resulting in electronic industry with considerable loss. ML algorithms plays significant role in automatic detection of counterfeit, testing of ICs [9, 10] circuit parameters, circuit structure and physical layout [11]. The computing power of computers has significantly risen due to the rapid advancements in semiconductor technology. ML approaches have gained popularity in several new applications as a result of this and subsequent advancements in theory. Researchers interested in the design and optimization of ICs have been motivated by this trend [12]. Additionally, the efficiency of analog/RF electronic design automation (EDA) tools is improved by the use of ML approaches. Furthermore, ML classification algorithm (KNN) is useful for VLSI test cost reduction [10]. Because AI/ML algorithms have so many benefits, VLSI design and technology have heavily used AI/ML techniques during the past decade. Therefore, it is worth knowing these algorithms.

AI uses a system that combines hardware and software. From a software standpoint, AI is especially focused on algorithms. In the recent years, AI has grown fast and advanced quickly in terms of intelligent algorithms, implementation of hardware, and other applications in a wide range of fields. In contrast to the intelligence of human beings or other living species, AI is the intelligence of machines. AI is just like the study of "intelligent agents", which refers to any agent or machine that can understand and comprehend its environment and take relevant actions to increase its chances of attaining its goals [13]. AI may also refer to scenarios in which machines can learn and analyze in the same way that humans do and thus be used to solve problems. ML is a subset of this type of intelligence. Numerous innovative automated methods for managing difficult and data-intensive jobs in VLSI design and testing are made possible by the unique learning techniques of AI [14]. The time and effort required to comprehend and process data inside and across various abstraction levels is decreased by using AI and ML techniques in VLSI design and manufacture. Modern circuits have a very large number of modules. The process of designing integrated circuits is challenging and encounters an increasing number of design issues. Partitioning of VLSI circuit into smaller subcircuit can be developed independently and effectively [15]. Therefore, there is a need to split the circuit into smaller and highly connected logical blocks [16]. VLSI physical design requires careful placement since it has a direct impact on the design cycle [17]. Partitioning of VLSI circuit plays an important role in VLSI physical design [18]. Partitioning of VLSI circuit is a concept of splitting circuit into a number of sub-circuits with minimal linkages between them [19]. A variety of clustering-based, i.e., bottom-up, partitioning methods have also been presented, with promising results. In addition to this, several supervised and unsupervised algorithms are used in optimized VLSI circuit partitioning [16].

Hardware-wise, the major focus of AI is the implementation of algorithms on a physical computing platform [13]. The AI algorithm may be implemented in a multithreaded or multicore configuration on a general-purpose central processing unit (CPU), which is the easiest method.

Additionally, for large-scale, it has been discovered that graphics processing units (GPUs), which are effective in convolutional calculations, are preferable than CPUs. Central processing unit (CPU) and graphics processing unit (GPU) co-processing has proven to be more effective than CPU alone. Additionally, some programmable or adaptable accelerator hardware platforms, such as field-programmable gate arrays (FPGAs) and application-specific integrated circuits (ASICs), can implement AIs towards a customized application in a more efficient manner in terms of power efficiency, computation capability, and form factor.

7.3 AI/ML Algorithms

In today's world, statistical learning is essential to almost every new area of research and technology. The word "mining" in computer science refers to extracting meaningful information from massive amounts of data.

Data can vary in terms of volume, variety, and velocity, which are known as the 3V's of big data. Data mining enables the search for important information in massive amounts of data. Data mining techniques are useful in various applications related to data such as classification, characterization, clustering, pattern recognition, pattern matching, association, and data visualization [41–46]. Data mining methods (DMT) have developed an emerging field of applied artificial intelligence (AI). To find patterns and relations among the features for additional research and predictions, it is possible to mine the enormous amounts of data generated within each field. There are several real-world problems that may be resolved using the statistical learning method. With the use of AI, a computer may mimic human behavior. The two primary subsets of AI are ML and deep learning (DL). One of the most suitable computing paradigms for integrating significant intelligence into different domains is ML [47–52].

ML is a collection of mathematical concepts and algorithms that enable a computer system to learn automatically from the past data without explicit programmed. DL is a subfield of ML that uses hierarchical structure to learn high-level abstractions from data. DL is currently growing for three key reasons: the considerably improved chip processing capabilities (such

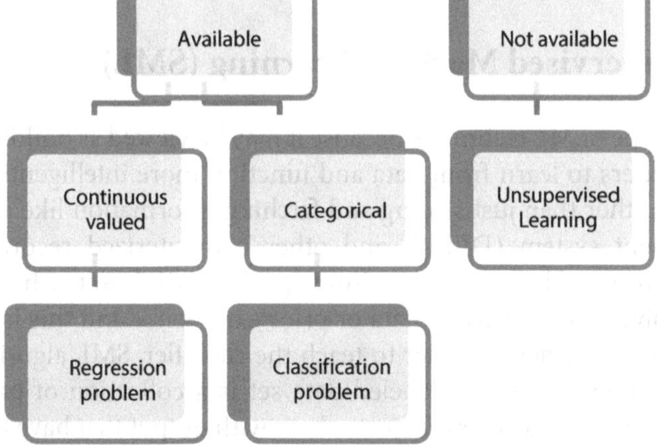

Figure 7.2 Identification of ML problem based on dataset output label.

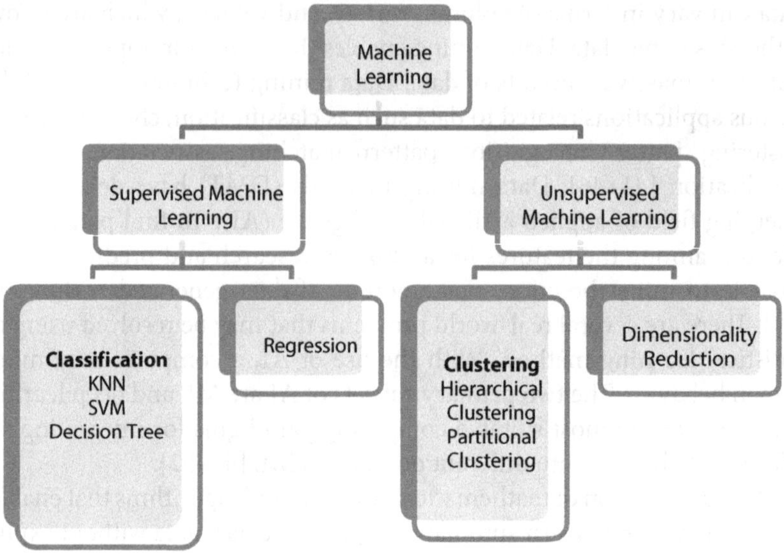

Figure 7.3 Classification of ML techniques.

as GPU units), the significantly reduced cost of computing hardware, and the significant developments in machine learning techniques [20]. In ML, a wide range of different algorithms can be used. Typically, a dataset may or may not have output label shown in Figure 7.2, ML algorithms belong to one of two categories of learning: either Supervised or Unsupervised learning shown in Figure 7.3.

7.4 Supervised Machine Learning (SML)

SML is a type of ML technique because it may be viewed as building blocks for computers to learn from data and function more intelligently by generalizing rather than just storing and fetching information like a database management system (DBMS) and other computerized record-keeping applications would. One of the primary goals of ML is to teach computers how to solve problems using data or prior experience and this is achieved by the good classifier. In order to teach the classifier, SML algorithms use labeled dataset [53–58]. A labeled data set is a collection of patterns or measurements of real-world objects along with output that have been used to train the model. A labeled data set provides the correct information of the patterns [59–65].

Every pattern uses same set of features in a certain dataset. These features might be categorical, continuous, or binary in nature. The mathematical representation of the labeled dataset is shown in Equation (7.1).

$$D = \{X_i, Y_i\}_{i=1}^{n} \tag{7.1}$$

Where D represents data set comprises n patterns and $X = \{x_1, x_2 \ldots x_n\}^{d}$: d – dimensional feature space. Y s$\{c_1, c_2, \ldots c_n\}$ represents set of class labels. Given a data set D, the primary goal of SML is to generate a prediction/classification model that connects the values of features in X to classes in Y; where X is an input and Y is an output. In SML, a mapping between a set of input feature set called X and an output label Y is learned, and the mapping is then utilized to predict the label for unseen data [66–72].

SML is mainly categorized into two classes: Classification and Regression. The label might be either continuous valued (for regression problems) or categorical (for classification problems). Some common SML algorithms are K-NN, SVM, and Decision tree that are explained in the upcoming sections [73–77].

7.5 Classification Techniques

The traditional ML algorithms, which are the most frequently used classification approaches in the literature, are presented in this section. Classification is the most common ML approach, which has many different domains of applications [78–81]. Classification techniques are useful in predictions. A machine learning classification technique has been developed to identify an appropriate patterns without requiring human participation or domain expertise from chip specialists [10]. In order to forecast the class-label of data instances, this technique tries to predict how a collection of features termed an attribute and a target (class-label) relate to one another. A general classification model used for prediction is shown in Figure 7.4.

Preparing and pre-processing the dataset is the initial stage in classification since it may contain noise, missing values, and impossible values. The training and test portions of this dataset will be randomly divided. Therefore, the quantity and quality of the dataset are essential requirements for any classifier's performance. It is a bit challenging to visualize

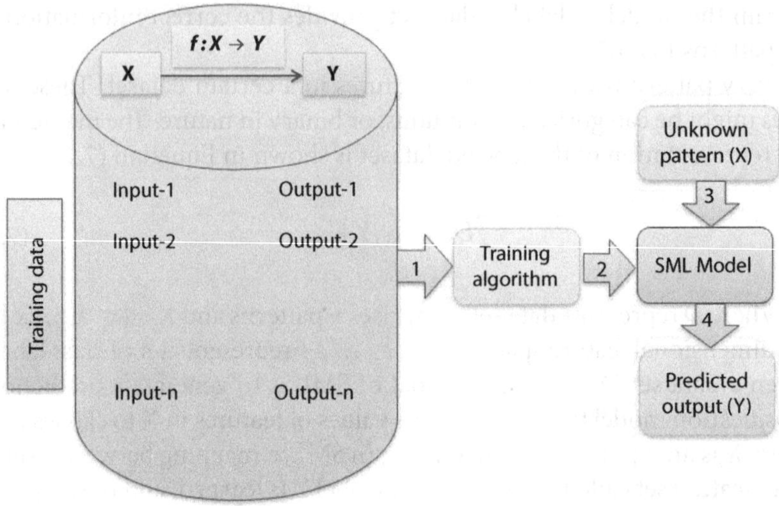

Figure 7.4 Supervised classification model.

and extract knowledge from high dimensional datasets. Therefore, dimensionality reduction is an important classification approach that removes redundant and irrelevant data [82–88]. A technique of dimensionality reduction known as "feature extraction" involves the process of extracting relevant information from raw data [21]. There are various methods available now for designing ML systems to handle classification and regression tasks [89–93].

7.6 K-Nearest Neighbors (KNN)

Cover and Hart first proposed the KNN method in 1968 [22]. Since the technique was first described in 1968, research on the KNN method has gained a lot of attention in the fields of data mining, pattern recognition and machine learning [23]. It is simple and easy to implement and can handle noisy data. Therefore, the KNN technique is frequently employed in pattern recognition to address problems of classification [10]. A non-parametric, supervised learning classifier, the k-nearest neighbor's method (also known as k-NN or KNN) is a successful lazy learning algorithm, which has been created for real-world applications [24]. Instance-based classifiers are sometimes known as lazy learners since they save all of the training data and do not generate a classifier until an unknown, unlabeled sample has to be categorized [94–98]. Each unlabeled sample (unknown

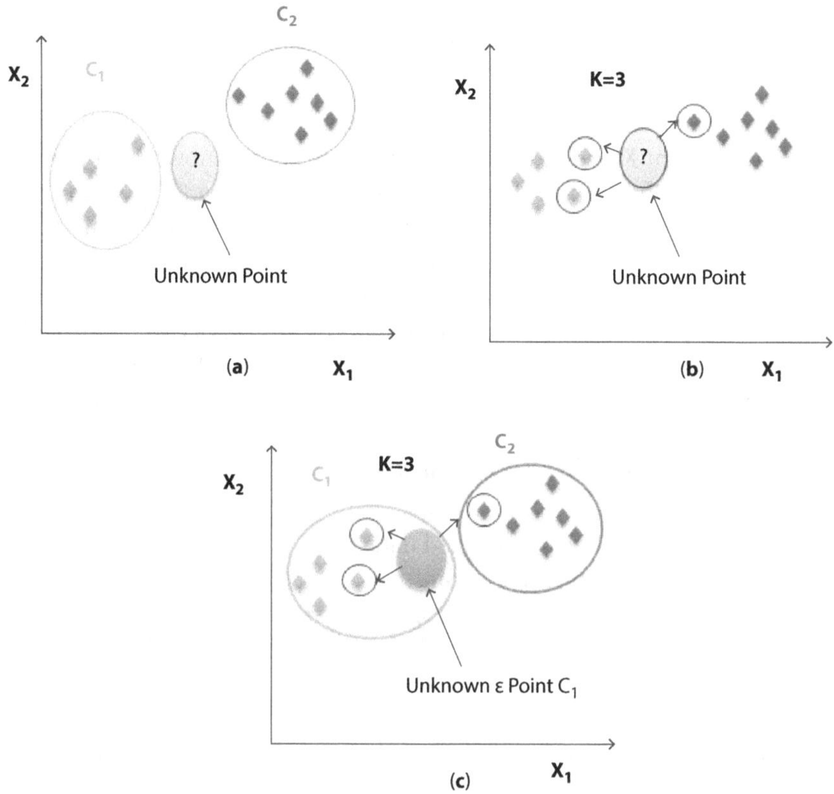

Figure 7.5 (a) Distribution of data points in 2-D feature space (x_1 and x_2). (b) K=3 closest data points from the unknown points based on Euclidean distance. (c) Classification of unknown data point based on majority of votes where c_1 and c_2 represent classes.

point) is categorized by this algorithm according to the label that is most commonly found among its K closest neighbors as shown in Figure 7.5. As a result, a key component of its performance is the ability to locate the closest neighbors using distance measures. The standard metric for numerical data is Euclidean Distance (ED). ED is the length of the straight line that connects the two points X and Y is calculated using the Pythagorean formula [99–103]. The ED between a test sample and the designated training samples is a typical basis for the k-nearest-neighbor classifier. The ED, which may be extended to higher dimensions as demonstrated in Equation (7.2), determines the square root of differences between the coordinates of two points [104–107].

$$Dist_{K,F} = \sqrt{\sum_{k=1}^{m}(X_k - Y_k)^2} \tag{7.2}$$

Where m represents dimensions (features) and X_k and Y_k are the k^h feature of data point X and Y. In KNN, a class label of unknown data point is chosen for classification problem based on a majority vote, meaning that the label that is most commonly occurred around an unknown data point is adopted. There is no optimal method for choosing "K" value it is based on the problem domain. The best value for "K" is the one that minimizes classification error and lends itself to generalization. It is a computationally efficient, easy-to-understand, and effective algorithm. The steps of K-NN classification algorithm includes following steps shown in Algorithm-7.1:

Algorithm 7.1: Steps of KNN classification algorithm

Input: $D = \{X_i, Y_i\}_{i=1}^{n}$ // D is a labeled dataset.
Output: Class of unknown data point.
(i) ***Initialization:*** *Initialize number of nearest neighbors (K).*
(ii) Calculate the distance between all of the sample points and the test sample that will be classified.
(iii) Sort the distances.
(iv) Select the k number of points with the smallest distance from the unknown data point.
(v) Calculate the frequency of occurrence of the majority of class label of the first k points.
(vi) Return the class label with the highest frequency of the occurrence of the first k points as the predicted class label of the unknown sample.

7.7 Support Vector Machine (SVM)

SVMs are the most effective and powerful ML techniques for supervised learning, introduced by Vapnik in 1992. It is scalable and effective in high dimensional spaces. It was employed in several applications including feature selection, regression, and classification. In classification, SVM is a supervised learning model that classifies objects of different classes using an optimal separating hyperplane [108–112].

A supervised learning model usually uses labeled data for learning and classifies the new sample based on that learning. The SVM method seeks to

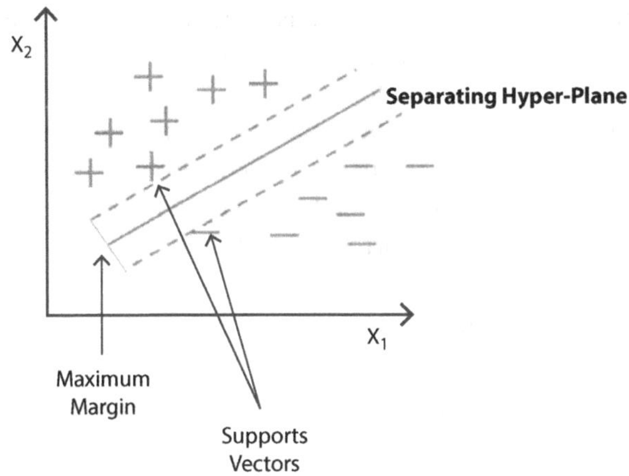

Figure 7.6 SVM classifier: $X_1$and X_2 are two arbitrary features, and positive (+) and negative (-) are two classes.

find the most suited decision boundary or borders known as a hyperplane that divides n-dimensional space into different classes. The main objective of the SVM classifier is to find the optimal hyperplane that maximizes the largest margin between the closest data points called support vectors and the hyperplane [113–116]. The number of dimensions of the decision boundary i.e. hyperplane are determined by the features in the data set; for example, a data set with two features uses line as hyperplane, whereas a data set with three features uses plane as hyperplane etc. The margin, which we aim to maximize for better generalization, is the distance between the support vectors on either side and separating hyperplane. The SVM method works best when there is a clear distinction between classes. SVM fails to work well when the data set contains a large amount of noise, such as overlapping target classes, which occurs frequently in real world. The pictorial representation of the SVM classifier is shown in Figure 7.6.

7.8 Linearly Separable Classification

Two classes C_1 and C_2 of d-dimensional feature space (d) considered to be linearly separable if there exists a hyperplane such that the data points of class (C_1) and class (C_2) lie on the two opposite sides of ^{d} classified by this hyperplane. SVM classifier can be of two types: linear and non- linear. In general, Non-linear classification problem is more challenging than linear one [25]. Assume we have N training instances that are completely separable linearly, each data point x_i has d attributes and belong to one of the two classes T and F represented by + 1 or − 1 symbol respectively. The give data set is of the following form $D = \{X_i, Y_i\}^{n}$ *where* x_i s$||^{d}_{i=1}$ and y_is {+ 1, −1}. The equation of decision plane in 2-dimensional feature space is given by Equation (7.3). Similarly, the equation of decision plane for n-Dimensional feature space is given by Equation (7.4). The Equation (7.4) can also be written like in the form of Equation (7.5).

$$w1\,x1 + w2x2 + w0 = 0 \tag{7.3}$$

$$w0 + w1\,x1 + w2x2 + \cdots w_n x_n = 0 \tag{7.4}$$

$$w_0 + \sum_{i=1}^{n} w_i x_i = 0 \quad \text{Or} \quad W^T X + w_0 = 0 \tag{7.5}$$

where $W = \begin{bmatrix} w_1 \\ w_2 \\ \vdots \\ w_n \end{bmatrix}$, $X = \begin{bmatrix} x_1 \\ x_2 \\ \vdots \\ x_n \end{bmatrix}$, w_0 is the bias which determines the point location relative to the origin and $W^T$is transpose of W. SVM has been widely employed in numerous fields due to its superior performance. The original SVM can only handle linearly or nearly linearly separable data. Many datasets are non-linear and difficult to separate, but they can be linearly separated by being mapped onto a higher dimensional space using a nonlinear mapping [26]. This mapping is defined by the kernel $K(x_i, x_j) = \Phi(x_i)^T\Phi(x_j)$ for this; different Kernel functions provide flexible means of representing data points for analyzing in a complex feature space. Kernel functions that are often used are linear, Gaussian Kernel, Radial basis function (RBF), polynomial and sigmoid [117–120].

7.9 Decision Tree Classifier (DTC)

One of the most used classifiers in statistics and ML is the decision tree. Decision tree classifier (DTC) is a tree structured supervised learning model based on divide and conquer algorithmic paradigm [24]. It is a nonparametric method used for both regression and classification. In real life, decision trees are frequently used for classification and less frequently for problems involving regression estimate. The main goal of employing decision trees is to develop a training model that can be used to predict class-label by learning decision rules inferred from training data shown in Algorithm 7.2. This algorithm classifies and predicts one or more discrete variables based on other variables in the dataset. It can easily be converted to a collection of basic **if-then** rules. The reader is able to comprehend and evaluate the results because to its clear portrayal. It is robust to outliers and missing values. The DTC is a tree type structure that has two types of nodes.

1. **Decision node**: Decision node specifies the direction based on decision test. Each decision node tests the value of an input feature.
2. **Leaf node**: Each leaf node match to a class label.

Algorithm 7.2 Decision tree

Step 1: Begin
Step 2: Input $D = \{X_i, Y_i\}_{i=1}^{n}$ // D is a labeled dataset.
Step 3: Use statistical approach for ordering the best feature. Step 4: Select the best feature as root node in each level.
Step 5: Dataset gets split recursively into smaller subsets on the basis of the feature's values.
Step 6: Repeat the step 2 and step 3 until the leaf node is found or when all the features are exhausted.
Step 7: End.

Quinlan presented the ID3 technique for inducing decision trees in 1986 [27]. In 1993, the ID3 algorithm was enhanced and updated to C4.5 [28]. There could be more than one decision tree that fix the training data in that case prefer low depth tree. To find the low depth decision tree that fixes the data is computationally hard problem. Choosing a feature from "n" distinct features to place at the root node at various levels of the tree is a computationally hard problem. Feature selection measure (FSM) is the collection of the best attribute for the root node as well as the sub-nodes. For solving this features selection problem, the best statistical methods are information gain (IG) and entropy. The term IG refers to the amount of information offered by an attribute about a class. The node is divided, and the tree is constructed based on information gain values. Entropy is a measure of the degree of arbitrariness in a dataset. The information gain is based on the decrease in entropy after a dataset is split on an attribute. The test set error may increase and the generalization may be impacted if we keep extending the tree until each node represents the lowest impurity.

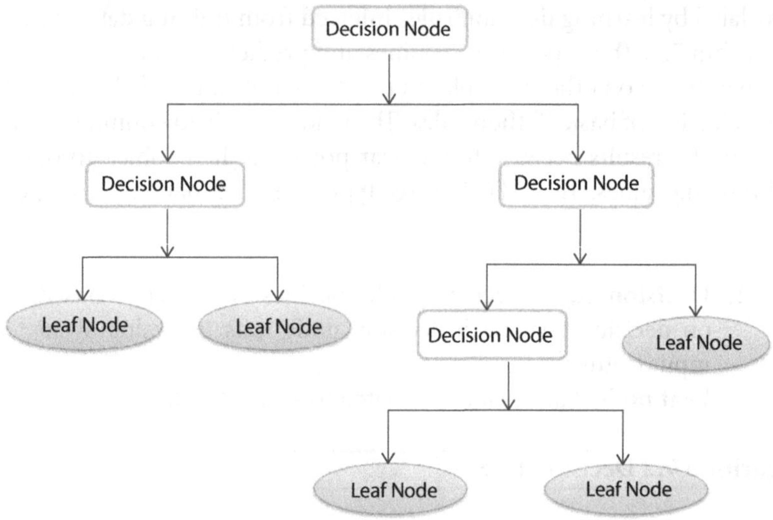

Figure 7.7 Decision tree classification model.

This is known as over fitting. Pruning the tree is one way to avoid over fitting. There are two techniques for pruning trees: and pre pruning and post pruning. Although pre-pruning is faster, post-pruning is really more effective since we cannot predict with certainty when to stop the tree's growth in advance.

DTC has tree-like graphical layout shown in Figure 7.7, which makes it simple to design and comprehend. The DTC is a versatile classifier that may be used with a variety of data types, including category, numerical, and textual data. The ability to deal with noisy datasets or missing values makes it useful for a variety of applications, including classification, clustering, regression, and feature selection [29]. In VLSI design, because of the increasing circuit intensity, chip temperature rises dramatically, affecting SoC reliability and performance. To solve these issues in the early stages of IC design, a C4.5 Decision Tree classifier technique is helpful.

7.10 Performance Measures in Classification

Classifier performance tests the predictive capabilities of ML models by using functions, score and loss of ML models. It measures the percentage of correct classifications using criteria such as accuracy, confusion matrix, precision, recall, and F1-score. Almost all metrics are depends on the concepts of true and false predictions created by the model and measured against the actual outcomes. If the model is error-free, all predictions will be either true positives (TP) or true negatives (TN) for binary classification, and predictions that are different from the true label of the pattern are annotated as false positives (FP) or false negatives (FN) [30]. The performance of the classifier may be assessed using a variety of metrics. We cannot fully understand the performance of the classifier from a single metric. Some of the commonly used performance measure metrics are discussed below.

(i) *Accuracy:* The following formula may be used to calculate a model's accuracy, which is a measurement of how accurately the algorithm classifies the unseen instances shown in Equation (7.6).

$$Accuracy = \frac{\text{Number of correct prediction}}{\text{Total number of points in Test Dataset}};\ 0 \leq \text{Accuracy} \leq 1 \tag{7.6}$$

(ii) *Confusion matrix*: There are other ways to assess performance other than accuracy, and sometimes we might require an additional analysis of the classifier's performance. Confusion Matrix is one such NXN matrix where N is number of target classes. It gives information about classifier errors and the various types of error that are being made. A good model will have large value across the diagonal and small values off the diagonal. Table 7.1 represents confusion matrix and Table 7.2 represents the most popular measures for binary classification based on the confusion matrix's values.

Table 7.1 Confusion matrix for binary class classification problem.

<table>
<tr><td rowspan="4">Predicted values</td><td rowspan="2"></td><td colspan="2">Actual values</td></tr>
<tr><td>0 (negative)</td><td>1 (positive)</td></tr>
<tr><td>0 (negative)</td><td>TN</td><td>FN</td></tr>
<tr><td>1 (positive)</td><td>FP</td><td>TP</td></tr>
</table>

Table 7.2 Popular measures for binary classification.

Measures	Formula	Description
Accuracy	$\frac{TP+TN}{TP+FN+FP+TP}$	Ratio between correctly classified data to total number of data.
Precision	$\frac{TP}{TP+FP}$	Ratio between true positive to total classified positive data.
Recall (Sensitivity)	$\frac{TP}{TP+FN}$	Ratio between true positive to actual positive data.
Specificity	$\frac{TN}{FP+TN}$	Ratio between true negative to total number of negative.

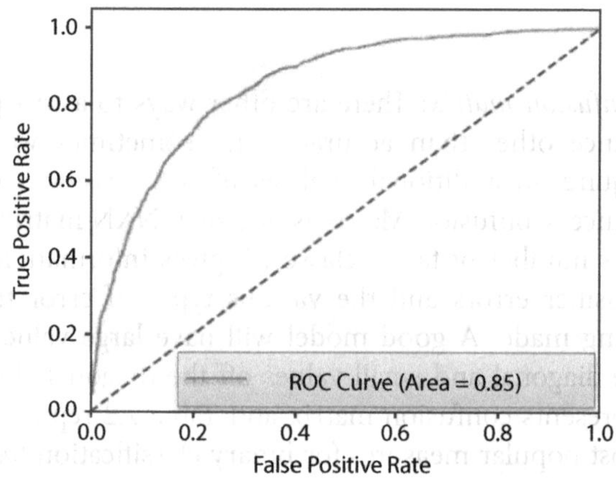

Figure 7.8 AUC-ROC curve.

(iii) *ROC curve*: A graph demonstrating the performance of a classification model is called a receiver operating characteristics (ROC) curve. Plotting the (TPR) versus the (FPR) at various threshold values results in the ROC curve. The full area under the ROC curve is known as AUC. Figure 7.8 represents the ROC curve and AUC. The perfect classifier has an AUC = 1.

7.11 Unsupervised Machine Learning (UML)

UML approaches facilitate the analysis of raw datasets, assisting in the generation of analytic discoveries from unlabeled data [31]. Unsupervised learning trains the model using unlabeled data. Unlabeled data set refers to data that has not been labeled. The general representation of unlabeled data set is shown in Equation 7.7.

$$D = \{X_i\}_{i=1}^{n} \tag{7.7}$$

Where D represents data set comprises n patterns and $X_i = \{x_1, x_2 \dots x_n\}^{d}$: d – dimensional feature space. In general, clustering methods split data set (D) into subsets $d_1, d_2 \dots d_n$ in such a way that each data point will be the member of only one set. All the data points in one set form cluster called cluster prototype and union of all the cluster prototypes form the original data set D as show in Figure 7.9.

Clustering is an example of an unsupervised machine learning approach that enables a model to learn from data without being assigned an output label. Data clustering is a popular data processing approach that groups objects into distinct clusters based on some level of similarity between data points. Unsupervised learning methods called clustering algorithms use a distance measure $d(x_i, x_j)$ between the data points to organize unlabeled data into predefined K number of clusters.

There are several mathematical similarity functions, such as Euclidean, Minkowski, Cosine Similarity, Manhattan Distance, and others, that have been used in diverse fields of research to form clusters of similar data points.

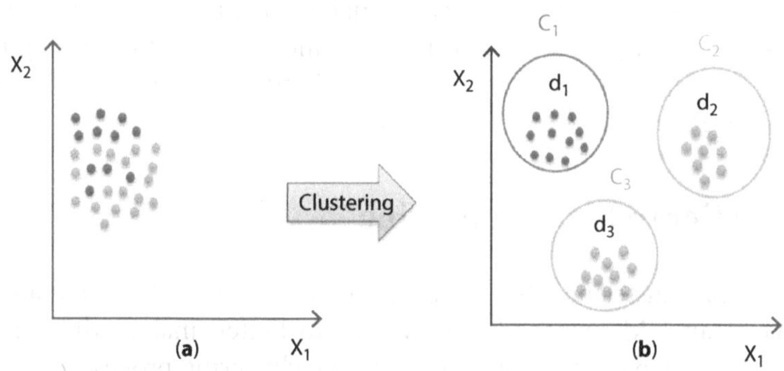

Figure 7.9 (a) Original data set where number of patterns (n) = 29. (b) Grouping of patterns into three clusters c_1, c_2, and c_3.

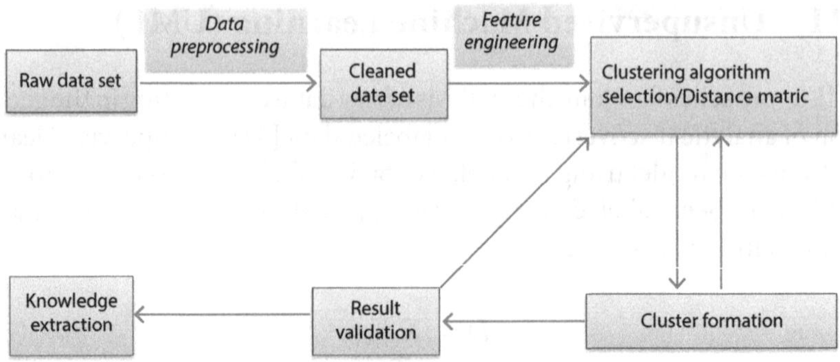

Figure 7.10 Common stages in clustering process.

These functions are essential for extracting information, pattern identification, classifications, and clustering from raw data. However, substantial work is required to determine the suitable similarity metrics depending on data point distribution and problem domains. Other than clustering, UML uses visualization, dimensionality reduction, and anomaly detection as examples [12]. Figure 7.10 shows the standard stages involved in data mining and knowledge extraction process.

The process of data clustering identifies a group of similar patterns within a dataset. In a circuit, a cluster is a collection of closely coupled components. Clustering techniques seek to locate the clusters within a circuit. Clustering methods can be employed in the design of VLSI layouts to create the circuit's natural hierarchy. According to recursive top-down partitioning, several known layout techniques create circuit hierarchies [32].

Various clustering paradigms provide a range of cluster methods and algorithmic approaches for cluster discovery. The clustering techniques can be broadly classified into two major classes: Hierarchical clustering and Partitional clustering. The Partitional algorithm includes two subdivisions: Hard/Crisp clustering techniques and Fuzzy/soft clustering techniques. The next sub- section presents and discusses the popular clustering techniques identified in the VLSI design.

7.12 Hierarchical Clustering

Hierarchical clustering also referred to as hierarchical cluster analysis (HCA), is an UML method that is used to better analyze the complex structure of the data. Without repeating the clustering process, hierarchical clustering algorithms can provide numerous consistent partitions of the data at different levels. Because all hierarchical clustering algorithms are

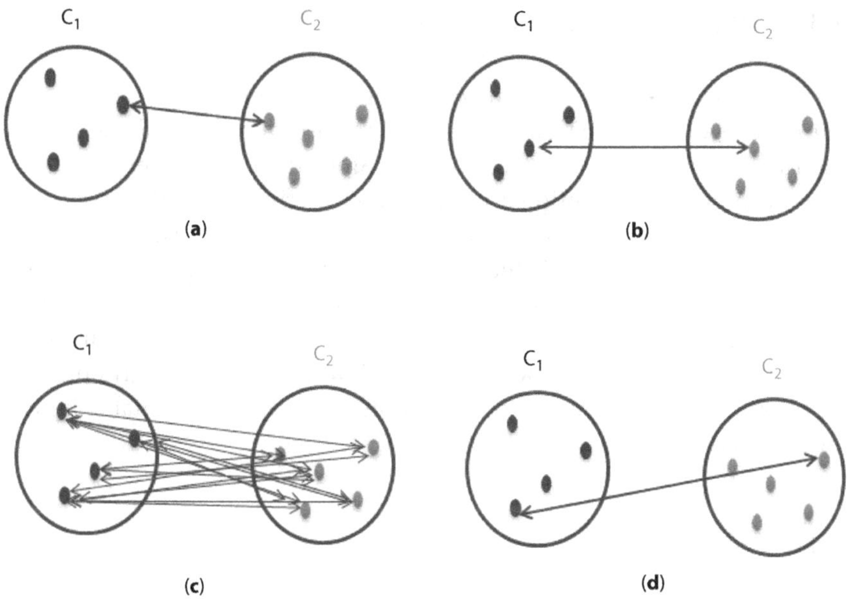

Figure 7.11 C_1 and C_2 represent different clusters. (a) Single linkage (b) centroid linkage (c) average linkage (d) complete linkage.

extremely sensitive to cutoff selection, a human expert is typically required to select the appropriate threshold for each clustering result.

Data objects are split into groups according to a hierarchy in hierarchical clustering. Iteratively creating clusters using a top-down or bottom-up approach reveals their hierarchical structure in the form of a dendrogram. A dendrogram is a tree diagram that depicts hierarchical clustering – linkages between similar groups of data. They don't need to know in advance how many clusters are there in the dataset. However, the algorithm's operation is believed to be stationary, and data points assigned to one group/cluster cannot be moved to another. Additionally, there are two categories for hierarchical clustering: (1) Agglomerative clustering algorithms (bottom-up method) and (2) Divisive clustering algorithms (top-down approach) [33]. Out of those two, latter is more popular. Linkages and a distance (proximity) measure are needed to determine how similar two objects are. ED is the most widely used distance measurement. Different links are employed in hierarchical algorithms to merge clusters. Common links include single, centroid, average and complete linkages shown in Figure 7.11.

7.13 Partitional Clustering

Partitional clustering is also known as the iterative relocation algorithm. The concept of partitional clustering was first introduced with the K-means algorithm, which was published in 1957. Since then, a number of studies have been done on traditional partitional clustering methods. Data clustering identifies a set of homogenous patterns in a dataset. Partitional clustering's objective is to divide the whole data set into homogeneous groups/clusters simultaneously without building a tree like structure (dendrogram) of clusters. Based on a certain criterion known as the fitness measure, partitional clustering groups the data points into homogeneous groups. The partitioning task becomes an optimization problem after a suitable fitness measure has been chosen [34].

7.14 K-Means

The easiest, most popular, and partition-based unsupervised learning approach is K-Means clustering (Hard C-means). It is a centroid-oriented partitioning strategy that partitions the provided data set points into a predetermined number of clusters known as *k* [35]. The K-means clustering algorithm initially selects *k* centroids by accepting the user's input for *k*. This algorithm will calculate the distance between each data point and each of the *k* numbers of selected centroids, and then it will assign each data point to its nearest centroid *k*. ED is typically used by K-Means to determine how far apart each pair of points and centroids are from one another. Each data point should be assigned to the closest centroids after once again calculating the real centroids from the previously generated clusters. The iterative K-mean clustering technique will keep recalculating the cluster until the centroids are fixed or the termination condition is not fulfilled. Time $O(nkl)$ is the running time of the conventional K-means clustering method, where "*n*" represents the total number of instances in the data set (D), "*k*" represents total cluster numbers, and "*l*" represents iteration numbers. An objective function (J) for K-means is shown in Equation (7.8).

$$J = min\sum_{j=1}^{k}\sum_{i=1}^{n}||x_i^{(j)} - c_j^2|| \tag{7.8}$$

Where $||x_i^{(j)} - c_j^2||$ is a chosen squared distance-measure between a_i data point $x^{(j)}$ and the cluster center c_j. The squared distance between each cluster's centroid and its data points should be minimized as possible.

Finding the intra-cluster similarity between the data points using the objective function is helpful, but computing its minimization for a smaller value of k ≥2 is challenging since it is an NP-hard problem [36]. In the 1950s and 1960s, a number of researchers separately proposed the K-means clustering approach, including Lloyd, MacQueen, Jancey, and Steinhaus from different fields. As a result, there are several variations of this method exists. The Lloyd's method, a simple and practically oriented local search approximation approach for clustering, was implemented by T. Kanungo *et al.* in 2002. The following steps are part of Lloyd's method for K-Means clustering shown in Algorithm-7.3.

Algorithm 7.3 K-Means

Input: $D = \{X_i\}_{i=1}^{n}$ *// D is a dataset.*
Output: *Group the data points into K clusters.*
Initialization: *Set the distance metric, centroids, and number of clusters (K).*
(ii) Select K number of centroids randomly.
(iii) Compute the ED between every data point and the (k) centroids.
(iv) Identify the closest centroid for each data point and assign to it.
(v) Iteratively compute the K centroids.
(vi) To get centroids stable, repeat steps (iii) and (IV).

It is a bit challenging task for this algorithm to determine the cluster-numbers a priori without having the good domain knowledge of the data sets [37]. One of two techniques may be used to identify the cluster numbers: either the algorithm is repeatedly run or the desired number of clusters is chosen based on some validity criterion, or the number of clusters is automatically decided using some relevant methods or criteria.

7.15 Fuzzy (soft) Clustering

Every data point in the dataset may only belong to one single cluster according to the hard (crisp) clustering techniques. In contrast, Fuzzy clustering is used when the elements do not belong completely to one group. It provides a better representation of the overlapping nature of clusters. Fuzzy clustering, which derives from fuzzy logic, shows the likelihood or degrees of a single data point simultaneously belonging to many groups or clusters. Fuzzy c-shells and Fuzzy c-means (FCM) algorithms are two of the widely used techniques. Fuzzy c-means works on an assumption that the number

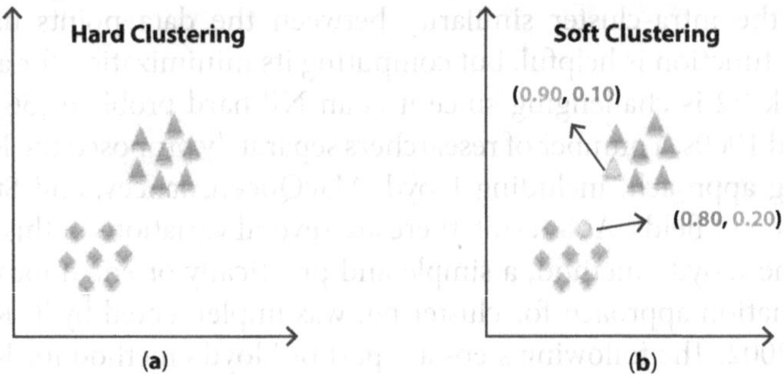

Figure 7.12 (a) Hard clustering (b) soft clustering with membership values between [0,1].

of clusters 'c' for the provided dataset is known and minimizes the objective function (J_{FCM}) given in Equation (7.9).

$$J_{FCM} = \sum_{k=1}^{c} \sum_{i=1}^{n} u_{ik}^{m} d_{ik}^{2} \tag{7.9}$$

Where u_{ik} represents the membership value of an instance x_i in cluster k and $d_{ik} = ||x_i - v_k||$ represent the Euclidean distance between x_i and v_k (cluster-centroid). m s $\{1, \infty\}$ is a control parameter that control the clustering result. u_{ik} fulfills the following relation shown in Equation (7.10) [38].

$$\sum_{k=1}^{c} u_{ik} = 1; \quad i = 1,2,3\ldots n \tag{7.10}$$

By assigning each instance to the cluster with the highest degree of membership value, the fuzzy clustering may be transformed into a hard clustering where each instance associated with only one cluster [34]. A hypothetical example of hard clustering and soft clustering is shown in Figure 7.12.

7.16 Cluster Validation Measures

Various clustering approaches for categorization, pattern analysis, decision making, image segmentation, information retrieval, and data mining have been developed; many important challenges remain in accurately detecting the clustering output [39]. Clustering validation is one of the primary issues for clustering applications. Without the use of any class labels, cluster validation is a technique for determining a collection of clusters that most closely fits natural partitions (number of clusters). For clustering algorithms, two types of validity measurements have been proposed: internal and external [33].

7.17 Internal Clustering Validation Measures

Internal validation criteria have been used to assess the outcomes of clustering to establish the ideal number of clusters. Internal validation of clustering based on separation and compactness. Commonly used internal validation criteria are discussed below.

(i) *Sum of square error:* Sum of Square Error (SSE), which is explained as follows and is presented in Equation (7.11), is the most commonly employed criterion metric for clustering.

$$SSE = \sum_{k=1}^{K} \sum * x_i g C_k \; ||x_i - \mu_k||^2 \tag{7.11}$$

where C_k is the collection of instances that constitute cluster k and μ_k is the cluster's vector mean.

(ii) *Silhouette index:* A compact and well-separated cluster is identified using the silhouette coefficient. The Silhouette index is defined as follows shown in Equation (7.12).

$$S(\mathrm{i}) = \frac{[b(\mathrm{i}) - a(\mathrm{i})]}{max\,[a(\mathrm{i}), b(\mathrm{i})]} \tag{7.12}$$

Where a(i) represents the average distance between the ith sample and all samples contained in X_j(j = 1, … C); (i) is the minimum average distance between the ith*s* and all of the samples clustered in $X_k(k = 1, .. c\,; k \neq j)$.

(iii) *Dunn index (DI):* The DI is a type of internal validation metric for cluster analysis. It determines the ratio between the smallest intracluster distance and the largest intercluster distance in a partitional clustering. The higher value of DI represents the good clustering result. The DI is defined as follows shown in Equation (7.13).

$$Dunn = \min_{1 \le i \le c} \left(min\left(\frac{d(c_i, c_j)}{\max_{1 \le k \le c} \; d(K_k)} \right) \right) \tag{7.13}$$

Where $d(c_i, c_j)$ is the distance between cluster X_i and X_j ; $d(X_k)$ represents the intracluster distance (X_k) and the total number of clusters present in a data set is represented by c.

7.18 External Clustering Validation Criteria

The basis for external clustering validation is a comparison between the partitions produced by the clustering methods/algorithms and the actual cluster-number of the datasets. The commonly used validation criteria under these categories are discussed below.

(i) *Entropy:* The entropy and purity metrics, which were developed by the information retrieval community, were often employed by the data mining community as external validation techniques. Entropy is used to evaluate the purity of the clusters' class labels. As a result, if each cluster is made up of objects with a single class label, the entropy is 0. However, when the number of class labels for the elements in a cluster increases, the entropy rises as well, as shown in Equation (7.14).

$$E_j = \Sigma_{i=1}^{n} \, p_{ij} \log(p_{ij}) \tag{7.14}$$

(ii) *Jaccard index (similarity)*: The number of observations in both sets is divided by the number of observations in either set to determine the Jaccard similarity. In other words, the Jaccard similarity can be calculated by dividing the size of the intersection by the size of the union of two sets as shown in Equation (7.15).

$$JJ(A,B) = \frac{A \cap B}{A \cup B} = \frac{TP}{TP + FP + FN} \tag{7.15}$$

If there are no values shared between the two sets, the Jaccard Similarity will be 0, and if two sets are identical, it will be 1.

(iii) *Rand measure:* It is also known as Rand index (RI) that calculates a similarity measure between two distinct clustering techniques by taking into account all sample pairings and counting pairs that are placed in the same or different clusters in the predicted and true clustering. The RI may also be viewed as a measure of how often the algorithm makes the right decision. The following formula may be used to calculate it shown in Equation (7.16).

$$RAND = \frac{TP + TN}{TP + FP + FN + TN} \tag{7.16}$$

Where TP is an acronym for "true positives," TN is an acronym for "true negatives," FP is an acronym for "false positives," and FN is an acronym for "false negatives". The RI scales is from 0 to 1, with value 0 indicates that there is no agreement between the two data clusterings by accessing all pairs of data points and value 1 shows perfect agreement between the two data clustering.

7.19 Limitation and Challenges - VLSI

Even though employing AI/ML techniques have been very beneficial for VLSI industry. However, there are still a number of drawbacks to AI/ML techniques are as follows:

(i) *Security*: AI/ML is an intelligent data-intensive technique. Adversarial examples are data points that are deliberately created by an adversary to trick a machine learning system into providing the wrong output [40]. Therefore, data security is seen as a serious issue while working with AI/ML techniques. Every data-oriented technology, including semiconductor production, has security as a key component. Very important data might be the target of attackers.

(ii) *High dimensional and redundant data:* The acquisition of unnecessary and excessive amounts of information might be detrimental since it can be difficult to use the data that is gathered.

(iii) *Missing and imbalanced data:* In AI/ML, missing and unbalanced data might be problematic. The majority of classification algorithms perform poorly when the data is unbalanced as a result.

(iv) *Preprocessing overhead:* Due to the fact that data preprocessing frequently uses significant amount of the time and effort required for the complete data analysis process. Certain ML methods could require too much computing power to be applied to large datasets with a high-dimensional feature map.

VLSI design is a specialized technology that deals with the design and implementation of integrated circuits (ICs) that comprise a large number of transistors on a single chip. AI and ML have several applications in VLSI design and optimization such as design optimization, layout optimization, and power optimization. AI/ML can help VLSI engineers and researchers develop and optimize circuits that are more efficient, reliable, and secure, while also cutting design and manufacturing costs and reducing time-to-market. Data mining methodologies such as clustering and classification may be included into VLSI partitioning, creating a new path for detecting hidden patterns in data and predicting the correlations between the variables that allow forecasting outcomes. The manufacturing of semiconductors is a very intricate process that involves several sub-processes and a wide variety of equipment. Understanding various models and algorithms might benefit the growth of theory and the comprehension of all the potential applications in various semiconductor production domains.

References

[1] S. B. V. Kumar, P. V Rao, H. A. Sharath, B. M. Sachin, U. S. Ravi, and B. V Monica, "Review on VLSI design using optimization and self-adaptive particle swarm optimization," *J. King Saud Univ. - Comput. Inf. Sci.*, vol. 32, no. 10, pp. 1095–1107, 2020, doi: 10.1016/j.jksuci.2018.01.001.

[2] P. Espadinha-cruz, R. Godina, and E. M. G. Rodrigues, "A Review of Data Mining Applications in Semiconductor Manufacturing," pp. 1–38, 2021.

[3] S. Madhura, "A Review on Low Power VLSI Design Models in Various Circuits," *J. Electron. Informatics*, vol. 4, no. 2, pp. 74–81, 2022.

[4] A. Malhotra, "Implementation of AI in the field of VLSI : A Review," *2022 Second Int. Conf. Power, Control Comput. Technol. IEEE*, pp. 1–5, 2022.
[5] H. Jung, S. Member, and M. Pedram, "Supervised Learning Based Power Management for Multicore Processors," *IEEE Trans. Comput. Des. Integr. Circuits Syst.*, vol. 29, no. 9, pp. 1395–1408, 2010.
[6] E. Poovannan and S. Karthik, "Pre Synthesis and Post Synthesis Power Estimation of VLSI Circuits Using Machine Learning Approach Pre Synthesis and Post Synthesis Power Estimation of VLSI," *Appl. Artif. Intell.*, vol. 36, no. 01, 2022, doi: 10.1080/08839514.2022.2145640.
[7] E. Poovannan and S. Karthik, "Role Of Machine Learning In Vlsi Ic Design," *Turkish J. Comput. Math. Educ.*, vol. 12, no. 10, pp. 2489–2494, 2021.
[8] A.K.Jain, "Data Clustering: A Review," *Adv. Mach. Learn. Data Min. Astron.*, vol. 31, no. 3, pp. 543–561, 1999, doi: 10.1201/b11822-19.
[9] B. Ahmadi, B. Javidi, and S. Shahbazmohamadi, "Microelectronics Reliability Automated detection of counterfeit ICs using machine learning," *Microelectron. Reliab.*, vol. 88–90, no. June, pp. 371–377, 2018, doi: 10.1016/j.microrel.2018.06.083.
[10] T. Song, Z. Huang, and A. Yan, "Machine learning classification algorithm for VLSI test cost reduction," *Integration*, vol. 87, no. June, pp. 40–48, 2022, doi: 10.1016/j.vlsi.2022.06.005.
[11] G. Thakur, S. Jain, and H. Sohal, "Measurement : Sensors Current issues and emerging techniques for VLSI testing - A review," *Meas. Sensors*, vol. 24, no. October, p. 100497, 2022, doi: 10.1016/j.measen.2022.100497.
[12] E. Afacan, N. Lourenço, R. Martins, and G. Dündar, "Review: Machine learning techniques in analog/RF integrated circuit design, synthesis, layout, and test," *Integration*, vol. 77, no. November 2020, pp. 113– 130, 2021, doi: 10.1016/j.vlsi.2020.11.006.
[13] G. Rong, A. Mendez, E. Bou, B. Zhao, and M. Sawan, "Artificial Intelligence in Healthcare : Review and Prediction Case Studies," *Engineering*, vol. 6, no. 3, pp. 291–301, 2020, doi: 10.1016/j.eng.2019.08.015.
[14] D. Amuru, H. V Vudumula, P. K. Cherupally, and S. R. Gurram, "AI/ML Algorithms and Applications in VLSI Design and Technology," *arXiv Prepr. arXiv*, pp. 1–41, 2022.
[15] J. Li and L. Behjat, "A Connectivity Based Clustering Algorithm With Application to VLSI Circuit Partitioning," *IEEE Trans. Circuits Syst.*, vol. 2, no. 5, pp. 384–388, 2006.
[16] K. A. Sumithradevi, "Optimization of Circuit Partitioning in VLSI through Classification Algorithms," *TENCON 2010-2010 IEEE Reg. 10 Conf.*, pp. 1554–1557, 2010.
[17] P. Design, "Progress of Placement Optimization for Accelerating VLSI Physical Design," *Electronics*, vol. 12, no. 2, pp. 1–23, 2023, doi: https://doi.org/10.3390/electronics12020337.
[18] R. Sujitha, "A Study on Various Data Mining Algorithms Pertaining to VLSI Cell Partitioning," vol. 7, no. 8, pp. 363–368, 2014.

[19] W. Deng, "VLSI Circuit Partitioning by Cluster-Removal Using Iterative Improvement Techniques," *Proc. Int. Conf. Comput. Aided Des. IEEE*, pp. 194–200, 1996.

[20] Y. Guo, Y. Liu, A. Oerlemans, S. Lao, S. Wu, and M. S. Lew, "Deep learning for visual understanding: A review," *Neurocomputing*, vol. 187, pp. 27–48, 2016, doi: 10.1016/j.neucom.2015.09.116.

[21] A. O. Salau and M. Min, "Feature Extraction : A Survey of the Types, Techniques, Applications," *2019 Int. Conf. signal Process. Commun. (ICSC). IEEE*, 2019.

[22] T. Cover and P. Hart, "Nearest neighbor pattern classification," *IEEE Trans. Inf. theory 13.1*, vol. I, pp. 21– 27, 1967, doi: http://dx.doi.org/10.1109/TIT.1967.1053964.

[23] Z. Deng, X. Zhu, D. Cheng, M. Zong, and S. Zhang, "Efficient kNN classification algorithm for big data," *Neurocomputing*, vol. 195, pp. 143–148, 2016, doi: 10.1016/j.neucom.2015.08.112.

[24] S. D. Jadhav and H. P. Channe, "Comparative Study of K-NN , Naive Bayes and Decision Tree Classification Techniques," *Int. J. Sci. Res.*, vol. 5, no. 1, pp. 1842–1845, 2016.

[25] D. A. Elizondo, R. Birkenhead, M. Gamez, N. Garcia, and E. Alfaro, "Linear separability and classification complexity," *Expert Syst. Appl.*, vol. 39, no. 9, pp. 7796–7807, 2012, doi: 10.1016/j.eswa.2012.01.090.

[26] T. Chen and G. L. Zhang, "Machine Learning in Advanced IC Design : A Methodological Survey," *IEEE Des. Test*, no. February, pp. 17–33, 2023.

[27] J. R. Quinlan, "Induction of Decision Trees," *Mach. Learn.*, pp. 81–106, 1986.

[28] H. Sharma and S. Kumar, "A Survey on Decision Tree Algorithms of Classification in Data Mining," *Int. J. Sci. Res.*, no. May, pp. 2094–2097, 2018.

[29] P. and D. Kumar, "Decision tree classifier : a detailed survey," *Int. J. Inf. Decis. Sci.*, vol. 12, no. 3, pp. 246– 269, 2020.

[30] M. Sokolova and G. Lapalme, "A systematic analysis of performance measures for classification tasks," *Inf. Process. Manag.*, vol. 45, no. 4, pp. 427–437, 2009, doi: 10.1016/j.ipm.2009.03.002.

[31] M. Usama *et al.*, "Unsupervised Machine Learning for Networking: Techniques, Applications and Research Challenges," *IEEE Access*, vol. 7, pp. 65579–65615, 2019, doi: 10.1109/ACCESS.2019.2916648.

[32] M. Lissa, "A parallel bottom-up clustering algorithm with applications to circuit partitioning in VLSI design," *Proc. 30th Int. Des. Autom. Conf.*, pp. 755–760, 1993, doi: 10.1145/157485.165119.

[33] A. E. Ezugwu, A. M. Ikotun, O. O. Oyelade, and L. Abualigah, "Engineering Applications of Artificial Intelligence Survey paper A comprehensive survey of clustering algorithms : State-of-the-art machine learning applications, taxonomy, challenges, and future research prospects," *Eng. Appl. Artif. Intell.*, vol. 110, no. January, p. 104743, 2022, doi: 10.1016/j.engappai.2022.104743.

[34] S. J. Nanda and G. Panda, "A survey on nature inspired metaheuristic algorithms for partitional clustering," *Swarm Evol. Comput.*, vol. 16, pp. 1–18, 2014, doi: 10.1016/j.swevo.2013.11.003.

[35] A. M. Ikotun, A. E. Ezugwu, L. Abualigah, B. Abuhaija, and J. Heming, "K-means clustering algorithms: A comprehensive review, variants analysis, and advances in the era of big data," *Inf. Sci. (Ny).*, vol. 622, pp. 178–210, 2023, doi: 10.1016/j.ins.2022.11.139.

[36] A. Saxena, M. Prasad, A. Gupta, N. Bharill, and O. Prakash, "A review of clustering techniques and developments," vol. 267, pp. 664–681, 2017.

[37] X. Yao, J. Wang, M. Shen, H. Kong, and H. Ning, "An improved clustering algorithm and its application in IoT data analysis," *Comput. Networks*, vol. 159, pp. 63–72, 2019, doi: 10.1016/j.comnet.2019.04.022.

[38] J. Nayak, B. Naik, and H. S. Behera, "Fuzzy C-Means (FCM) Clustering Algorithm : A Decade Review from 2000 to 2014," *Comput. Intell. Data Min.*, vol. 2, pp. 133–149, 2014, doi: 10.1007/978-81-322-2208-8.

[39] P. Mitra, "A survey of density based clustering algorithms," *Front. Comput. Sci.*, vol. 15, no. 1, pp. 1–27, 2021.

[40] E. Alhajjar, P. Maxwell, and N. Bastian, "Adversarial machine learning in Network Intrusion Detection Systems," *Expert Syst. Appl.*, vol. 186, no. May, 2021.

[41] N. Jain, and B. Raj, "Parasitic Capacitance and Resistance Model Development and Optimization of Raised Source/Drain SOI FinFET Structure for Analog Circuit Applications", *J. Nanoelectron. Optoelectron.*, ASP, USA, Vol. 13, pp. 531–539, 2018.

[42] S. Singh, S. K.Vishvakarma, and B.Raj, "Analytical Modeling of Split-Gate Junction-Less Transistor for a Biosensor Application", *Sens. Bio-Sens. Res.*, Elsevier, Vol. 18, pp. 31–36, 2018.

[43] M. Gopal, and B. Raj, "Low Power 8T SRAM Cell Design for High Stability Video Applications", *ITSI Trans. Electr. Electron. Eng.*, Vol.1, No. 5, pp. 91–97, 2013.

[44] B. Raj, J. Mitra, D. K. Bihani, V. Rangharajan, A. K. Saxena, and S. Dasgupta, "Analysis of Noise Margin, Power and Process Variation for 32 nm FinFET Based 6T SRAM Cell", *J. Comput. (JCP)*, Academy Publisher, FINLAND, Vol. 5, No. 6, PP.1-6, 2010.

[45] D. Sharma, R. Mehra, and B. Raj, "Comparative Analysis of Photovoltaic Technologies for High Efficiency Solar Cell Design", *Superlattices and Microstruct.*, Elsevier, Vol. 153, pp. 106861, 2021.

[46] P. Kaur, A. S. Buttar, and B. Raj, "A Comprehensive Analysis of Nanoscale Transistor Based Biosensor: A Review", *Indian J. Pure Appl. Phys.*, Vol. 59, pp. 304–318, 2021.

[47] D.Yadav, B.Raj, and B.Raj, "Design and Simulation of Low Power Microcontroller for IoT Applications", *J. Sens. Lett.*, ASP USA,Vol. 18, pp. 401–409, 2020.

[48] S.Singh, and B.Raj, "A 2-D Analytical Surface Potential and Drain Current Modeling of Double-Gate Vertical t-Shaped Tunnel FET", *J. Comput. Electron.*, Springer,Vol. 19, pp. 1154–1163, 2020.

[49] J.Singh, and B.Raj, "An Accurate and Generic Window function for Nonlinear Memristor Model" *J. Comput. Electron.*, Springer, Vol. 18, No. 2, pp. 640–647, 2019.

[50] M.Kaur, N.Gupta, S.Kumar, B.Raj, and A. K.Singh, "Comparative RF and Crosstalk Analysis of Carbon Based Nano Interconnects", *IET Circuits, Devices Syst.*, Vol. 15, No. 6, pp. 493–503, 2021.

[51] N.Kandasamy, F.Ahmad, D.Ajitha, B.Raj, and N.Telagam, "Quantum Dot Cellular Automata Based Scan Flip Flop and Boundary Scan Register", *IETE J. Res.*, Vol.69, PP. 535-548, 2023.

[52] S. K.Sharma, B.Raj, and M.Khosla, "Enhanced Photosensitivity of Highly Spectrum Selective Cylindrical Gate In1-xGaxAs Nanowire MOSFET Photodetector", *Mod. Phys. Lett. B*, Vol. 33, No. 12, p. 1950144, 2019.

[53] J.Singh, and B.Raj, "Design and Investigation of 7T2M NVSARM with Enhanced Stability and Temperature Impact on Store/Restore Energy", *IEEE Trans. Very Large Scale Integr. VLSI Syst.*, Vol. 27, No. 6, pp. 1322–1328, 2019.

[54] A. K.Bhardwaj, S.Gupta, B.Raj, and A.Singh,"Impact of Double Gate Geometry on the Performance of Carbon Nanotube Field Effect Transistor Structures for Low Power Digital Design", *Comput. Theor. Nanosci.*, ASP USA, Vol. 16, pp. 1813–1820, 2019.

[55] N.Jain, and B.Raj, "Thermal Stability Analysis and Performance Exploration of Asymmetrical Dual-k Underlap Spacer (ADKUS) SOI FinFET for Security and Privacy Applications", *Indian J. Pure Appl. Phys. (IJPAP)*, Vol. 57, pp. 352–360, 2019.

[56] A.Singh, M.Khosla, and B.Raj, "Design and Analysis of Dynamically Configurable Electrostatic Doped Carbon Nanotube Tunnel FET", *Microelectronics J.*, Elsevier, Vol. 85, pp. 17–24, 2019.

[57] N.Jain, and B.Raj, "Dual-k Spacer Region Variation at the Drain Side of Asymmetric SOI FinFET Structure: Performance Analysis towards the Analog/RF Design Applications", *J. Nanoelectron. Optoelectron.*, American Scientific Publishers, Vol. 14, pp. 349–359, 2019.

[58] J.Singh, S.Sharma, B.Raj, and M.Khosla, "Analysis of Barrier Layer Thickness on Performance of In1-xGaxAs Based Gate Stack Cylindrical Gate Nanowire MOSFET", *JNO*, ASP USA, Vol. 13, pp. 1473–1477, 2018.

[59] N.Jain, and B.Raj, "Analysis and Performance Exploration of High-k SOI FinFETs over the Conventional Low-k SOI FinFET toward Analog/RF Design", *J. Semicond. (JoS)*, IOP Science, Vol. 39, No. 12, pp. 124002-1-7, 2018.

[60] C.Goyal, J. S.Ubhi, and B.Raj, "A Reliable Leakage Reduction Technique for Approximate Full Adder with Reduced Ground Bounce Noise", *J. Math. Probl. Eng.*, Hindawi, Vol. 2018, Article ID 3501041, PP.1-16, 2018.

[61] Anuradha, J. Singh, B.Raj, and M.Khosla, "Design and Performance Analysis of Nano-Scale Memristor-Based Nonvolatile SRAM", *J. Sens. Lett.*, American Scientific Publishers USA, Vol. 16, pp. 798–805, 2018.

[62] G.Wadhwa, and B.Raj, "Parametric Variation Analysis of Charge-PlasmaBased Dielectric Modulated JLTFET for Biosensor Application", *IEEE Sens. J.*, Vol. 18, No. 15, PP. 6070 – 6077, 2018.

[63] J.Singh, and B.Raj, "Comparative Analysis of Memristor Models for Memories Design", *JoS*, IoP, Vol. 39, No. 7, PP. 074006-1-12, 2018.

[64] D.Yadav, S. S.Chouhan, S. K.Vishvakarma, and B.Raj, "Application Specific Microcontroller Design for IoT Based WSN", *Sens. Lett.*, ASP, Vol. 16, pp. 374–385, 2018.

[65] G.Singh, R. K.Sarin, and B.Raj, "Fault-Tolerant Design and Analysis of Quantum-Dot Cellular Automata Based Circuits", *IEEE/IET Circuits, Devices Syst.*, Vol. 12, pp. 638–664, 2018.

[66] J.Singh, and B.Raj, "Modeling of Mean Barrier Height Levying Various Image Forces of Metal Insulator Metal Structure to Enhance the Performance of Conductive Filament Based Memristor Model", *IEEE Nanotechnol.*, Vol. 17, No. 2, pp. 268–275, 2018.

[67] A.Jain, S.Sharma, and B.Raj, "Analysis of Triple Metal Surrounding Gate (TM-SG) III-V Nanowire MOSFET for Photosensing Application", *OptoElectron. J.*, Elsevier, Vol. 26, No. 2, PP. 141–148, 2018.

[68] A.Jain, S.Sharma, and B.Raj, "Design and Analysis of High Sensitivity Photosensor Using Cylindrical Surrounding Gate MOSFET for Low Power Sensor Applications", *Eng. Sci. Technol. Int. J.*, Elsevier's, Vol. 19, No. 4, pp. 1864–1870, 2016.

[69] A.Singh, M.Khosla, and B.Raj, "Analysis of Electrostatic Doped Schottky Barrier Carbon Nanotube FET for Low Power Applications," *J Mater. Sci. Mater. Electron.*, Springer,Vol. 28, PP. 1762–1768, 2017.

[70] G.Saiphani Kumar, A.Singh, B.Raj, "Design and Analysis of Gate All around CNTFET Based SRAM Cell Design", *J. Comput. Electron.*, Springer, Vol. 17, No.1, PP. 138–145, 2018.

[71] G. P.Singh, B. S.Sohi, and B.Raj, "Material Properties Analysis of Graphene Base Transistor (GBT) for VLSI Analog Circuits", *Indian J. Pure Appl. Phys., (IJPAP)*, Vol. 55, pp. 896–902, 2017.

[72] S.Kumar and B.Raj, "Estimation of Stability and Performance Metric for Inward Access Transistor Based 6T SRAM Cell Design Using n-type/p-type DMDG-GDOV TFET", *IEEE VLSI Circuits and Systems Letter*, Vol. 3, No. 2, PP. 25–39, 2017.

[73] S.Sharma, A.Kumar, M.Pattanaik, and B.Raj, "Forward Body Biased Multimode Multi-Threshold CMOS Technique for Ground Bounce Noise Reduction in Static CMOS Adders", *Int. J. Inf. Electron. Eng.*, Vol. 3, No. 3, PP. 567–572, 2013.

[74] H.Singh, P.Kumar, and B.Raj, "Performance Analysis of Majority Gate SET Based 1-bit Full Adder", *International Journal of Computer and Communication*

Engineering (IJCCE), IACSIT Press Singapore, ISSN: 20103743, Vol. 2, No. 4, PP.1-8, 2013.

[75] A. K.Bhardwaj, S.Gupta, and B.Raj, "Investigation of Parameters for Schottky Barrier (SB) Height for Schottky Barrier Based Carbon Nanotube Field Effect Transistor Device", *J. Nanoelectron. Optoelectron.*, ASP, Vol. 15, PP. 783–791, 2020.

[76] P.Bansal, and B.Raj, "Memristor: A Versatile Nonlinear Model for Dopant Drift and Boundary Issues", *JCTN*, American Scientific Publishers, Vol. 14, No. 5, PP. 2319–2325, 2017.

[77] N.Jain, and B.Raj, "An Analog and Digital Design Perspective Comprehensive Approach on Fin-FET (Fin-Field Effect Transistor) Technology - A Review", *Reviews in Advanced Sciences and Engineering (RASE)*, ASP, Vol. 5, PP. 1–14, 2016.

[78] S.Sharma, B.Raj, and M.Khosla, "Subthreshold Performance of In1-xGaxAs Based Dual Metal with Gate Stack Cylindrical/Surrounding Gate Nanowire MOSFET for Low Power Analog Applications", *J. Nanoelectron. Optoelectron.*, American Scientific Publishers, USA, Vol. 12, PP. 171- 176, 2017.

[79] B.Raj, A. K.Saxena, and S.Dasgupta, "Analytical Modeling for the Estimation of Leakage Current and Subthreshold Swing Factor of Nanoscale Double Gate FinFET Device" *Microelectron. Int.*, UK, Vol. 26, PP. 53–63, 2009.

[80] S. S.Soniya, G.Wadhwa, and B.Raj, "An Analytical Modeling for Dual Source Vertical Tunnel Field Effect Transistor", *International Journal of Recent Technology and Engineering (IJRTE)*, Vol. 8, No. 2, 2019.

[81] S.Singh, and B.Raj, "Design and Analysis of Hetrojunction Vertical T-Shaped Tunnel Field Effect Transistor", *J. Electron. Mater.*, Springer, Vol. 48, No. 10, PP. 6253–6260, 2019.

[82] C.Goyal, J. S.Ubhi, and B.Raj, "A Low Leakage CNTFET Based Inexact Full Adder for Low Power Image Processing Applications", *Int. J. Circuit Theory Appl.*, Wiley, Vol. 47, No. 9, PP. 1446–1458, 2019.

[83] B.Raj, A. K.Saxena, and S.Dasgupta, "A Compact Drain Current and Threshold Voltage Quantum Mechanical Analytical Modeling for FinFETs", *J. Nanoelectron. Optoelectron. (JNO)*, USA, Vol. 3, No. 2, PP. 163–170, 2008.

[84] G.Wadhwa, and B.Raj, "An Analytical Modeling of Charge Plasma Based Tunnel Field Effect Transistor with Impacts of Gate Underlap Region", *Superlattices Microstruct.*, Elsevier, Vol. 142, PP. 106512-22, 2020.

[85] S.Singh, and B.Raj, "Modeling and Simulation Analysis of SiGe Heterojunction Double GateVertical t-Shaped Tunnel FET", *Superlattices Microstruct.*, Elsevier,Vol. 142, PP. 106496-104, 2020.

[86] A.Singh, D. K.Saini, D.Agarwal, S.Aggarwal, M.Khosla, and B.Raj, "Modeling and Simulation of Carbon Nanotube Field Effect Transistor and Its Circuit Application", *J. Semicond. (JoS)*, IOP Science, Vol. 37, PP. 074001–074006, 2016.

[87] N.Jain, and B.Raj, "Device and Circuit Co-Design Perspective Comprehensive Approach on FinFET Technology - A Review", *Journal of Electron Devices*, Vol. 23, No. 1, PP. 1890–1901, 2016.

[88] S.Kumar and B.Raj, "Analysis of ION and Ambipolar Current for DualMaterial Gate-Drain Overlapped DG-TFET", *J. Nanoelectron. Optoelectron.*, American Scientific Publishers, USA, Vol. 11, PP. 323–333, 2016.

[89] N.Anjum, T.Bali, and B.Raj, "Design and Simulation of Handwritten Multiscript Character Recognition", *Int. J. Adv. Res. Comput. Commun. Eng.*, Vol. 2, No. 7, PP. 2544–2549, 2013.

[90] S.Sharma, B.Raj, and M.Khosla, "A Gaussian Approach for Analytical Subthreshold Current Model of Cylindrical Nanowire FET with Quantum Mechanical Effects", *Microelectronics J.*, Elsevier, Vol. 53, PP. 65–72, 2016.

[91] K.Singh, and B.Raj, "Performance and Analysis of Temperature Dependent Multi-Walled Carbon Nanotubes as Global Interconnects at Different Technology Nodes", *J. Comput. Electron.*, Springer, Vol. 14, No. 2, PP. 469–476, 2015.

[92] S.Kumar and B.Raj, "Compact Channel Potential Analytical Modeling of DG-TFET Based on Evanescent–Mode Approach", *J. Comput. Electron.*, Springer, Vol. 14, No. 2, PP. 820–827, 2015.

[93] K.Singh, and B.Raj, "Temperature Dependent Modeling and Performance Evaluation of Multi-Walled CNT and Single-Walled CNT as Global Interconnects", *J. Electron. Mater.*, Springer, Vol. 44, No. 12, PP. 4825–4835, 2015.

[94] V. K. Sharma, M. Pattanaik, and B. Raj, "INDEP Approach for Leakage Reduction in Nanoscale CMOS Circuits", *Int. J. Electron.*, Taylor & Francis, Vol. 102, No. 2, PP. 200–215, 2014.

[95] K.Singh, and B.Raj, "Influence of Temperature on MWCNT Bundle, SWCNT Bundle and Copper Interconnects for Nanoscaled Technology Nodes", *J. Mater. Sci. Mater. Electron.*, Springer, Vol. 26, No. 8, PP. 6134–6142, 2015.

[96] N.Anjum, T.Bali, and B.Raj, "Design and Simulation of Handwritten Gurumukhi and Devanagri Numerical Recognition", *International Journal of Computer Applications, Published by Foundation of Computer Science*, New York, USA,Vol. 73, No. 12, PP. 16–21, 2013.

[97] S.Khandelwal, V.Gupta, B.Raj, and R. D.Gupta, "Process Variability Aware Low Leakage Reliable Nano Scale DG-FinFET SRAM Cell Design Technique" , *J. Nanoelectron. Optoelectron.*, Vol. 10, No. 6, PP. 810–817, 2015.

[98] V. K. Sharma, M. Pattanaik, and B.Raj, "ONOFIC Approach: Low Power High Speed Nanoscale VLSI Circuits Design", *Int. J. Electron.*, Taylor & Francis, Vol. 101, No. 1, PP. 61–73, 2014.

[99] S.Khandelwal, B.Raj, and R. D.Gupta, "FinFET Based 6T SRAM Cell Design: Analysis of Performance Metric, Process Variation and Temperature Effect", *J. Comput. Theor. Nanosci.*, ASP, USA, Vol. 12, PP. 2500–2506, 2015.

[100] S.Singh, Y.Shekhar, R.Jagdeep, S.Anurag, and B.Raj, "Impact of HfO2 in Graded Channel Dual Insulator Double Gate MOSFET", *J. Comput. Theor. Nanosci., American Scientific Publishers*, Vol. 12, No. 6, PP. 950–953, 2015.

[101] V. K.Sharma, M.Pattanaik, and B.Raj, "PVT Variations Aware Low Leakage INDEP Approach for Nanoscale CMOS Circuits", *Microelectron. Reliab.*, Elsevier, Vol. 54, PP. 90–99, 2014.

[102] B.Raj, A. K.Saxena and S.Dasgupta, "Quantum Mechanical Analytical Modeling of Nanoscale DG FinFET: Evaluation of Potential, Threshold Voltage and Source/Drain Resistance", *Elsevier's J. Mater. Sci. Semicond. Process.*, Elsevier, Vol. 16, No. 4, PP. 1131–1137, 2013.

[103] M.Gopal, S. S. D.Prasad, and B.Raj, "8T SRAM Cell Design for Dynamic and Leakage Power Reduction", *Int. J. Comput. Appl.*, Published by Foundation of Computer Science, New York, USA, Vol. 71, No. 9, PP. 43–48, 2013.

[104] M.Pattanaik, B.Raj, S.Sharma, and A.Kumar, "Diode Based Trimode MultiThreshold CMOS Technique for Ground Bounce Noise Reduction in Static CMOS Adders", *Advanced Materials Research, Trans Tech Publications*, Switzerland, Vol. 548, PP. 885–889, 2012.

[105] B.Raj, A. K.Saxena, and S.Dasgupta, "Nanoscale FinFET Based SRAM Cell Design: Analysis of Performance Metric, Process Variation, Underlapped FinFET and Temperature Effect", *IEEE Circuits Syst. Mag.*, Vol. 11, No. 2, PP. 38–50, 2011.

[106] V. K. Sharma, M. Pattanaik, and B.Raj, "Leakage Current ONOFIC Approach for Deep Submicron VLSI Circuit Design", *International Journal of Electrical, Computer, Electronics and Communication Engineering*, World Academy of Sciences, Engineering and Technology, Vol. 7, No. 4, PP. 239–244, 2013.

[107] T.Chawla, M.Khosla, and B.Raj, "Design and Simulation of Triple Metal Double-Gate Germanium on Insulator Vertical Tunnel Field Effect Transistor", *Microelectron. J.*, Elsevier, Vol. 114, PP. 105125-32, 2021.

[108] P.Kaur, S. S.Gill, and B.Raj, "Comparative Analysis of OFETs Materials and Devices for Sensor Applications", *Journal of Silicon*, Springer, Vol. 14, PP. 4463–4471, 2022.

[109] S. K.Sharma, P.Kumar, B.Raj, and Balwinder Raj, "In1-xGaxAs Double Metal Gate-Stacking Cylindrical Nanowire MOSFET for Highly Sensitive Photo Detector", *Journal of Silicon*, Springer, Vol. 14, PP. 3535–3541,2022.

[110] B.Raj, A. K.Saxena, and S.Dasgupta, "Analytical Modeling of Quasi Planar Nanoscale Double Gate FinFET with Source/Drain Resistance and Field Dependent Carrier Mobility: A Quantum Mechanical Study", *J. Comput. (JCP)*, Academy Publisher, FINLAND, Vol. 4, No. 9, PP.1-8, 2009.

[111] S.Bhushan, S.Khandelwal, and B.Raj, "Analyzing Different Mode FinFET Based Memory Cell at Different Power Supply for Leakage Reduction", *Seventh International Conference on Bio-Inspired Computing: Theories and Application*, (BIC-TA 2012), organized IIITM Gwalor, PP. 89–100, 14-16 Dec 2012.

[112] J.Singh, and B.Raj, "Temperature Dependent Analytical Modeling and Simulations of Nanoscale Memristor", *Eng. Sci. Technol. Int. J.*, Elsevier's, Vol. 21, PP. 862–868, 2018.
[113] S.Singh, S.Bala, B.Raj, and Balwinder Raj, "Improved Sensitivity of Dielectric Modulated Junctionless Transistor for Nanoscale Biosensor Design", *Sens. Lett.*, ASP,Vol. 18, PP. 328–333, 2020.
[114] V.Kumar, S. K.Vishvakarma, and B.Raj, "Design and Performance Analysis of ASIC for IoT Applications", *Sens. Lett.*, ASP, Vol. 18, PP. 31–38, 2020.
[115] A.Jaiswal, R. K.Sarin, B.Raj, and S.Sukhija, "A Novel Circular Slotted Microstrip-fed Patch Antenna with Three Triangle Shape Defected Ground Structure for Multiband Applications", *Adv. Electromagn. (AEM)*, Vol. 7, No. 3, PP. 56–63, 2018.
[116] G.Wadhwa, and BRaj, "Label Free Detection of Biomolecules Using ChargePlasma-Based Gate Underlap Dielectric Modulated Junctionless TFET", *J. Electron. Mater. (JEMS)*, Springer, Vol. 47, No. 8, PP. 4683– 4693, 2018.
[117] G.Singh, R. K.Sarin, and B.Raj, "Design and Performance Analysis of a New Efficient Coplanar Quantum-Dot Cellular Automata Adder", *Indian J. Pure Appl. Phys. (IJPAP)*, Vol. 55, PP. 97–103, 2017.
[118] A.Singh, M.Khosla, and B.Raj, "Design and Analysis of Electrostatic Doped Schottky Barrier CNTFET Based Low Power SRAM", *Int. J. Electron. Commun.*, (AEÜ), Elsevier,Vol. 80, PP. 67–72, 2017.
[119] P.Kaur, V.Pandey, and B.Raj, "Comparative Study of Efficient Design, Control and Monitoring of Solar Power Using IoT", *Sens. Lett.*, ASP, Vol. 18, PP. 419–426, 2020.
[120] G.Wadhwa, P.Kamboj, J.Singh, B.Raj, "Design and Investigation of Junctionless DGTFET for Biological Molecule Recognition", *Trans. Electr. Electron. Mater.*, Springer, Vol. 22, PP. 282–289, 2020.

8

Advancement of Neuromorphic Computing Systems with Memristors

Jeetendra Singh[1], Shailendra Singh[2]*, Balwant Raj[3], Vikas Patel[1] and Balwinder Raj[4]

[1]*Dept. of ECE, NIT Sikkim, Sikkim, India*
[2]*Dept. of ECE, Pranveer Singh Institute of Technology Kanpur, Kanpur, Uttar Pradesh, India*
[3]*Dept. of ECE, NIT Jalandhar, UIET, Panjab University SSG Regional Centre Hoshiarpur, Punjab, India*
[4]*Dept. of ECE, NIT Jalandhar, Punjab, India*

Abstract

A computing device is deemed rudimentary if its operational scope is confined to elementary high-school-level mathematical tasks. Conversely, a mechanism that incorporates components of biotic origin, like as DNA units, as opposed to conventional electrical elements, closely mirrors the intricacies of the human brain. The instantiation of a computation system that operates neurons and synapses in parallel, exhibiting high capacity and low power consumption, constitutes a neuromorphic computing system (NCS). Neurons communicate through the transmission of both chemical and electrical signals. A neural network is established through minuscule junctions, termed "synapses," where each neuron is intricately linked to others, facilitating signal propagation. Neuromorphic systems seamlessly integrate computing systems and memories, thereby mitigating the separation between them and overcoming the challenges associated with the "memory wall." This chapter provides an exhaustive perspective on neuromorphic computing, commencing with an examination of recent challenges in achieving effective computing systems capable of supporting diverse novel applications. Subsequently, device-level perspectives are scrutinized to enhance the efficacy of neuromorphic solutions. The comprehensive applications of neuromorphic computing are

**Corresponding author*: shailendras.nitj@gmail.com

Balwinder Raj, Suman Lata Tripathi, Tarun Chaudhary, K. Srinivasa Rao and Mandeep Singh (eds.) Integrated Devices for Artificial Intelligence and VLSI, (193–216) © 2024 Scrivener Publishing LLC

delineated to maximize exposure for research groups and the scientific community engaged in this field.

Keywords: DNA (Deoxyribonucleic acid), neuromorphic computing system (NCS), synapses, phase change memory (PCM), memory wall, optical devices, image processing

8.1 Introduction

A computing device is basic if it can only do elementary school math calculations. However, a machine that mimics neurobiological architectures using electronic circuits would function similarly to the human brain. A computer system that imitates the neurons and synapses of the human brain and nervous system with high power and low energy consumption is called a neuromorphic computing system.

Neurons communicate with each other by sending chemical and electrical signals. Each neuron is connected to other neurons through small junctions called synapses (the end of a neuron allows a signal to be transmitted from one to another). The tightly integrated computing systems and memories of neuromorphic systems have reduced the distance between them and thus can break through the so-called "memory wall", the processor-memory performance gap that has steadily increased in recent decades. This chapter provides a detailed overview of neuromorphic computing. We begin by reviewing recent challenges in implementing efficient computing systems that enable a variety of new applications. We then consider hardware-level perspectives for better neuromorphic solutions. Broad applications of neuromorphic computing are presented here to provide maximum exposure to research groups and scientific communities working with this technology. Finally, the role of new memory devices, including memristors, conductive bridge memories, phase change memory (PCM) and spin devices, as well as floating gate transistors and optical devices, in the development of advanced neuromorphic computing systems will be explored. This chapter is useful for undergraduate students, graduate students and various research groups to better implement and effectively apply neuromorphic systems in the fields of social learning, teaching, speech, image processing and statistical analysis.

8.1.1 Evolution in Neural Networks

The computer's evolution accelerated over the last 50 years due to the von Neumann architecture presented in the 1940s [1]. This model of a stored-program digital computer stores both program instructions and data in the same read-write, random-access memory and runs for high data communication needs. According to Moore's predictions, the number of transistors will double every 18–24 months, and this represented massive evolution for the semiconductor industry in the 1960s [2]. Higher growth allowed increased processing speed, and lower power consumption meant more efficient computing systems were developed. However, in the semiconductor industry, current fabrication technology proposes a limit to Moore's law [3], so many approaches have been suggested [4] to cope with the expected growth of processing systems. Some approaches recommend new processing models and novel devices beyond conventional, complementary metal-oxide-semiconductor (CMOS) technologies.

During the 1960s, scientists used memorization techniques such as EET (emerging electronics technology) to model the nervous system, which is more complex than neurons [5–10], to pass from person to person. Extending the appearance to the structure and fundamentals of the behavior of the human brain shows a very diverse processing paradigm requiring an advanced performance from older computer architecture. [11]. Nowadays, we have many more supercomputers as compared to the last decade, with faster access, speed, and more powerful performance. However, when attempting to model the human brain in neuromorphic electronic systems, it is a much more complex problem to do through computers all that the human brain can do. Yes, it can be done [12]. This linkage between the standard computer and the human brain started the expansion of neuromorphic computing [13, 14].

This merger of the function of biological neurons and the function of transistors in the sub- threshold regime began in the 1980s.Computational neuroscience has several components: one is computational modeling and another is the philosophical study of biological neural networks. Computational modeling aims to obtain data collected from biological behavior, potentially functional tools of biological systems, through data and statistics [15]. These data may describe different levels of thought and the dynamics of neural modes, including their short-term interactions in individual neurons and the development of neural circuits through interactions with other neurons. These also include short- and long-term neuronal flow models and various overall systemic levels at the individual neuronal level. Thus a plausible resolution may be based on the relationship between performance and ambiguous neural mechanisms in learning and recognition [16].

Faced with the rapidly increasing computational demands of neural networks, many neuromorphic ASICs have been developed and demonstrated over conventional CMOS technology. For example, Google designed their Tensor Processing Unit to accelerate neural network processing, and IBM designed the True North neuromorphic IC [17]. This design, the fundamental operation, will fulfill the high performance of MAC (multiplication-and- accumulation). The nanoscale memristor is one of the most promising devices for this application available today. The two-terminal device can be adjusted by acting as a variable blocker whose value specific voltage or currents. Metals used are dielectric base dielectric film nanometers that can be smaller. Application by creating a filament or other mechanism such as resistor modulation controlled the electric field and current across a dielectric, causing a change in its resistance. In neuromorphic engineering, the memristor has drawn particular interest due to its unique plasticity behavior that parallels the optimization rules observed in the biological synapse. Memories can modify and change their performance over time in response to the human brain's different stimulus patterns. Vision restoration [18] and convolutional neural networks (CNNs) [19] are also fields that can use memristors to connect memory and data processing.

A number of emerging technologies are outlined in Figure 8.1 that have the potential to reshape the artificial intelligence research environment in the future. The computer's development has increased throughout the past 70 years due to the von Neumann architecture. From 1950 to 2020, we have seen many advances in neuromorphic computing systems. During 2006, several organizations began to focus on developing neuromorphic hardware devices using memristors.

Since the resistance of memristors is linked to past device inputs, they could be used to replace the high energy storage logic in neurons. When effectively applied, they can be significantly smaller than the digital circuitry necessary to achieve the same operation, allowing for greater network density.

8.1.2 Study Plan and Difficulties in Exhibiting Effective Neuromorphic Computing Systems

Models, algorithms, complete hardware systems, device-level components, new materials, and communication infrastructure are discussed in this section, along with helpful software operations. This chapter introduces hardware reference systems as neuromorphic applications. Thus far there have been many obstacles in neuromorphic computing and hardware neural networks. Many intriguing new research avenues are emerging in a range of fields, which will assist to revolutionize the neuromorphic computers in the future scenario. The exciting question is neuromorphic system training

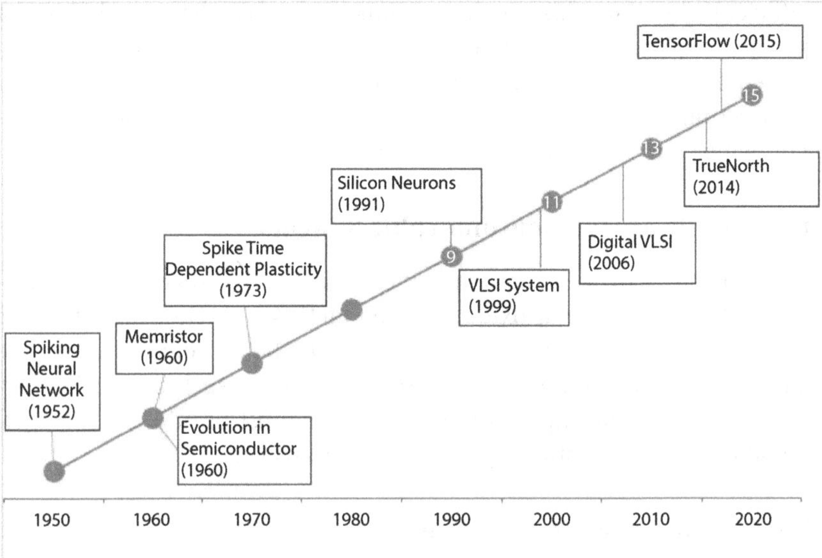

Figure 8.1 Evolution in neuromorphic computing systems.

and whether it is suitable for machine learning algorithms for learning. If used properly, we expect the effect of neuromorphic computing to enhance the representation of racemic neural networks to be similar to the effect of GPUs in extended training networks [20]. In other words, when algorithm developers do not rely on passive-interface simulation and training programs, the response time will be too rapid to realize efficient methods. However, to make these innovation possible, algorithm developers must be ready to go beyond conventional algorithms and to think outside the von Neumann architecture. This is not an insignificant assignment. However, if it is successful, it may change our understanding of machine learning and the methodologies used by neuromorphic robots. Many machine-learning methods for Big Data have been developed to date, with the CPU power of server-class machines being used for optimization [21]. However, offline storage space is frequently required for huge amounts of data, although the storage duration and viability complexity of methodologies used in such systems in both cases. For example, new methods for learning algorithms that use a stochastic approach allow us to deal with huge amounts of data. Given the luxury of substantial computational resources, it has allowed us to follow the views of hard times. Natural obstacles need to be addressed in order to understand the practical applications, such as the viability of the hardware implementation of these algorithms and approaches for embedded platforms. On such platforms, storage and computational capacity are insufficient; therefore this research points to the fact that translating the algorithm requires less computational complexity and low-power hardware requirements, where neuromorphic engineering seeks to provide a new direction. Some necessary improvements in neuromorphic engineering drawings have been compiled into one [21].

8.1.3 Hardware for Neuromorphic Systems

Neuromorphic hardware equipment covers the full spectrum, including complementary metal-oxide semiconductors (CMOS), digital signal processor hardware, graphics processing units, field-programmable gate arrays (FPGAs), combined with unique equipment such as accelerators (Figure 8.2). The challenges of large data sizes must be addressed, which demand innovative solutions, particularly actions [22].

8.1.4 Device-Level Perspective

Neuromorphic devices are among the most exciting components of the emerging technologies and elements of current neuromorphic computing analysis. With the nano-scale material structures of today, many novel devices can be produced as crucial elements of neuromorphic computing. In neuromorphic computing and modeling, researchers compete with materials scientists at all levels, including algorithms, as the new elements are adapted to different neuromorphic applications. It is not only the tiny, ultra-capacity rapid neuromorphic computers making use of modern technologies, but these technologies can also support constructing new composite elements to trigger actions designed explicitly for neuromorphic computing. From a software engineering standpoint, neuromorphic computer support software systems and represent a new challenge to be met to develop neuromorphic computers to be used by non-specialists. As

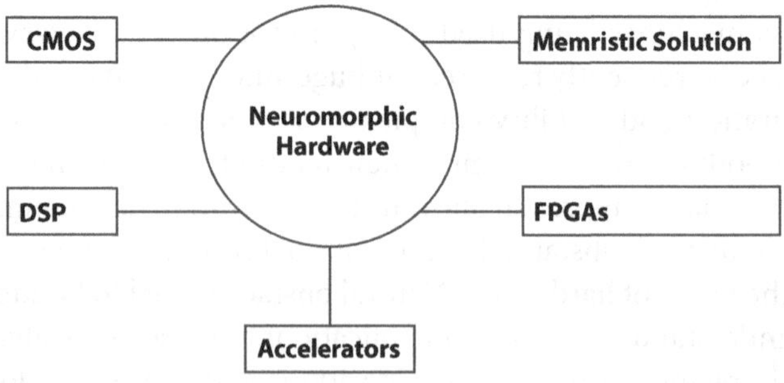

Figure 8.2 Neuromorphic system hardware.

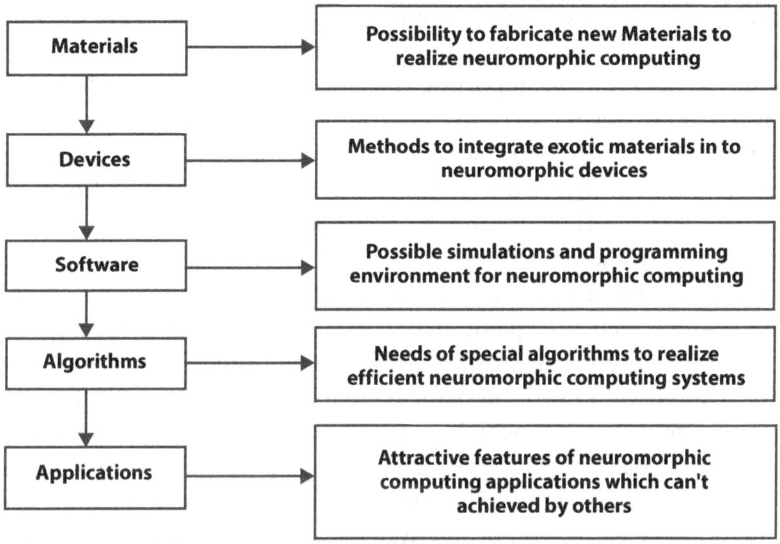

Figure 8.3 Design strategy and challenges in developing efficient neuromorphic computing systems.

neuromorphic computing has significant advantages, we must provide the data to create support software to develop new neuromorphic devices and train software engineers developing these systems and designs. A design strategy for developing efficient neuromorphic computing systems, and the challenges faced, are shown in Figure 8.3. Again, neuromorphic computers need a completely different path of learning from the traditional von Neumann architecture. A programming language particularly for neuromorphic devices is needed, where the device is not used as a neural network simulator but as a particular kind of computer with specific features (for example, for mass correspondence and computing and memory parts are kept) is constructed the original. While attracting new users, we believe that these languages will help us move forward [23]. Much work is needed, and the neuromorphic community should be seeking to widen the range of applications of neuromorphic systems. These applications include the future dissipation at low power irregularities detectors, autonomous vehicle neuromorphic coprocessors, intelligent sensors, Internet of Things (IoT) independent computer down intelligent controllers and deploy systems sputnik, and many other application areas. Analysis platform included [24]. Real-time spatiotemporal (belonging to both space and time) data analysis or real-time instructions must be reported to the potential wider community of neuromorphic systems, to see the computer in this that people with such applications as a solution. Neuromorphic computing is an exciting growth area and neuromorphic supercomputers are the future of neuromorphic computing devices (algorithms and models), which can play an essential role in supporting research at each computing level.

Over the last several decades, neuromorphic computing has arisen as a way to build a "thinking machine." Whereas specific machine-learning algorithms are bio-inspired to some extent, DNNs and convolutional neural networks (CNNs) focus more on optimal signal processing rather than biomimicry. On the other hand, neuromorphic systems generally portray brain like activity in more detail, frequently imitating numerous complex characteristics seen in organic neural systems. Of course, the degree to which neuromorphic designs resemble biology differs from one system to the next. Neuromorphic systems handle data in the form of streams of electric pulses or spikes, which is a common trait, as far as we can tell. This spiking data structure is primarily based on how the brain represents information. As a result, as compared to more traditional machine learning methodologies, neuromorphic systems are often targeted at more brain like processing, even brain mimicry.

8.1.5 Electrical Circuits to Realize Neurons

A significant change exists in the planning phase of neuromorphic systems which are realized by electrical circuits. An electrical network which realized neurons is shown in Figure 8.4. The communication connections are relied on classical von Neumann computing transistors, resistors, capacitors, inductors, and essential equipment [79–83]. These traditional tools have distinct characteristics like velocity, volume, control range and also they are defined in terms of other essential aspects such as energy

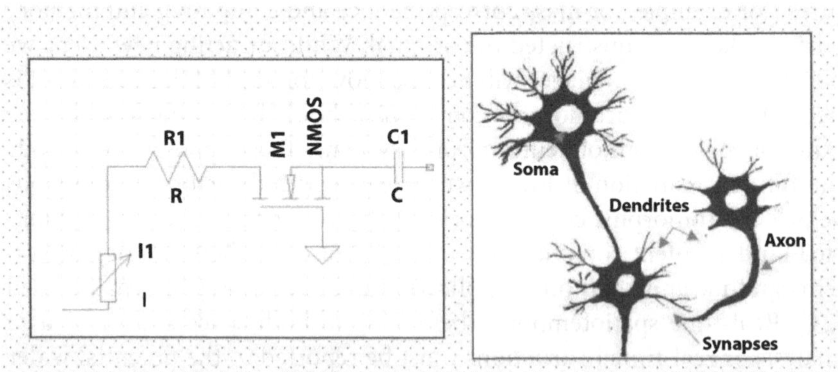

Figure 8.4 Electrical circuits to realize neurons.

dissipation, inability to determine good design, functionality, faults allowed and poor connectivity. On the other hand, the brain is based on an elaborate collection of neurons, each consisting of a body (soma), synapse, axons, and dendrites that are multifunctional and incomplete [25].

8.1.6 Broad Applications of Neuromorphic Computing

Most of the killing applications show that the neuromorphic computers and the equipment needed for developing the neuromorphic computing systems are remain to be determined. Memory [26] speech [27], statistical analysis [28], control [29], and many applications, including the detection of irregularities [30], have been applied to various neural network types. Figure 8.5 shows wide applications of the neuromorphic in the regime of medical, sensors, data and image processing etc. These applications are implemented directly on neural network hardware, which can have lesser power, rapid computation and a smaller yield footprint than the von Neumann architecture providing [84–90].

The researchers believe in better algorithms and more extensive reinforcement of computer circuits for neuromorphic computers. However, the use of the information processing capability-driven processors which behave like the human brain is expected to determine as well. With these improvements, let us examine their goals for upgrading neuromorphic computers and AI application areas. While existing CPU and GPU are

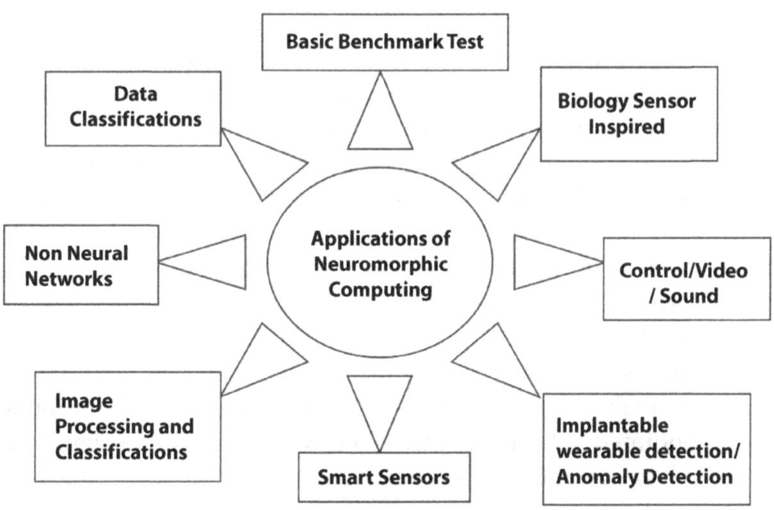

Figure 8.5 Applications of neuromorphic computing.

supercomputers are taking on an excellent level and have tremendous development in terms of AI applications, researchers are still faced with some shortcomings [91–95].

Furthermore, spiking neural networks to their complete capacity has now no longer been studied the manner synthetic neural networks were, and it could be that physiological neuromorphic hardware is wanted to develop neuromorphic structures to decide killer programs. This is going together with suitable algorithms being evolved for neuromorphic structures, as mentioned in phase IV. We speak numerous programs of neuromorphic structures. We pass over one primary utility area: the use of neuromorphic structures to take a look at neuroscience thru faster, extra green simulations than conventional computing platforms. Instead, we consciousness on were applied neuromorphic structures in different real-international programs [96–100]. The reason of this phase is to effectively offer scope for the sort of attempted issues tackled via way of means of neuromorphic structures and programs them to encourage the reader to use of their very own set. Figure five programs of neuromorphic structures and a way to provide a top-level view of the sort of well-known as they were. A complete set of neuromorphic structures which have been completely evolved is primarily based totally on a selected sensory gadget and implemented to the one's areas. The "sensing of them" fashion is primarily based totally on famous vision-primarily based totally gadget implementations, which regularly structure absolutely organic visible structures [31, 32] and on numerous functions of somatosensory or tactile-primarily based totally structures copy [33–39].

Another class of applications for neuromorphic system intelligent vehicle owners with high expectations: They work like self-parking and command facilities, but they also want an easy way to control their communication with the public and want infinite interfaces. Voice-based controls, identify the underlying intent is more likely to provide more personalization and seamless connectivity [101–106]. They also help in displaying the clutter of dashboards and cars. Why don't today's vehicles make widespread use of this approach? This requires a thorough evaluation. The term traditional AI hardware is draining the battery, and there is a constant hunger to ship only to use without cloud-based AI, and the wake creates many lags, leading to a poor experience. Neuromorphic technologies make AI hardship possible [107–112]. In collaboration with a newly appointed automotive customer, we demonstrated that a thousand times less power could recognize commands for 0.2 seconds using spiking neural networks. Embedded GPU accelerator running on neuromorphic processors can make simple sound faster. This gives the game enlightening, low-latency conversations but power-limiting obstructions, sideways, even within a parked vehicle [40].

Neuromorphic systems are another class of human-inspired brain application designed to model neuromorphic hardware driving behavior [41–44]. This entails researching brain signals acquired by various modalities and building systems to understand how they change in response to a specific stimulus, paving the way for the advent of the brain-computer interface. Producing in accordance with is a component of paying taxes. Many problems exist, including the noise and inconsistency inherent in these data collecting modalities, the scale of the dataset, and the limits imposed by the acquisition modalities. These biological cues are frequently easy to get, and multi-modal materials are frequently beneficial but more difficult to implement on typical hardware platforms. Corradi *et al.* [45] work in this field to construct an event-based neuromorphic electroencephalogram (EEG) record-keeping system [46]. Memristive synapse neural networks recognize human emotions epistolizing to three-tone hypothetical English alphabet speech [47] revealed a structure for spatiotemporal modeling of intelligence data called NewCube. Record EEG by ear is aided by the characterization of tapes composed using this process by Mikkelsen *et al.* [48]. Barzegarjalli has received a neuromorphic system impersonating schizophrenia [49]. Kudithipudi [50], anywhere he generates and probes a neuromagnetic receptacle computing architecture for this plan [113–116].

Interfaces with biological systems either directly additional a group of appeals to neuromorphic systems or constituted or worn as a part of other objects ordinarily used for medical treatment or observance [51]. a necessary characteristic of those applications is that they need devices designed to be terribly little and have low power requirements. Brain-machine neuromorphic system interfaces became more current in recent years [52–63]. By its very nature, spike-based Neuromorphic models victimization communication of an equivalent type, equivalent to a biological system, are a natural selection for the brain-machine or brain-computer interface. Pacemaker or defibrillator system [64–68], retinal implants [69], prosthesis [71] elderly users [70], Neuromorphic system, and other wearable or wearable implant developed by fall detector. Robotics applications are primarily neuromorphic systems. Autonomous robots often need remote and power-efficient systems. Many requirements and applications for robotics, including motor control, have been demonstrated successfully in neural networks. Any approved applications of neuromorphic systems for robotics involve acquiring a specific behavior [72] and ensuing a specific movement, social learning, learning, and measuring common goals or walls. So far, the most common use autonomous sailing functions of neuromorphic implementation [73] to robotics. Like robotics, space utilization generates speed through a central model generator (CPG). CPGs originate oscillatory acts as if they were formed by swimming motion gait. Neuromorphic [74] CPG is executing numerous systems. Neuromorphic control tasks have become popular since they typically need real-time performance, are many times employed in natural systems that require low volume and less power and have a temporal processing element. They will profit from the use of figures that are steadfast associations or suspensions in synapses. Several specific control applications used neuromorphic systems [75], but the common control test car-pole problem or inverted pendulum work. Neuromorphic structures have additionally been implemented to video games along with Pong, Pacman, and Flappy Bird. It has been done by the various image-based applications using neural networks and neuromorphic implementation, including detection widely edge; other pattern recognition, such as an application to simple shapes or pixel patterns using neuromorphic systems, has been done.

Explore other image analyses, including real-world classified images, such as traffic signals, facial recognition or recognition, vehicle identification or detection, Air pollutants in images, production defects or omissions detection, hand gesture popularity—human identification, item texture analysis, and different real-global picture popularity tasks [76].

Many standard real-world image data set Caltech -101 data sets, Google Street View house numbers (SVHN) dataset, CIFAR 10 dataset, and is rated at neuromorphic systems, including ImageNet. It is a well-known fixture for evaluating neural network architecture, whether to implement image classification, AlexNet. Various voice recognition functions are also considered along with the neuromorphic system. Speech recognition, for example, is a combined treatment for the neuromorphic system. The functions of the neuromorphic system of music perception have been implemented. The temporal processing element may require both functions for speech and music recognition and may require real-time.

Thus, neuromorphic constructs based on neural network systems with temporal processing components as recurrent or spiking are a natural fit for these applications. Other voice-based neuromorphic system functions include speech recognition, noise, voice activity separation, and voice analysis of identified motives. Neuromorphic systems have also been implemented to change noisy speech or different signals into smooth signals for noise filtering applications. Video access applications have also been an everyday use of neuromorphic systems. The several basic examples for video are thing attention within the video section. This statement does not need a volatile element, as it can separate video stages as images [77]. However, there is a need for temporary factors, such as motion detectors and some other video applications, motion prediction, motion tracking, and activity recognition. The neuromorphic system requires some recurrent network applied to the task of natural language processing (NLP). Sentence construction in the example applications in NLP displayed using Neuromorphic system, sentence completion, questions concerning classification, sentiment analysis, and document resemblance. Tasks with real-time performance, the capability to employ in the area with a tiny footprint and matters of general use less power require neuromorphic systems.

Intelligent sensors interest, humidity sensors, intensity sensor lighting and are classified as mobile devices and can also include being used to authenticate users. Similarly, exception detectors also include peculiarities detection in traffic, biological data, industrial data, and fault exposure applications in cyber security purposes in diesel engines and analog chips for neuromorphic systems. Usual data classification using neuromorphic systems has also become well-known. This capital includes suitable value forecasting, wind velocity estimation, soldering, including space accident diagnosis, grain identification, driver temporary state detection, computer user analysis, gas detection or detection, product categorization, as well as hyper spectral information categorization, and neuromorphic systems have several rubbing areas, mixed classification, radiation detection, climate prognostication and applications at intervals high potential physics. One set of standards, the UCI Machine Learning Repository neural networks has become popular in neuromorphic systems including the following data sets: build, connect4, gene, glass, heart, manosphere; however, some propose to use the neuromorphic system for non-neural network applications. A typical application area has been graphed algorithms that may be expressed as neuromorphic architecture plots—for example, graph algorithms, including minimum graph color, maximum cut, and smallest route. The neuromorphic system is also utilized to affect action in droves and is used to solve compliance problems. Going ahead, many neuromorphic architectures are expected and use instances for non-neural community applications, as neuromorphic computer systems emerge as greater typically available and ideas with specific features, such as a new type of computer [78].

8.2 Summary

This chapter provides an overview of the realization and implementation of neuromorphic systems. The current challenges and wide application of neuromorphic computing systems are considered in full detail. Also, models, device-level components, algorithms, hardware in the context of complete hardware systems, new materials, communication infrastructure, and supporting systems such as supporting software systems and systems of neuromorphic applications are discussed here. Contributions of memristive devices along with traditional devices are also explained in detail, as they are helpful to gear up the current neuromorphic systems. To replicate the nonlinear behavior of the memristor, we have proposed a viable memristor emulator circuit creation approach. Two alternative emulator circuits for a flux-controlled memristor have been presented. Our numerical analysis and simulation using software, as well as experimental testing, show that the suggested circuits can precisely simulate the behavior of a memristor and satisfy all three memristor fingerprints. Furthermore, in the absence of an actual solid-state device for the memristor, these emulator circuits could be highly beneficial in researching the characteristics and applications of memristors. As a result, our emulator circuits have the potential to be employed in a wide range of analog and digital applications. Compared to many previous emulator circuits developed by other organizations, the proposed circuits are practical and straightforward to create. As a result, the suggested emulator circuit construction approach would substantially influence the research's developmental and instructional elements. Because of their quick processing pace and reduced energy consumption, memristors are helpful in neural network systems. At the device layer, memristors have high energy efficiency and a high on-chip density. At the architectural level, concurrent processing is carried out at the same time as information is stored, eliminating the need for frequent data movement and memory wall issues. Based on electron effect and non-ideal features in the manufacture of nano-scale memristors, the robust performance of memristor neural networks still needs to be improved. Meanwhile, many memristor types have limited flexibility, and chip designers have not thoroughly investigated this. At the present time, chip designers do not have access to complete unified memristor models. High-density memristor crossbar array integration is additionally complicated by wire resistance, stealth path current and half-select difficulties. The joint efforts of academia and industry will result in revolutionary studies in memristor-based ANNs.

References

[1] Von Neumann, J., 1993. First Draft of a Report on the EDVAC. *IEEE Ann. Hist. Comput.*, *15*(4), pp.27-75.
[2] Moore, G.E., 1965. Cramming more components onto integrated circuits.
[3] Waldrop, M.M., 2016. The chips are down for Moore's law. *Nature News,* 530(7589), p.144.
[4] Kaur, J., 2016. Life Beyond Moore: More Moore or More than Moore—A *Review. Int. J. Comput. Sci. Mob. Comput,* 5, pp.233-237.
[5] Hodgkin, A.L. and Huxley, A.F., 1952. Currents carried by sodium and potassium ions through the membrane of the giant axon of Loligo. *J. Physiol.*, 116(4), pp.449-472.
[6] FitzHugh, R., 1961. Impulses and physiological states in theoretical models of nerve membrane. *Biophys. J.*, 1(6), pp.445-466.
[7] Morris, C. and Lecar, H., 1981. Voltage oscillations in the barnacle giant muscle fiber. *Biophys. J.*, 35(1), pp.193-213.
[8] Brette, R. and Gerstner, W., 2005. Adaptive exponentials integrate-and-fire model as an effective description of neuronal activity. *J. Neurophysiol.*, 94(5), pp.3637-3642.
[9] A.Bhardwaj, P.Kumar, B.Raj, and S.Anand, "Design and Performance Optimization of Doping-Less Vertical Nanowire TFET Using Gate Stack Technique," *J. Electron. Mater. (JEMS),* Springer, Vol. 41, No. 7, pp. 4005–4013, 2022.
[10] J. Singh, and B. Raj, "Tunnel Current Model of Asymmetric MIM Structure Levying Various Image Forces to Analyze the Characteristics of Filamentary Memristor", *Appl. Phys. A,* Springer, Vol. 125, No. 3, pp. 203.1 to 203.11, 2019.
[11] C. Goyal, J. S. Ubhi, and B. Raj, "Low Leakage Zero Ground Noise Nanoscale Full Adder Using Source Biasing Technique", *J. Nanoelectron. Optoelectron.*, American Scientific Publishers,Vol. 14, pp. 360–370, 2019.
[12] G.Singh, R. K.Sarin, and B.Raj, "A Novel Robust Exclusive-OR Function Implementation in QCA Nanotechnology with Energy Dissipation Analysis", *J. Comput. Electron.*, Springer, Vol. 15,No. 2, pp. 455–465, 2016.
[13] T.Wadhera, D.Kakkar, G.Wadhwa, and B.Raj, " Recent Advances and Progress in Development of the Field Effect Transistor Biosensor: A Review", *J. Electron. Mater.*, Springer,Vol. 48, No. 12, pp. 7635–7646, 2019.
[14] G.Wadhwa, P.Kamboj, and B.Raj, "Design Optimisation of Junctionless TFET Biosensor for High Sensitivity", *Adv. Nat. Sci.: Nanosci. Nanotechnol.*, Vol. 10, No. 7, p. 045001, 2019.
[15] P.Bansal, and B.Raj, "Memristor Modeling and Analysis for Linear Dopant Drift Kinetics", *J. Nanoeng. Nanomanuf.*, American Scientific Publishers, Vol. 6, pp. 1–7, 2016.

[16] A.Singh, M.Khosla, and B.Raj, "Circuit Compatible Model for Electrostatic Doped Schottky Barrier CNTFET," *J. Electron. Mater.*, Springer,Vol. 45, No. 12, pp. 4825–4835, 2016.

[17] Ashima, D.Vaithiyanathan, and B.Raj, "Performance Analysis of Charge Plasma induced Graded Channel Si Nanotube", *J. Eng. Res. (JER)*, EMSME Special Issue, Vol.9, pp. 146–154, 2021.

[18] A. S.Tomar, V. K.Magraiya, and B.Raj, "Scaling of Access and Data Transistor for High Performance DRAM Cell Design", *Quantum Matter*, Vol. 2, pp. 412–416, 2013.

[19] N.Jain, and B.Raj, "Parasitic Capacitance and Resistance Model Development and Optimization of Raised Source/Drain SOI FinFET Structure for Analog Circuit Applications", *J. Nanoelectron. Optoelectron.*, ASP, USA, Vol. 13, pp. 531–539, 2018.

[20] S.Singh, S. K.Vishvakarma, and B.Raj, "Analytical Modeling of Split-Gate Junction-Less Transistor for a Biosensor Application", *Sensing and BioSensing, Elsevier*, Vol. 18, pp. 31–36, 2018.

[21] M.Gopal, and B.Raj, "Low Power 8T SRAM Cell Design for High Stability Video Applications", *ITSI Transaction on Electrical and Electronics Engineering*, Vol.1, No. 5, pp. 91–97, 2013.

[22] B.Raj, J.Mitra, D. K.Bihani, V.Rangharajan, A. K. Saxena, and S. Dasgupta, "Analysis of Noise Margin, Power and Process Variation for 32 nm FinFET Based 6T SRAM Cell", *J. Comput. (JCP)*, Academy Publisher, FINLAND, Vol. 5, No. 6, PP.1-6, 2010.

[23] D.Sharma, R.Mehra, and B.Raj, "Comparative Analysis of Photovoltaic Technologies for High Efficiency Solar Cell Design", *Superlattices Microstruct.*, Elsevier, Vol. 153, pp. 106861, 2021.

[24] P.Kaur, A. S.Buttar, and B.Raj, "A Comprehensive Analysis of Nanoscale Transistor Based Biosensor: A Review", *Indian J. Pure Appl. Phys.*, Vol. 59, pp. 304–318, 2021.

[25] D.Yadav, B.Raj, and B.Raj, "Design and Simulation of Low Power Microcontroller for IoT Applications", *J. Sens. Lett.*, ASP USA,Vol. 18, pp. 401–409, 2020.

[26] S.Singh, and B.Raj, "A 2-D Analytical Surface Potential and Drain Current Modeling of Double-Gate Vertical t-Shaped Tunnel FET", *J. Comput. Electron.*, Springer,Vol. 19, pp. 1154–1163, 2020.

[27] J.Singh, and B.Raj, "An Accurate and Generic Window function for Nonlinear Memristor Model" *J. Comput. Electron.*, Springer, Vol. 18, No. 2, pp. 640–647, 2019.

[28] M.Kaur, N.Gupta, S.Kumar, B.Raj, and A. K.Singh, "Comparative RF and Crosstalk Analysis of Carbon Based Nano Interconnects", *IET Circuits Devices Syst.*, Vol. 15, No. 6, pp. 493–503, 2021.

[29] N.Kandasamy, F.Ahmad, D.Ajitha, B.Raj, and N.Telagam, "Quantum Dot Cellular Automata Based Scan Flip Flop and Boundary Scan Register", *IETE J. Res.*, Vol.69, PP. 535-548, 2023.

[30] S. K.Sharma, B.Raj, and M.Khosla, "Enhanced Photosensitivity of Highly Spectrum Selective Cylindrical Gate In1-xGaxAs Nanowire MOSFET Photodetector", *Mod. Phys. Lett. B*, Vol. 33, No. 12, p. 1950144, 2019.

[31] J.Singh, and B.Raj, "Design and Investigation of 7T2M NVSARM with Enhanced Stability and Temperature Impact on Store/Restore Energy", *IEEE Trans. Very Large Scale Integr. Syst.*, Vol. 27, No. 6, pp. 1322–1328, 2019.

[32] A. K.Bhardwaj, S.Gupta, B.Raj, and A.Singh,"Impact of Double Gate Geometry on the Performance of Carbon Nanotube Field Effect Transistor Structures for Low Power Digital Design", *Comput. Theor. Nanosci.*, ASP USA, Vol. 16, pp. 1813–1820, 2019.

[33] N.Jain, and B.Raj, "Thermal Stability Analysis and Performance Exploration of Asymmetrical Dual-k Underlap Spacer (ADKUS) SOI FinFET for Security and Privacy Applications", *Indian J. Pure Appl. Phys. (IJPAP)*, Vol. 57, pp. 352–360, 2019.

[34] A.Singh, M.Khosla, and B.Raj, "Design and Analysis of Dynamically Configurable Electrostatic Doped Carbon Nanotube Tunnel FET", *Microelectron. J.*, Elsevier, Vol. 85, pp. 17–24, 2019.

[35] N.Jain, and B.Raj, "Dual-k Spacer Region Variation at the Drain Side of Asymmetric SOI FinFET Structure: Performance Analysis towards the Analog/RF Design Applications", *J. Nanoelectron. Optoelectron.*, American Scientific Publishers, Vol. 14, pp. 349–359, 2019.

[36] J.Singh, S.Sharma, B.Raj, and M.Khosla, "Analysis of Barrier Layer Thickness on Performance of In1-xGaxAs Based Gate Stack Cylindrical Gate Nanowire MOSFET", *JNO*, ASP USA, Vol. 13, pp. 1473–1477, 2018.

[37] N.Jain, and B.Raj, "Analysis and Performance Exploration of High-k SOI FinFETs over the Conventional Low-k SOI FinFET toward Analog/ RF Design", *J. Semicond. (JoS)*, IOP Science,Vol. 39, No. 12, pp. 124002-1-7, 2018.

[38] C.Goyal, J. S.Ubhi, and B.Raj, "A Reliable Leakage Reduction Technique for Approximate Full Adder with Reduced Ground Bounce Noise", *J. Math. Probl. Eng.*, Hindawi, Vol. 2018, Article ID 3501041, PP.1-16, 2018.

[39] Anuradha, J. Singh, B.Raj, and M.Khosla, "Design and Performance Analysis of Nano-Scale Memristor-Based Nonvolatile SRAM", *J. Sens. Lett.*, American Scientific Publishers USA, Vol. 16, pp. 798–805, 2018.

[40] G.Wadhwa, and B.Raj, "Parametric Variation Analysis of Charge-PlasmaBased Dielectric Modulated JLTFET for Biosensor Application", *IEEE Sens. J.*, Vol. 18, No. 15, PP. 6070 – 6077, 2018.

[41] J.Singh, and B.Raj, "Comparative Analysis of Memristor Models for Memories Design", *JoS*, IoP, Vol. 39, No. 7, PP. 074006-1-12, 2018.

[42] D.Yadav, S. S.Chouhan, S. K.Vishvakarma, and B.Raj, "Application Specific Microcontroller Design for IoT Based WSN", *Sens. Lett.*, ASP, Vol. 16, pp. 374–385, 2018.

[43] G.Singh, R. K.Sarin, and B.Raj, "Fault-Tolerant Design and Analysis of Quantum-Dot Cellular Automata Based Circuits", *IEEE/IET Circuits Devices Syst.*, Vol. 12, pp.638–664, 2018.

[44] J.Singh, and B.Raj, "Modeling of Mean Barrier Height Levying Various Image Forces of Metal Insulator Metal Structure to Enhance the Performance of Conductive Filament Based Memristor Model", *IEEE Nanatechnol.*, Vol. 17, No. 2, pp. 268–275, 2018.

[45] A.Jain, S.Sharma, and B.Raj, "Analysis of Triple Metal Surrounding Gate (TM-SG) III-V Nanowire MOSFET for Photosensing Application", *Optoelectron. J.*, Elsevier, Vol. 26, No. 2, PP. 141–148, 2018.

[46] A.Jain, S.Sharma, and B.Raj, "Design and Analysis of High Sensitivity Photosensor Using Cylindrical Surrounding Gate MOSFET for Low Power Sensor Applications", *Eng. Sci. Technol. Int. J.*, Elsevier's, Vol. 19, No. 4, pp. 1864–1870, 2016.

[47] A.Singh, M.Khosla, and B.Raj, "Analysis of Electrostatic Doped Schottky Barrier Carbon Nanotube FET for Low Power Applications," *J. Mater. Sci.: Mater. Electron.*, Springer,Vol. 28, PP. 1762–1768, 2017.

[48] G.Saiphani Kumar, A.Singh, B.Raj, "Design and Analysis of Gate All around CNTFET Based SRAM Cell Design", *J. Comput. Electron.*, Springer, Vol. 17, No.1, PP. 138–145, 2018.

[49] G. P.Singh, B. S.Sohi, and B.Raj, "Material Properties Analysis of Graphene Base Transistor (GBT) for VLSI Analog Circuits", *Indian J. Pure Appl. Phys. (IJPAP)*, Vol. 55, pp. 896–902, 2017.

[50] S.Kumar and B.Raj, "Estimation of Stability and Performance Metric for Inward Access Transistor Based 6T SRAM Cell Design Using n-type/p-type DMDG-GDOV TFET", *IEEE VLSI Circuits and Systems Letter*, Vol. 3, No. 2, PP. 25–39, 2017.

[51] S.Sharma, A.Kumar, M.Pattanaik, and B.Raj, "Forward Body Biased Multimode Multi-Threshold CMOS Technique for Ground Bounce Noise Reduction in Static CMOS Adders", *Int. J. Inf. Electron. Eng.*, Vol. 3, No. 3, PP. 567–572, 2013.

[52] H.Singh, P.Kumar, and B.Raj, "Performance Analysis of Majority Gate SET Based 1-bit Full Adder", *Int. J. Comput. Commun. Eng. (IJCCE)*, IACSIT Press Singapore, ISSN: 20103743, Vol. 2, No. 4, PP.1-8, 2013.

[53] A. K.Bhardwaj, S.Gupta, and B.Raj, "Investigation of Parameters for Schottky Barrier (SB) Height for Schottky Barrier Based Carbon Nanotube Field Effect Transistor Device", *J. Nanoelectron. Optoelectron.*, ASP, Vol. 15, PP. 783–791, 2020.

[54] P.Bansal, and B.Raj, "Memristor: A Versatile Nonlinear Model for Dopant Drift and Boundary Issues", *JCTN*, American Scientific Publishers, Vol. 14, No. 5, PP. 2319–2325, 2017.

[55] N.Jain, and B.Raj, "An Analog and Digital Design Perspective Comprehensive Approach on Fin-FET (Fin-Field Effect Transistor) Technology - A Review", *Rev. Adv. Sci. Eng. (RASE)*, ASP, Vol. 5, PP. 1–14, 2016.

[56] S.Sharma, B.Raj, and M.Khosla, "Subthreshold Performance of In1-xGaxAs Based Dual Metal with Gate Stack Cylindrical/Surrounding Gate Nanowire MOSFET for Low Power Analog Applications", *J. Nanoelectron. Optoelectron.*, American Scientific Publishers, USA, Vol. 12, PP. 171– 176, 2017.

[57] B.Raj, A. K.Saxena, and S.Dasgupta, "Analytical Modeling for the Estimation of Leakage Current and Subthreshold Swing Factor of Nanoscale Double Gate FinFET Device" *Microelectron. Int.*, UK, Vol. 26, PP. 53–63, 2009.

[58] S. S.Soniya, G.Wadhwa, and B.Raj, "An Analytical Modeling for Dual Source Vertical Tunnel Field Effect Transistor", *Int. J. Recent Technol. Eng. (IJRTE)*, Vol. 8, No. 2, 2019.

[59] S.Singh, and B.Raj, "Design and Analysis of Hetrojunction Vertical T-Shaped Tunnel Field Effect Transistor", *J. Electron. Mater.*, Springer, Vol. 48, No. 10, PP. 6253–6260, 2019.

[60] C.Goyal, J. S.Ubhi, and B.Raj, "A Low Leakage CNTFET Based Inexact Full Adder for Low Power Image Processing Applications", *Int. J. Circuit Theory Appl.*, Wiley, Vol. 47, No. 9, PP. 1446– 1458, 2019.

[61] B.Raj, A. K.Saxena, and S.Dasgupta, "A Compact Drain Current and Threshold Voltage Quantum Mechanical Analytical Modeling for FinFETs", *J. Nanoelectron. Optoelectron. (JNO)*, USA, Vol. 3, No. 2, PP. 163–170, 2008.

[62] G.Wadhwa, and B.Raj, "An Analytical Modeling of Charge Plasma Based Tunnel Field Effect Transistor with Impacts of Gate Underlap Region", *Superlattices Microstruct.*, Elsevier, Vol. 142, PP. 106512-22, 2020.

[63] S.Singh, and B.Raj, "Modeling and Simulation Analysis of SiGe Heterojunction Double GateVertical t-Shaped Tunnel FET", *Superlattices Microstruct.*, Elsevier,Vol. 142, PP. 106496-104, 2020.

[64] A.Singh, D. K.Saini, D.Agarwal, S.Aggarwal, M.Khosla, and B.Raj, "Modeling and Simulation of Carbon Nanotube Field Effect Transistor and Its Circuit Application", *J. Semicond. (JoS)*, IOP Science, Vol. 37, PP. 074001-074006, 2016.

[65] N.Jain, and B.Raj, "Device and Circuit Co-Design Perspective Comprehensive Approach on FinFET Technology - A Review", *J. Electron Devices,* Vol. 23, No. 1, PP. 1890–1901, 2016.

[66] S.Kumar and B.Raj, "Analysis of ION and Ambipolar Current for DualMaterial Gate-Drain Overlapped DG-TFET", *J. Nanoelectron. Optoelectron.*, American Scientific Publishers, USA, Vol. 11, PP. 323–333, 2016.

[67] N.Anjum, T.Bali, and B.Raj, "Design and Simulation of Handwritten Multiscript Character Recognition", *Int. J. Adv. Res. Comput. Commun. Eng.*, Vol. 2, No. 7, PP. 2544–2549, 2013.

[68] S.Sharma, B.Raj, and M.Khosla, "A Gaussian Approach for Analytical Subthreshold Current Model of Cylindrical Nanowire FET with Quantum Mechanical Effects", *Microelectron. J.*, Elsevier, Vol. 53, PP. 65–72, 2016.

[69] K.Singh, and B.Raj, "Performance and Analysis of Temperature Dependent Multi-Walled Carbon Nanotubes as Global Interconnects at Different

Technology Nodes", *J. Comput. Electron.*, Springer, Vol. 14, No. 2, PP. 469–476, 2015.

[70] S.Kumar and B.Raj, "Compact Channel Potential Analytical Modeling of DG-TFET Based on Evanescent-Mode Approach", *J. Comput. Electron.*, Springer, Vol. 14, No. 2, PP. 820–827, 2015.

[71] K.Singh, and B.Raj, "Temperature Dependent Modeling and Performance Evaluation of Multi-Walled CNT and Single-Walled CNT as Global Interconnects", *J. Electron. Mater.*, Springer,Vol. 44, No. 12, PP. 4825–4835, 2015.

[72] V. K. Sharma, M. Pattanaik, and B. Raj, "INDEP Approach for Leakage Reduction in Nanoscale CMOS Circuits", *Int. J. Electron.*, Taylor & Francis, Vol. 102, No. 2, PP. 200–215, 2014.

[73] K.Singh, and B.Raj, "Influence of Temperature on MWCNT Bundle, SWCNT Bundle and Copper Interconnects for Nanoscaled Technology Nodes", *J. Mater. Sci.: Mater. Electron.*, Springer, Vol. 26, No. 8, PP. 6134–6142, 2015.

[74] N.Anjum, T.Bali, and B.Raj, "Design and Simulation of Handwritten Gurumukhi and Devanagri Numerical Recognition", *Int. J. Comput. Appl.*, Published by Foundation of Computer Science, New York, USA,Vol. 73, No. 12, PP. 16–21, 2013.

[75] S.Khandelwal, V.Gupta, B.Raj, and R. D.Gupta, "Process Variability Aware Low Leakage Reliable Nano Scale DG-FinFET SRAM Cell Design Technique", *J. Nanoelectron. Optoelectron.*, Vol. 10, No. 6, PP. 810–817, 2015.

[76] V. K. Sharma, M. Pattanaik, and B.Raj, "ONOFIC Approach: Low Power High Speed Nanoscale VLSI Circuits Design", *Int. J. Electron.*, Taylor & Francis, Vol. 101, No. 1, PP. 61–73, 2014.

[77] S.Khandelwal, B.Raj, and R. D.Gupta, "FinFET Based 6T SRAM Cell Design: Analysis of Performance Metric, Process Variation and Temperature Effect", *J. Comput. Theor. Nanosci.*, ASP, USA, Vol. 12, PP. 2500–2506, 2015.

[78] S.Singh, Y.Shekhar, R.Jagdeep, S.Anurag, and B.Raj, "Impact of HfO2 in Graded Channel Dual Insulator Double Gate MOSFET", *J. Comput. Theor. Nanosci.*, American Scientific Publishers, Vol. 12, No. 6, PP. 950–953, 2015.

[79] V. K.Sharma, M.Pattanaik, and B.Raj, "PVT Variations Aware Low Leakage INDEP Approach for Nanoscale CMOS Circuits", *Microelectron. Reliab.*, Elsevier, Vol. 54, PP. 90–99, 2014.

[80] B.Raj, A. K.Saxena and S.Dasgupta, "Quantum Mechanical Analytical Modeling of Nanoscale DG FinFET: Evaluation of Potential, Threshold Voltage and Source/Drain Resistance", Elsevier's *J. Mater. Sci. Semicond. Process.*, Elsevier, Vol. 16, No. 4, PP. 1131–1137, 2013.

[81] M.Gopal, S. S. D.Prasad, and B.Raj, "8T SRAM Cell Design for Dynamic and Leakage Power Reduction", *Int. J. Comput. Appl.*, Published by Foundation of Computer Science, New York, USA,Vol. 71, No. 9, PP. 43–48, 2013.

[82] M.Pattanaik, B.Raj, S.Sharma, and A.Kumar, "Diode Based Trimode MultiThreshold CMOS Technique for Ground Bounce Noise Reduction in

Static CMOS Adders", *Advanced Materials Research, Trans Tech Publications,* Switzerland, Vol. 548, PP. 885–889, 2012.

[83] B.Raj, A. K.Saxena, and S.Dasgupta, "Nanoscale FinFET Based SRAM Cell Design: Analysis of Performance Metric, Process Variation, Underlapped FinFET and Temperature Effect", *IEEE Circuits Syst. Mag.*, Vol. 11, No. 2, PP. 38–50, 2011.

[84] V. K. Sharma, M. Pattanaik, and B.Raj, "Leakage Current ONOFIC Approach for Deep Submicron VLSI Circuit Design", *International Journal of Electrical, Computer,* Electronics and Communication Engineering, World Academy of Sciences, Engineering and Technology, Vol. 7, No. 4, PP. 239–244, 2013.

[85] T.Chawla, M.Khosla, and B.Raj, "Design and Simulation of Triple Metal Double-Gate Germanium on Insulator Vertical Tunnel Field Effect Transistor", *Microelectron. J.*, Elsevier, Vol. 114, PP. 105125-32, 2021.

[86] P.Kaur, S. S.Gill, and B.Raj, "Comparative Analysis of OFETs Materials and Devices for Sensor Applications", *J. Silicon,* Springer,Vol. 14, PP. 4463–4471, 2022.

[87] S. K.Sharma, P.Kumar, B.Raj, and Balwinder Raj, "In1-xGaxAs Double Metal Gate-Stacking Cylindrical Nanowire MOSFET for Highly Sensitive Photo Detector", *J. Silicon,* Springer, Vol. 14, PP. 3535–3541,2022.

[88] B.Raj, A. K.Saxena, and S.Dasgupta, "Analytical Modeling of Quasi Planar Nanoscale Double Gate FinFET with Source/Drain Resistance and Field Dependent Carrier Mobility: A Quantum Mechanical Study", *J. Comput. (JCP)*, Academy Publisher, FINLAND, Vol. 4, No. 9, PP.1-8, 2009.

[89] S.Bhushan, S.Khandelwal, and B.Raj, "Analyzing Different Mode FinFET Based Memory Cell at Different Power Supply for Leakage Reduction", *Seventh International Conference on Bio-Inspired Computing: Theories and Application,* (BIC-TA 2012), organized IIITM Gwalor, PP. 89–100, 14-16 Dec 2012.

[90] J.Singh, and B.Raj, "Temperature Dependent Analytical Modeling and Simulations of Nanoscale Memristor", *Eng. Sci. Technol. Int. J.*, Elsevier's, Vol. 21, PP. 862–868, 2018.

[91] S.Singh, S.Bala, B.Raj, and Balwinder Raj, "Improved Sensitivity of Dielectric Modulated Junctionless Transistor for Nanoscale Biosensor Design", *Sens. Lett.*, ASP,Vol. 18, PP. 328–333, 2020.

[92] V.Kumar, S. K.Vishvakarma, and B.Raj, "Design and Performance Analysis of ASIC for IoT Applications", *Sens. Lett.*, ASP,Vol. 18, PP. 31–38, 2020.

[93] A.Jaiswal, R. K.Sarin, B.Raj, and S.Sukhija, "A Novel Circular Slotted Microstrip-fed Patch Antenna with Three Triangle Shape Defected Ground Structure for Multiband Applications", *Advanced Electromagnetic (AEM),* Vol. 7, No. 3, PP. 56–63, 2018.

[94] G.Wadhwa, and BRaj, "Label Free Detection of Biomolecules Using ChargePlasma-Based Gate Underlap Dielectric Modulated Junctionless TFET", *J. Electron. Mater. (JEMS),* Springer, Vol. 47, No. 8, PP. 4683– 4693, 2018.

[95] G.Singh, R. K.Sarin, and B.Raj, "Design and Performance Analysis of a New Efficient Coplanar Quantum-Dot Cellular Automata Adder", *Indian J. Pure Appl. Phys. (IJPAP),* Vol. 55, PP. 97–103, 2017.

[96] A.Singh, M.Khosla, and B.Raj, "Design and Analysis of Electrostatic Doped Schottky Barrier CNTFET Based Low Power SRAM", *Int. J. Electron. Commun.,* (AEÜ), Elsevier,Vol. 80, PP. 67–72, 2017.

[97] P.Kaur, V.Pandey, and B.Raj, "Comparative Study of Efficient Design, Control and Monitoring of Solar Power Using IoT", *Sens. Lett.,* ASP, Vol. 18, PP. 419–426, 2020.

[98] G.Wadhwa, P.Kamboj, J.Singh, B.Raj, "Design and Investigation of Junctionless DGTFET for Biological Molecule Recognition", *Trans. Electr. Electron. Mater.,* Springer, Vol. 22, PP. 282–289, 2020.

[99] T.Chawla, M.Khosla, and B.Raj, "Optimization of Double-Gate Dual Material GeOI-Vertical TFET for VLSI Circuit Design", *IEEE VLSI Circuits and Systems Letter,* Vol. 6, No. 2, PP. 13–25, 2020.

[100] S. K.Verma, S.Singh, G.Wadhwa and B.Raj, "Detection of Biomolecules Using Charge-Plasma Based Gate Underlap Dielectric Modulated Dopingless TFET", *Trans. Electr. Electron. Mater. (TEEM),* Springer, Vol. 21, PP. 528–535, 2020.

[101] N.Jain, and B.Raj, "Impact of Underlap Spacer Region Variation on Electrostatic and Analog/RF Performance of Symmetrical High-k SOI FinFET at 20 nm Channel Length", *J. Semicond. (JoS),* IOP Science, Vol. 38, No. 12, PP. 122002-10, 2017.

[102] S.Singh, and B.Raj, "Analytical Modeling and Simulation Analysis of T-Shaped III-V Heterojunction Vertical T-FET", *Superlattices Microstruct.,* Elsevier, Vol. 147, PP. 106717-25, 2020.

[103] G.Singh, R. K.Sarin, and B.Raj, "Design and Analysis of Area Efficient QCA Based Reversible Logic Gates", *J. Microprocess. Microsyst.,* Elsevier, Vol. 52, PP. 59–68, 2017.

[104] A.Singh, M.Khosla, and B.Raj, "Compact Model for Ballistic Single Wall CNTFET under Quantum Capacitance Limit", *J. Semicond. (JoS),* IOP Science,Vol. 37, PP. 104001–104008, 2016.

[105] S.Singh, M.Khosla, G.Wadhwa, and B.Raj, "Design and Analysis of DoubleGate Junctionless Vertical TFET for Gas Sensing Applications", *Appl. Phys. A,* Springer,Vol. 127, No. 16, PP.883-890, 2021.

[106] I.Singh, B.Raj, M.Khosla, and B. K.Kaushik, "Potential MRAM Technologies for Low Power SoCs", *SPIN World Scientific Publisher,* SCIE; Vol. 10, No. 04, PP. 2050027-35, 2020.

[107] S.Singh, and B.Raj, "Parametric Variation Analysis on Hetero-Junction Vertical t-Shape TFET for Suppressing Ambipolar Conduction", *Indian J. Pure Appl. Phys.,* Vol. 58, PP. 478–485, 2020.

[108] S. S.Soniya, G.Wadhwa, and B.Raj, "Design and Analysis of Dual Source Vertical Tunnel Field Effect Transistor for High Performance", *Trans. Electr. Electron. Mater.,* Springer, Vol. 21, PP. 74–82, 2019.

[109] M.Kaur, N.Gupta, S.Kumar, B.Raj, and A. K.Singh, "RF Performance Analysis of Intercalated Graphene Nanoribbon Based Global Level Interconnects", *J. Comput. Electron.*, Springer, Vol. 19, PP.1002–1013, 2020.

[110] G.Wadhwa, B.Raj, "Design and Performance Analysis of Junctionless TFET Biosensor for High Sensitivity" *IEEE Nanatechnol.*, Vol.18, PP. 567–574, 2019.

[111] J.Singh, and B.Raj, "Enhanced Nonlinear Memristor Model Encapsulating Stochastic Dopant Drift", *JNO*, ASP, Vol.14, PP. 958–963, 2019.

[112] Izhikevich, E.M., 2003. Simple model of spiking neurons. *IEEE Trans. Neural Networks*, 14(6), pp.1569-1572.

[113] Runge, R.G.; Uemura, M.; Viglione, S.S. Electronic synthesis of the avian retina. *IEEE Trans. Biomed. Eng.* 1968, 15, 138–151. [CrossRef] [PubMed]

[114] Douglas, R., Mahowald, M. and Mead, C., 1995. Neuromorphic analogue VLSI. *Annu. Rev. Neurosci.*, 18(1), pp.255-281.

[115] Mead, C., 1990. Neuromorphic electronic systems. *Proceedings of the IEEE*, 78(10), pp.1629-1636

[116] Mahowald, M.A., Douglas, R.J., LeMoncheck, J.E. and Mead, C.A., 1992, February. An introduction to silicon neural analogs. *In Seminars in Neuroscience* (Vol. 4, No. 1, pp. 83-92). Academic Press.

9

Neuromorphic Computing and Its Application

Tejasvini Thakral, Lucky Lamba, Manjeet Singh, Tarun Chaudhary and Mandeep Singh*

Department of Electronics and Communication Engineering, Dr. BR Ambedkar National Institute Of Technology, Jalandhar, Punjab, India

Abstract

Neuromorphic computing is a rapidly developing field that seeks to emulate the neural structure and function of the human brain using hardware and software technologies. In recent years, the development of neuromorphic computing has been fueled by advancements in semiconductor technology and the need for more efficient and intelligent computing systems. This review chapter provides an overview of the state-of-the-art in neuromorphic computing, including the principles and concepts that underlie the technology, its key applications, and the challenges and opportunities that lie ahead. In this chapter discussion about the potential of neuromorphic computing to enable a wide range of applications, including sensory processing, robotics, machine learning, and cognitive computing. In this chapter discussions are made on some of the key challenges associated with the development of neuromorphic computing systems, including scalability, power consumption, and programming models. Overall, this review chapter provides a comprehensive overview of the current state of the art in neuromorphic computing and its potential impact on the future of computing and artificial intelligence.

***Keywords*:** Neuromorphic, cognitive, nano-neurons, synapses, algorithm, semiconductor, human neuron, human neuron

**Corresponding author*: mandeeps.ec.21@nitj.ac.in

Balwinder Raj, Suman Lata Tripathi, Tarun Chaudhary, K. Srinivasa Rao and Mandeep Singh (eds.) Integrated Devices for Artificial Intelligence and VLSI, (217–246) © 2024 Scrivener Publishing LLC

9.1 Introduction

With technology that is both energy-efficient and capable of performing very complex tasks, neuromorphic computing draws inspiration from the brain. In this chapter, discussion is made on redesigning electronics is necessary in order to produce new technology. show that research in physics and materials science will be crucial in creating artificial nano-neurons and synapses, as well as for density them, putting them together, and making them work well for computing. We talk about how some scientists choose to move forward this way by looking to artificial intelligence for notions, while others elect to move in this way by glancing to neuroscience for ideas. Discussions are made about the impressive results that how these approaches have led to relatively recent times [1].

9.2 Evolution of Neuroinspired Computing Chips

For decades, researchers have worked to build machines that can think like people. The development of this field has been strongly related to that of mathematics and neuroscience. Our knowledge of the physical makeup and functional operations of the brain has grown significantly over the past century. The increased understanding Some early successes include the following of neurobiology inspires the concept of developing technological devices that mimic the brain's neural network. Some early successes include the following:

In 1936 A mathematical argument was created by mathematician and computer scientist Alan Turing to demonstrate that, given any mathematical computation and an algorithm, a computer is capable of performing it. In his 1948 publication "Intelligent Machinery," Turing proposed a cognitive modelling tool based on human neurons. In 1949, Canadian psychologist Donald Hebb made a key contribution to neuroscience by hypothesizing a link between synaptic plasticity and learning. Hebb is frequently referred regarded as the "father of neuropsychology." In 1950 The Turing Test, developed by Turing, is still viewed as the gold standard for evaluating AGI [2]. The Perceptron, a dedicated analogue computer that have used motorized variable resistor as the weights, was unveiled in 1957 [3], introducing the concept of neuro-inspired computing hardware at that time. In 1961, ADALINE [4], an adjustable hardware perceptron with learning capability, was built using discrete variable resistors. However, because to the impossibility of high-density device integration, even the first sets of circuitry were very colossal in size and included only a small number of synaptic devices. With the debut of silicon-based integration technologies for VLSI and the emergence of significant demands for computing capabilities, including studies on analogue spiking integrated neural networks, interest in neuro-inspired computer chips quickly rose during the 1980s. The spiking method was founded on an analogy between the physical behaviors of integrated circuits and or the biological spiking behaviors in the brain. a concrete instance The spiking technique was created using an analogy between biological brain spiking behaviors and the physical behaviors of silicon chips. a powerful example Mead's work [5], which imitated brain behaviors, was built on metal-oxide-semiconductor field-effect transistor (MOSFET) circuits. The non-spiking method, on the other hand, used physical rules to speed up matrix multiplication in neural network processes.

For example, in order to multiply vector matrices inside a neural network, a group of researchers at Bell Laboratories developed a 12*12 unprogrammable synapse resistive array [6]. However, these ground-breaking chips could only be shown on a limited scale because of the practical limitations of the semiconductor technology that was available at the time. These limitations included a lack of miniaturized devices that could simulate the actions of living organisms. When HP Laboratories were first connected the resistor changing behavior of a solid-state device to the compositional memristor in 2008 [7], a fundamental revolution was initiated. This revolution was brought about by leveraging the underlying device characteristics in order to fabricate computer chips that were inspired by neuronal circuitry. In an endeavor to study NVM-based neuro-inspired technology, an analogue memristive device was built in 2010 to demonstrate synaptic functions [8].

The first attempt of its type, this. Since that time, a growing number of researchers have been concentrating their efforts on developing NVM-based neuro-inspired computing hardware, which includes chips, architectures, and algorithms.2013. With the intention of developing a synthetic human brain, Henry Markram started the HBP. With a 10-year time horizon, HBP aims to better understand the human brain and use this understanding in technology and medicine. It also incorporates cutting-edge neuromorphic computers. Working on the initiative are more than 500 scientists from 140 universities in Europe.2014 saw the True North neuromorphic chip's development by IBM. Compared to conventional von Neumann hardware, the device is utilized for optical object detection and uses less power. In 2018The Loihi neuromorphic microprocessor was created by Intel and has been used for robotics, gesture recognition, and smell analysis [2].

An exascale supercomputer would be equivalent to the human brain and at least fifty times quicker compared to the most advanced supercomputer in the United States (the Titan at Oak Ridge National Laboratory). The DOE has established a goal date of 2021 for the first exascale system; it has also mandated that one of the exascale supercomputers it is sponsoring employ a unique design, and neuromorphic computing is a good option to offer that new technology. Based on our investigation, It has been anticipated that First, neuromorphic processors will be incorporated in smart phones by 2025.

9.3 Science Behind Brain Physics

Our knowledge of the physical makeup and functional operations of the brain has grown significantly over the past century. The idea of creating electronic gear that mimics the brain's neural network was inspired by the advancement of neuroscience. When it comes to activities requiring recognition and decision-making, the human brain is far more efficient than classic Von Neumann computers, and it only consumes 20 watts of power [9]. More than 10^11 neurons and 10^15 synapses connect each other in the human brain. In addition to this, it demonstrates a number of exciting phenomena for physicists to study, including as self-oscillations, chaos, synchronization, unpredictable resonance, and also the reduction of energy or entropy. Its ability to learn through experience sets it distinct from standard instruction-based algorithms in certain significant ways.

From the very beginning, physicists have participated in theoretical research programs to gain a better understanding of the computing processes that occur in the brain. As a result, they have contributed to both the discipline of computational comp sci and the field of computational neuroscience. Statistical physics, non-linear dynamics, and complex systems concept have all contributed to the categorization of the neural processes that underlie learning [1].

The Figure 9.1 includes computing concepts and architectures. (a) A "memory constraint" between the CPU and RAM is depicted in this diagram of von Neumann architecture. (b) Storage-based-program-computing as a heuristic paradigm. The processing unit retrieves data from memory, manipulates it, and then saves the result back into memory. (c) A plan depicting the brain-inspired design philosophy. A paradigm shift toward computers that is inspired by the brain's wiring (d).

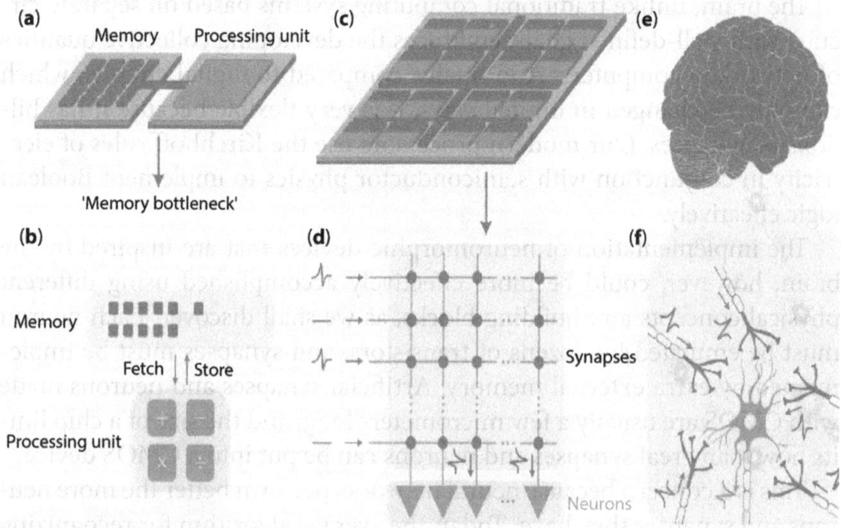

Figure 9.1 Architecture and paradigm of computing [10].

The granular synapse-neurons allow for free-form connections between electronic neurons, both in terms of input and output. (e) A schematic representation of the brain's structural components. (f) Conceptualization of neuronal connections and synapses in biology. When a neuron "fires," it means it has processed the incoming spikes impulses from the other neurons with which it is synaptically linked and has reached its own threshold for firing [11].

9.4 Limitations of Semiconductor Devices

Our general-purpose computers are built on digital circuits, unlike the physical substrate of the brain, which differs fundamentally from them. It is not based on electronic components like memory cells or transistors, but rather on biological components like neurons and synapses. To operate at low energy levels, this system takes use of the randomness of cells rather than highly exact circuits, as it is in digital systems. Aside from binary encoding, it also makes use of analogue encoding (soma dynamics, synaptic values) (single spikes). To conserve energy, the brain uses asynchronous communication rather than the single clock used by computers for synchronous communication.

The brain, unlike traditional computing systems based on separate circuits with well-defined goals, embraces the developing collective qualities of networked computers. And finally, compared to digital circuits, which can only be changed in limited ways, it is very flexible because it has billions of synapses. Our modern processors use the Kirchhoff rules of electricity in conjunction with semiconductor physics to implement Boolean logic effectively.

The implementation of neuromorphic devices that are inspired by the brain, however, could be more effectively accomplished using different physical concepts and building blocks, as we shall discover. Each neuron must be emulated by dozens of transistors, and synapses must be implemented by extra external memory. Artificial synapses and neurons made with CMOS are usually a few micrometers long, and the size of a chip limits how many real synapses and neurons can be put into a CMOS device.

This is a concern because neural networks perform better the more neurons and synapses they have; Today, the average algorithm for recognizing images has large numbers of neurons and connections between them. By connecting chips together, it is possible to make a lot of neurons and synapses, but the system as an entire gets bigger and loses a large amount of efficiency throughout the process. The same as Spinnaker or Brainscales, the high tech neuromorphic systems can occupy up a few cubic meters of space.

To make chips of lower power with millions of neurons and synapses, we need nanodevices that can precisely replicate important characteristics of synapses and neurons of the nanometer scale, such as absence linearity, memory, and learning. Lastly, CMOS technology alone makes it hard to

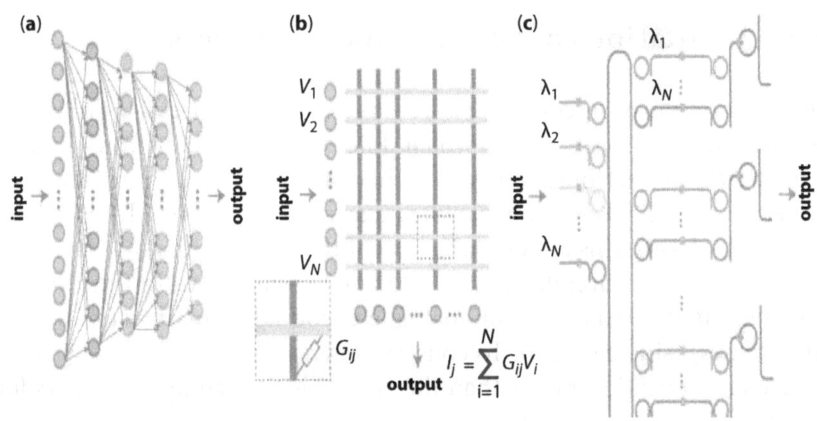

Figure 9.2 Hardware of deep neural network [1].

connect neurons in a way that is very strong. Each neuron in the brain is connected by an overall mean of 10,000 synapses. Modern electronics are mostly restricted to two dimensions (2D), fan-out is limited, and it is hard to power circuit components evenly and efficiently, so this kind of connection can't be made Axons and dendrites of neurons have a high fan-in/fan-out ratio, and blood effectively transports energies across the vast system in the brain due to being three-dimensional (3D) [1]. Figure 9.3 shows neuromorphic computing physics.

Hardware for deep neural networks (a) Deep neural network with circles of neurons and synapses (arrows). (b) Memristive crossbar array between neuron layers. The inset shows a vertically connected memristor cell. Red pre-synaptic CMOS neurons apply voltages to rows. All input voltages Vi weighted by memristor conductances Gij determine each column's output current Ij. Column amplifiers power post-synaptic CMOS neurons (in blue). (c) Circular resonators connect wavelength I inputs (in different hues) to a grey neuron in an optical neural network. Phase change materials implement neurons (green squares) and synapses (orange squares).

(a-f) Synaptic structures. Top row is extremely conductive, lower row is low conductive. (a) A filamentary device (oxide or conductive bridge): the resistance is dependent on the diameter of the filament connecting the top and bottom electrodes. (b) When it comes to chalcogenide-based phase change memories, the resistance is determined by the size of the amorphous

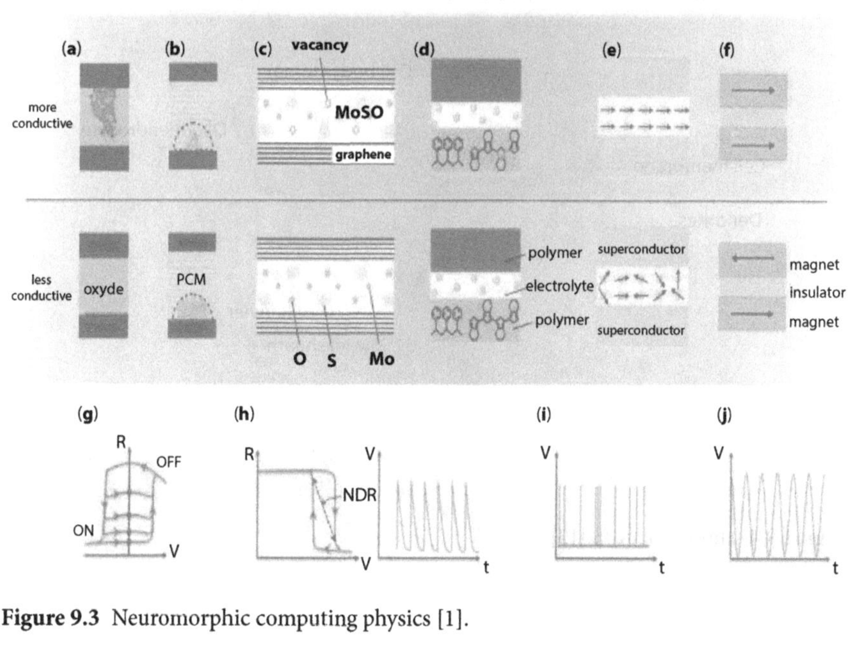

Figure 9.3 Neuromorphic computing physics [1].

region. (c) Oxygen vacancy concentration determines resistance in Van der Waals heterostructure graphene/MoSO/graphene. (d) Positive ions in the electrolyte layer conduct the organic electrochemical device. Polymer reduction at the postsynaptic electrode absorbs some conducting ions, decreasing synaptic conductivity. (e) Magnetic barrier Josephson junction: resistance depends on magnetic order. (f) Magnetic tunnel junction: magnetic layer orientation determines resistance. (g) A non-volatile memristor's resistance versus applied voltage. (h) Voltage surges caused by a variable memristive switch and its electrical resistance function which is the cause of the voltage applied. (i) Josephson junction voltage spikes caused by resistive and capacitive shunting and (j) Superparamagnetic tunnel junction electrical resistance as a function of time [30–36].

In essence, an artificial neuron is a biological neuron with engineering modifications. The device has a single output and many inputs. An ANN [12, 14] is composed of a substantial number of fundamental processing units connected to one another and layered together.

A hybrid neural network combines different types of neural network architectures or components to leverage the strengths of each approach and address specific challenges or tasks more effectively. It refers to the integration of multiple neural network models, such as combining artificial neural networks (ANNs) with spiking neural networks (SNNs), or combining different types of ANNs [37–41]. Figures 9.4, 9.5 and 9.6 represents human neuron, artificial neuron and multilayered ANN respectively.

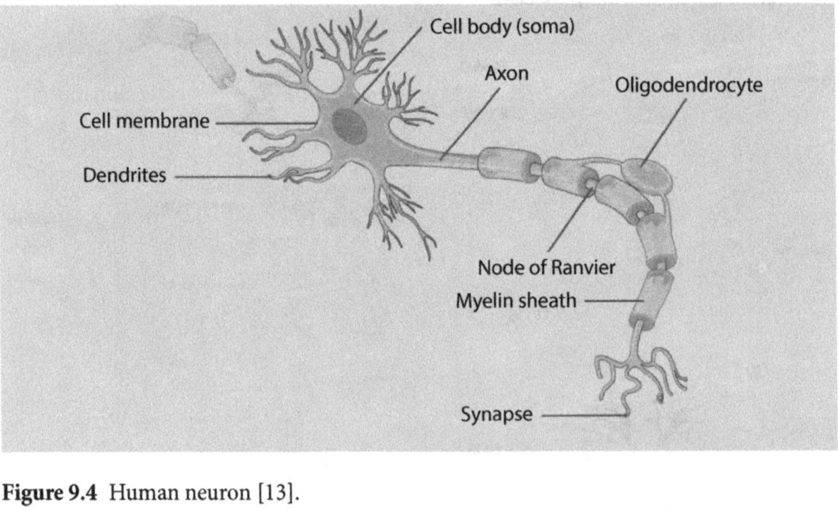

Figure 9.4 Human neuron [13].

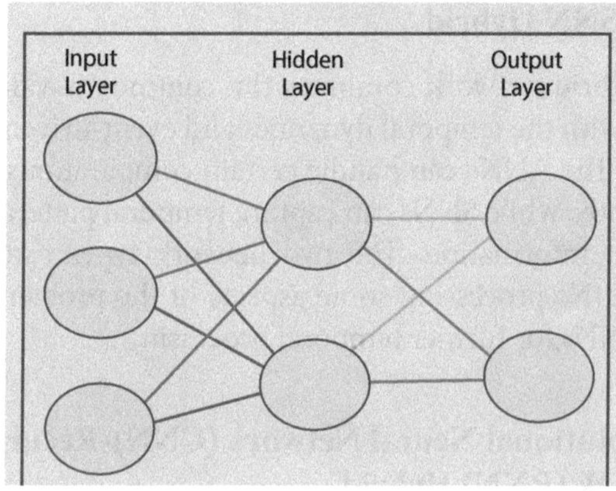

Figure 9.5 Artificial neuron [12].

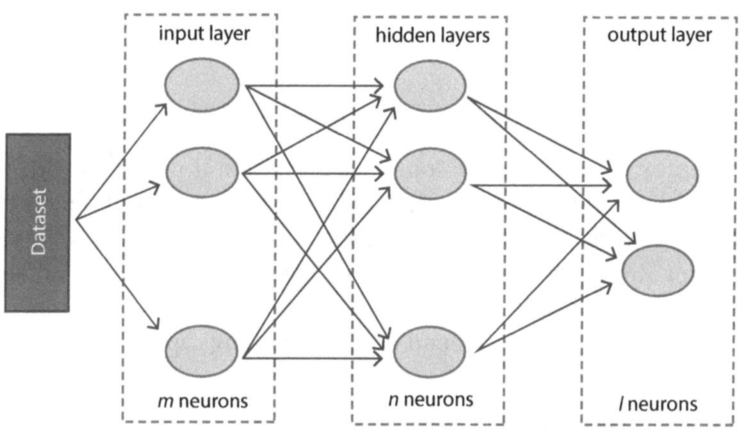

Figure 9.6 Multilayered ANN [15].

9.5 Various Combination of Networks

There are various ways to construct a hybrid neural network, and the specific approach depends on the problem at hand and the desired advantages of each neural network type. Here are a few examples:

9.5.1 ANN-SNN Hybrid

This type of hybrid network combines the continuous-valued computations of ANNs with the temporal dynamics and event-driven processing of SNNs [42–47]. The ANNs can handle certain computations and transformations efficiently, while SNNs can capture temporal patterns and process time-dependent information. The two network types can be interconnected, with ANNs processing some aspects of the problem and passing the results to SNNs for further temporal processing

9.5.2 Convolutional Neural Network (CNN)-Recurrent Neural Network (RNN) Hybrid

In computer vision tasks, a hybrid network can combine the spatial hierarchy and feature extraction capabilities of CNNs with the temporal modeling and sequence processing abilities of RNNs. The CNN component can extract relevant visual features from input images, and the RNN component can capture sequential dependencies and long-term context.

9.5.3 Deep Reinforcement Learning (DRL) Hybrid

Deep reinforcement learning combines deep neural networks with reinforcement learning algorithms. In a hybrid DRL network, different types of neural networks can be integrated, such as combining CNNs for visual perception, RNNs for sequential decision making, and ANNs for value estimation. This allows the network to learn complex tasks by combining perception, temporal processing, and decision-making capabilities [48–53].

9.5.4 Ensemble Hybrid

In an ensemble hybrid network, multiple neural network models of the same type (e.g., multiple ANNs) are combined to improve performance. Each model in the ensemble can have different architectures, initial conditions, or training data. By aggregating the predictions or decisions of multiple models, the ensemble hybrid network can achieve better accuracy, robustness, or generalization [54–58].

The design and training of hybrid neural networks require careful consideration of how to combine the different components effectively. It involves developing appropriate architectures, designing information flow between the components, and training the network in a way that leverages the strengths of each neural network type.

Hybrid neural networks can offer enhanced capabilities compared to using a single neural network type. They have been applied in various domains, including robotics, computer vision, natural language processing, and reinforcement learning, to address complex problems that require both spatial and temporal processing, or combination of different learning and inference mechanisms.

9.5.5 Different Types of Neural Networks

Parameter	ANN	CNN	RNN
Architecture	Composed of multiple layers of interconnected nodes, with each node performing a simple computation	Contains convolutional layers, pooling layers, and fully connected layers, with each layer performing a specific operation	Contains recurrent layers, with each layer processing sequential input and sharing parameters across time
Data type	Suitable for structured data and tabular data	Sui table for image and video data	Suitable for sequential data and time-series data
Learning efficiency	Requires large amounts of training data and time to learn complex patterns	Learns spatial features efficiently and is faster to train than ANN for image classification tasks	Can learn long-term dependencies in sequential data, but can be slow to train

(Continued)

(*Continued*)

Parameter	ANN	CNN	RNN
Memory usage	Memory-efficient compared to CNN and RNN	Memory-intensive due to the need to store large amounts of image data	Memory-intensive due to the need to store the previous hidden state
Input size	Can handle variable input size	Requires fixed input size	Can handle variable input size
Output size	Can produce a single scalar value or a vector of values	Can produce a 2D or 3D output, such as an image or a video	Can produce a sequence of output values
Use cases	Suitable for a wide range of tasks, such as regression, classification, and pattern recognition	Suitable for image and video tasks, such as object detection and image classification	Suitable for natural language processing tasks, such as language translation and speech recognition

9.6 Artificial Intelligence

In Artificial intelligence (AI) already can beat human brains in several restricted groups of issues, like as image/speech classification and games [16]. However, computer equipment is hampered by AI applications that need the processing of huge amounts of data, putting great demands on computing power and speed efficiency [17]. Traditional computer equipment is constructed on the von Neumann design, which physically divides processor and memory units (see Figure 9.1.a,b above) [10]. This architecture is well-suited for scientific computing due to its accurate Boolean calculations and high-precision data representations, but it is ineffective for AI applications [10]. As a result, using AI technology in power-constrained situations like edge computing and the Internet of Things (IoT) necessitates the creation of entirely new chip designs and device technologies [10]. The adoption of neuro-inspired microchips, which replicate the structure and functioning of the human mind, is one potential technique for the evolution of intelligent computing [17]. When compared to traditional systems, these neuro-inspired computer chips are expected to deliver gains in both terms of economy and processing power while executing AI workloads. Several neuro-inspired computer chip types have indeed been developed in recent years, with numerous neuro-inspired components included into each wafer at different levels the device to the circuit and architecture. Even so, given the technology for neuro-inspired microchips is still at its infancy, it is critical to assess the industry's challenges and opportunities [10].

9.7 A Summary of Neuromorphic Hardware Methodologies

In the conventional von-Neumann modern computer, the memory wall problem, which affects how data moves between the CPU and memory/storage off-chip, has evolved into the framework constraint [18]. The problems are getting even worse when a vast neural network needs a large amount of information to do calculations when it is being trained or tested. Even though large-scale matrix operations are a big part of neuro-inspired learning strategies, computing methodologies that use parallel processing at a finer granularity level directly on the chip are enticing. The neuro-inspired design, which takes advantage of the way neurons do distributed computing and synapses record information close to where it is used, is an interesting choice [59–63].

The neuroinspired architecture [19] stores weights locally in the synapses and does computing in a distributed way in the neurons. The shift from the computation-focused (von-Neumann) architecture to the data-focused (neuro-inspired) architecture in computing is shown in (Figure 9.2). The synapses are native memories that are linked to a lot of other memories through the channels of communication. The neurons, on the other hand, are just simple computing components (for nonlinear activation or thresholding function). The ultimate objective of putting neural circuits processing into hardware is to support (but not replace) the existing von-Neumann design for application-specific intelligent tasks like speech and image retrieval, self-driving cars, etc. [64–69].

So far, neuro-inspired learning techniques have been tested on different embedded systems with some concurrency. In general, there are two ways to make neuromorphic hardware, or philosophies, depending how the info is encoded. The first method stays with the digital (nonspiking) implementation of machine learning, deep learning, or artificial neural networks (ANNs), while the second method gets ideas from the way the brain works to make the most of parallel or dispersed computation. Figure 9.7 shows a dramatic change in the way technology works.

In digital deployments, the value of a neuron is encoded by the number of pulses or voltage levels, or by a combination of the two. GPUs [20] or field-programmable gate arrays (FPGAs) [21] have been used a lot to speed up hardware for machine learning and deep learning. CMOS ASIC accelerators [22–24] have already been made to save even more energy. For example, Google implemented their custom-built TPU framework to speed up the complicated intelligent computing tasks that were employed to make AlphaGo [20].

These are just a few examples of how neuromorphic computing can enhance natural language processing tasks. The ability of neuromorphic systems to capture the complexities and nuances of human language can lead to significant advancements in NLP, enabling more accurate, context-aware, and natural language understanding and generation [70–74].

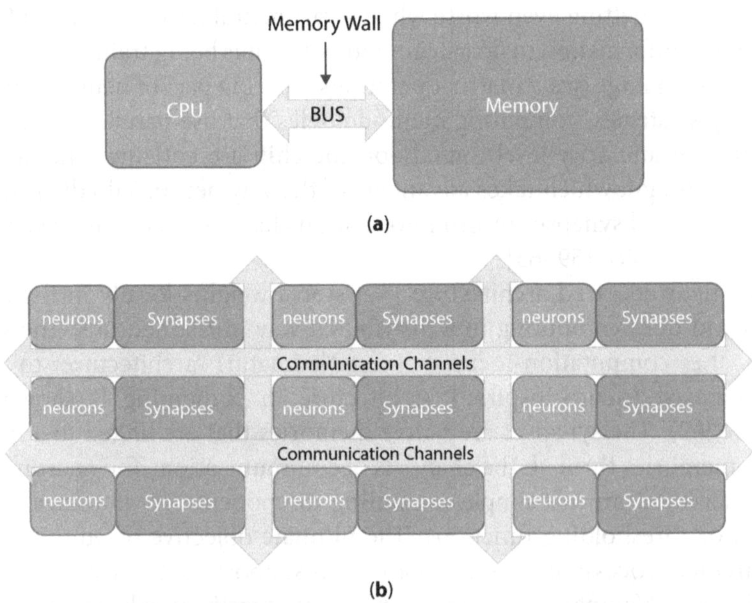

Figure 9.7 A dramatic change in the way technology works, as from computation-focused (Von-Neumann) design to the information (neuro-inspired) architecture [25]. (a) The von Neumann layout (b) The neuro-inspired layout.

9.8 Neuromorphic Computing in Robotics

Neuromorphic computing holds great potential for revolutionizing robotics by emulating the computational principles and capabilities of the human brain. Here are some applications of neuromorphic computing in robotics:

9.8.1 Sensor Processing and Perception

Neuromorphic systems can enhance sensor processing and perception in robots. By mimicking the brain's ability to process sensory information in parallel and in real-time, these systems can enable robots to efficiently process and interpret data from various sensors, such as cameras, tactile sensors, and inertial sensors, allowing for more accurate and responsive perception [75–78].

9.8.2 Motor Control and Movement

Neuromorphic computing can improve motor control and movement capabilities in robots. By emulating the Neuromorphic computing is a rapidly evolving field with several exciting future directions and emerging trends. Here are some key areas of development and research in the field:

9.8.3 Neuromorphic Hardware Advances

There is ongoing research and development in the design and fabrication of advanced neuromorphic hardware. This includes the exploration of novel device technologies, such as memristors, spintronics, and photonic devices that can enable more efficient and scalable neuromorphic systems. Additionally, the development of neuromorphic chips with increased computational power and energy efficiency is a crucial focus [79–85].

9.8.4 Brain-Inspired Learning Algorithms

The development of new learning algorithms inspired by the brain is a key research direction. The Neuromorphic computing holds immense promise for the future of computing and is poised to bring about transformative advancements in various domains. By drawing inspiration from the architecture and functioning of the human brain, neuromorphic computing offers unique capabilities and potential solutions to complex computational problems.

With ongoing research and development, the field of neuromorphic computing is addressing key challenges such as hardware variability, fault tolerance, power efficiency, and robust learning. Advances in neuromorphic hardware design, brain-inspired learning algorithms, and cognitive architectures are driving the field forward [86–92].

The future of neuromorphic computing encompasses exciting possibilities. As neuromorphic hardware continues to advance, we can expect more efficient and powerful neuromorphic chips, enabling real-time and energy-efficient processing. The development of hybrid approaches that integrate neuromorphic computing with traditional architectures will unlock new avenues for flexible and high-performance computing systems.

The integration of neuromorphic computing into edge devices will enable on-device intelligence and empower a wide range of applications in IoT, wearables, and mobile devices. Moreover, the combination of neuromorphic computing with AI hardware will enhance the capabilities of both fields, leading to more intelligent and efficient computing systems.

Ethical considerations and explainability in neuromorphic computing are receiving increasing attention, ensuring responsible and transparent use of these technologies. By addressing privacy concerns and developing ethical guidelines, the deployment and adoption of neuromorphic systems can proceed in a trustworthy and accountable manner.

As researchers and developers continue to push the boundaries of neuromorphic computing, the potential applications are vast. From natural language processing and robotics to cognitive computing and edge intelligence, neuromorphic computing has the potential to revolutionize various fields, enabling more efficient, adaptive, and human-like systems [93–97].

In conclusion, the future of neuromorphic computing looks bright. As the field advances, we can expect to witness groundbreaking innovations and solutions that will shape the next generation of computing systems, leading us toward a more intelligent and interconnected world.

9.9 Challenges in Neuromorphic Computing

Neuromorphic computing has the potential to significantly impact natural language processing (NLP) tasks by leveraging the brain-inspired architecture and principles. Here are some applications of neuromorphic computing in NLP [103–105]:

9.9.1 Language Understanding and Interpretation

Neuromorphic systems can be used to develop models that understand and interpret natural language more effectively. By emulating the parallel and distributed nature of the human brain, these systems can capture the semantic relationships, context, and subtle nuances in language, leading to improved language understanding.

9.9.2 Sentiment Analysis and Emotion Recognition

Neuromorphic computing can enhance sentiment analysis and emotion recognition tasks by enabling more nuanced understanding of human emotions and expressions. By modeling the complex interactions between words, phrases, and context, neuromorphic systems can capture the subtleties of sentiment and emotion, resulting in more accurate and contextually aware analysis [98–102].

9.9.3 Natural Language Generation

Generating natural language that is coherent, contextually appropriate, and expressive is a challenging task. Neuromorphic computing can aid in developing models that generate more human-like and contextually relevant text. By leveraging the principles of neural networks and distributed representations, these systems can generate more fluent and coherent text outputs.

9.9.4 Language Translation and Multilingual Processing

Neuromorphic computing can facilitate improvements in machine translation and multilingual processing tasks. By capturing the structural and semantic similarities between languages, neuromorphic systems can better handle complex linguistic phenomena, resulting in more accurate and context-aware translations [102–106].

9.9.5 Dialogue Systems and Conversational Agents

Developing intelligent dialogue systems and conversational agents that can understand and respond to natural language inputs requires robust and contextually aware models. Neuromorphic computing can enable more efficient and interactive dialogue systems by capturing the dynamics of human conversation, modeling context, and generating appropriate responses.

9.9.6 Language Modeling and Prediction

Neuromorphic systems can be leveraged for language modeling tasks, such as predicting the next word in a sentence or generating coherent text. By learning from large-scale language corpora, these systems can capture statistical regularities and dependencies in language, enabling better prediction and generation capabilities.

9.9.7 Text Summarization and Information Extraction

Extracting key information and generating concise summaries from large volumes of text is essential in various applications. Neuromorphic computing can aid in developing models that can automatically summarize text and extract relevant information by capturing the salient features and context in the text data.

These are just a few examples of how neuromorphic computing can enhance natural language processing tasks. The ability of neuromorphic systems to capture the complexities and nuances of human language can lead to significant advancements in NLP, enabling more accurate, context-aware, and natural language understanding and generation [107–109].

9.10 Applications of Neuromorphic Computing

9.10.1 Medicines

The ability of neuromorphic devices to receive and process information from their surroundings is incredibly effective. These gadgets can work with the human body when combined with organic components. In the not too distant future, neuromorphic technologies might make drug delivery methods even more efficient. Because of their extraordinary reactivity, they could potentially release a drug whenever they detected a change in the conditions of the body (i.e. varying insulin and glucose levels.)

There is also the possibility of implementing neuromorphic computing technologies into prostheses. This technology has the capacity to accept and process an external signal in an effective manner, which is another of its many advantages. People who have prostheses may find that by employing neuromorphic devices instead of conventional ones, they are able to have an experience that is more natural and does not cause any disruptions [26].

9.10.2 Artificial Intelligence [AI]

From definition, the goal of neuromorphic is to replicate how the brain of human works. Neurons in the brain quickly and efficiently receive, process, and send electrical impulses. So, it makes more sense that tech experts, especially those who work in the artificial intelligence field (AI), would consider this branch of computers to be highly fascinating. Even as name implies, cerebral intelligence is the main focus of AI scientists. The capacity of the brain to learn and apply what it has learned is known as intelligence. The two fields should work together in the future because this concept is so similar to neuromorphic computing [2]. Copying brain activity to the level of a single artificial neuron is the key to creating extremely intelligent computers that can contend with humans in intelligence [110–112].

9.10.3 Imaging

Similar to how the human eye creates images, neuromorphic vision sensors do the same. They take pictures "based on events" [27]. This shows that they make images based on how bright the light is, which is an external signal, not an internal one [28]. Furthermore, they move at a faster pace independent of conventional frame rates. In a neuromorphic sensor, each pixel functions independently of its neighbors. Additionally, the gadget almost instantly communicates changes to each pixel [27]. These strategies work together to make data utilization significantly more effective. Like their traditional equivalents, These sensors don't blur motion or take too long to react to their surroundings. Incorporating neuromorphic vision sensors into virtual and augmented reality systems may be advantageous given these qualities.

9.10.4 Sensor Processing and Perception

Neuromorphic systems can enhance sensor processing and perception in robots. By mimicking the brain's ability to process sensory information in parallel and in real-time, these systems can enable robots to efficiently process and interpret data from various sensors, such as cameras, tactile sensors, and inertial sensors, allowing for more accurate and responsive perception.

9.10.5 Motor Control and Movement

Neuromorphic computing can improve motor control and movement capabilities in robots. By emulating the brain's control mechanisms, neuromorphic systems can enable robots to generate smooth and adaptive movements, adjust to changing environmental conditions, and execute complex motions with greater agility and precision.

9.10.6 Autonomous Navigation and Mapping

Neuromorphic computing can enhance autonomous navigation and mapping in robots. By leveraging the brain-inspired principles of spatial awareness, memory, and learning, neuromorphic systems can enable robots to build and update maps of their environment, localize themselves accurately, and plan optimal paths for navigation in dynamic and unknown environments.

9.10.7 Human-Robot Interaction and Collaboration

Neuromorphic systems can facilitate more natural and intuitive human-robot interaction. By modeling human-like cognitive processes and understanding human gestures, expressions, and speech, these systems can enable robots to perceive and respond to human cues, fostering effective collaboration and communication between humans and robots.

9.10.8 Adaptive and Learning Capabilities

Neuromorphic computing can enhance the adaptive and learning capabilities of robots. By incorporating neural networks and synaptic plasticity mechanisms, robots can learn from experience, adapt to new tasks and environments, and continuously improve their performance over time. This enables robots to become more versatile and adaptable to a wide range of applications.

9.10.9 Task Planning and Decision Making

Neuromorphic systems can aid in task planning and decision making for robots. By leveraging the brain's ability to integrate information, evaluate alternatives, and make efficient decisions, these systems can enable robots to plan and execute complex tasks, optimize resource allocation, and adapt their strategies based on real-time feedback.

9.10.10 Robustness and Fault Tolerance

Neuromorphic computing can enhance the robustness and fault tolerance of robots. By incorporating redundancy, distributed processing, and self-healing capabilities inspired by the brain, robots can recover from hardware or software failures, adapt to unforeseen situations, and maintain reliable operation in challenging environments.

These applications demonstrate the potential of neuromorphic computing to enhance various aspects of robotics, ranging from perception and motor control to decision making and human-robot interaction. By leveraging the brain-inspired computational principles, neuromorphic systems can enable robots to exhibit more intelligent, adaptive, and human-like behavior, leading to advancements in autonomous systems, industrial automation, healthcare robotics, and many other domains.

9.10.11 Some More Applications

Neuromorphic computing may be excellent for implementation at the network's perimeter due to its low power consumption [29]. The periphery is the part of a network where a user's device can communicate with a cloud service. Since that driverless cars would be operating in this environment, neuromorphic computing could help them respond to risks more quickly. When these vehicles lose internet access, a system using neuromorphic computing may take control of them. This could improve the safety and environmental compatibility of driverless vehicles. The expanded sensory capabilities of neuromorphic computing may also improve present "smart technologies" [2]. This improvement, comparable to autonomous cars, may improve the effectiveness of the devices in a broader range of situations. Also, neuromorphic computing may be utilized to improve communication.

9.11 Conclusion

In conclusion, neuromorphic computing is an exciting and rapidly developing field that holds great promise for the future of computing and artificial intelligence. By emulating the neural structure and function of the human brain, neuromorphic computing systems have the potential to revolutionize a wide range of applications, from sensory processing to cognitive computing. While there are still many challenges to be addressed, such as scalability and power consumption, significant progress has been made in recent years, and the field is poised for continued growth and innovation. As we continue to develop and refine neuromorphic computing systems, we can expect to see new breakthroughs in artificial intelligence, robotics, and other fields, opening up new possibilities for solving complex problems and improving the human experience. Overall, neuromorphic computing represents a major advance in our ability to create intelligent machines, and its impact on society is likely to be profound in the years ahead.

References

[1] D. Markovic, A. Mizrahi, D. Querlioz, and J. Grollier, "Physics for Neuromorphic Computing."

[2] "What is Neuromorphic Computing?" https://www.techtarget.com/searchenterpriseai/definition/neuromorphic-computing (accessed Feb. 01, 2023).

[3] "The Perceptron, a Perceiving and Recognizing Automaton Project Para - F. Rosenblatt - Google Books." https://books.google.co.in/books/about/The_ Perceptron_a_Perceiving_and_Recogniz.html?id=P_XGPgAACAAJ&redir_esc=y (accessed Feb. 01, 2023).

[4] B. Widrowt, W. H. Piercet, and J. B. Angellt, "IRE Transactions on Military Electronics 191 Birth, Life, and Death in Microelectronic Systems*."

[5] C. Mead, "Neuromorphic Electronic Systems," *Proceedings of the IEEE*, vol. 78, no. 10, pp. 1629– 1636, 1990, doi: 10.1109/5.58356.

[6] L. D. Jackel, R. E. Howard, H. P. Graf, B. Straughn, and J. S. Denker, "Artificial Neural Networks for Computing.," *J. Vac. Sci. Technol. B.: Microelectron. Nanometer Struct.*, vol. 4, no. 1, pp. 61–63, Jan. 1986, doi: 10.1116/1.583351.

[7] D. B. Strukov, G. S. Snider, D. R. Stewart, and R. S. Williams, "The missing memristor found," *Nature*, vol. 453, no. 7191, pp. 80–83, May 2008, doi: 10.1038/nature06932.

[8] S. H. Jo, T. Chang, I. Ebong, B. B. Bhadviya, P. Mazumder, and W. Lu, "Nanoscale memristor device as synapse in neuromorphic systems," *Nano Lett*, vol. 10, no. 4, pp. 1297–1301, Apr. 2010, doi: 10.1021/nl904092h.

[9] H. Dureja, Y. Garg, R. Chaujar, and B. Kumar, "Review Research Paper Neuromorphic Computing and Applications", *International Research Journal of Modernization in Engineering Technology and Science (IRJMETS)*, Volume 03, Issue 11, November 2021.

[10] W. Zhang *et al.*, "Neuro-inspired computing chips," *Nature Electronics*, vol. 3, no. 7. *Nature Research*, pp. 371–382, Jul. 01, 2020. doi: 10.1038/s41928-020-0435-7.

[11] C. Prakash, L. R. Gupta, A. Mehta, H. Vasudev, R. Tominov, E. Korman, A. Fedotov, V. Smirnov and K. K.,Kesari, "Computing of neuromorphic materials: an emerging approach for bioengineering solutions", *J. Mater. Adv.*, Vol. 4, PP. 5882-5919, 2023.

[12] C. Stergiou and D. Siganos, "Neural Networks." [Online]. Available: http://www.doc.ic.ac.uk/~nd/...

[13] Science based - Where can the brain be arranged inside the torso? - Worldbuilding Stack Exchange

[14] E. Y. Li, "Artificial neural networks and their business applications," 1994.

[15] A. Abraham, "Artificial Neural Networks," *Handbook of Measuring System Design*, Jul. 2005, doi: 10.1002/0471497398.MM421.

[16] Y. Lecun, Y. Bengio, and G. Hinton, "Deep learning," *Nature*, vol. 521, no. 7553. Nature Publishing Group, pp. 436–444, May 27, 2015. doi: 10.1038/nature14539.

[17] IEEE Solid-State Circuits Society., *2019 IEEE International Solid-State Circuits Conference: Digest of Technical Papers: ISSCC: February 2019.*

[18] S. A. McKee, "Reflections on the memory wall," in *2004 Computing Frontiers Conference*, 2004, pp. 162–167. doi: 10.1145/977091.977115.

[19] C. S. Poon and K. Zhou, "Neuromorphic silicon neurons and large-scale neural networks: Challenges and opportunities," *Front. Neurosci.*, no. SEP. 2011. doi: 10.3389/fnins.2011.00108.

[20] S. Chetlur*et al.*, "cuDNN: Efficient Primitives for Deep Learning," Oct. 2014, [Online]. Available: http://arxiv.org/abs/1410.0759

[21] X. Chen, Y. Duan, R. Houthooft, J. Schulman, I. Sutskever, and P. Abbeel, "InfoGAN: Interpretable Representation Learning by Information Maximizing Generative Adversarial Nets," Jun. 2016, [Online]. Available: http://arxiv.org/abs/1606.03657

[22] N. P. Jouppi *et al.*, "In-Datacenter Performance Analysis of a Tensor Processing Unit TM," 2017.

[23] Y. H. Chen, T. Krishna, J. S. Emer, and V. Sze, "Eyeriss: An Energy-Efficient Reconfigurable Accelerator for Deep Convolutional Neural Networks," *IEEE J. Solid-St. Circ.*, vol. 52, no. 1, pp. 127–138, Jan. 2017, doi: 10.1109/JSSC.2016.2616357.

[24] B. Moons, R. Uytterhoeven, W. Dehaene, and M. Verhelst, "Envision: A 0.26-to-10TOPS/W subword-parallel dynamic-voltage-accuracy-frequency-scalable Convolutional Neural Network processor in 28nm FDSOI," in *Digest of Technical Papers - IEEE International Solid-State Circuits Conference*, Mar. 2017, vol. 60, pp. 246–247. doi: 10.1109/ISSCC.2017.7870353.

[25] S. Yu, "Neuro-Inspired Computing with Emerging Nonvolatile Memorys," *Proceedings of the IEEE*, vol. 106, no. 2, pp. 260–285, Feb. 2018, doi: 10.1109/JPROC.2018.2790840.

[26] Y. Tuchman *et al.*, "Organic neuromorphic devices: Past, present, and future challenges," *MRS Bull.*, vol. 45, no. 8, pp. 619–630, Aug. 2020, doi: 10.1557/mrs.2020.196.

[27] "Neuromorphic Vision Sensors Eye the Future of Autonomy - EE Times." https://www.eetimes.com/neuromorphic-vision-sensors-eye-the-future-of-autonomy/ (accessed Feb. 01, 2023).
[28] J. P. Boettiger, "AFIT Scholar AFIT Scholar Theses and Dissertations Student Graduate Works A Comparative Evaluation of the Detection and Tracking A Comparative Evaluation of the Detection and Tracking Capability Between Novel Event-Based and Conventional Frame-Capability Between Novel Event-Based and Conventional Frame-Based Sensors Based Sensors." [Online]. Available: https://scholar.afit.edu/etd/3154
[29] S. Greengard, "Neuromorphic chips take shape," *Commun. ACM*, vol. 63, no. 8, pp. 9–11, Jul. 2020, doi: 010.1145/3403960.
[30] N.Jain, and B.Raj, "Parasitic Capacitance and Resistance Model Development and Optimization of Raised Source/Drain SOI FinFET Structure for Analog Circuit Applications", *J. Nanoelectron. Optoelectron.*, ASP, USA, Vol. 13, pp. 531–539, 2018.
[31] S.Singh, S. K.Vishvakarma, and B.Raj, "Analytical Modeling of Split-Gate Junction-Less Transistor for a Biosensor Application", *Sens. Bio-Sens.*, Elsevier, Vol. 18, pp. 31–36, 2018.
[32] M.Gopal, and B.Raj, "Low Power 8T SRAM Cell Design for High Stability Video Applications", *ITSI Trans. Electr. Electron. Eng.*, Vol.1, No. 5, pp. 91–97, 2013.
[33] B.Raj, J.Mitra, D. K.Bihani, V.Rangharajan, A. K. Saxena, and S. Dasgupta, "Analysis of Noise Margin, Power and Process Variation for 32 nm FinFET Based 6T SRAM Cell", *J. Comput. (JCP)*, Academy Publisher, FINLAND, Vol. 5, No. 6, PP.1-6, 2010.
[34] D.Sharma, R.Mehra, and B.Raj, "Comparative Analysis of Photovoltaic Technologies for High Efficiency Solar Cell Design", *Superlattices Microstruct.*, Elsevier, Vol. 153, pp. 106861, 2021.
[35] P.Kaur, A. S.Buttar, and B.Raj, "A Comprehensive Analysis of Nanoscale Transistor Based Biosensor: A Review", *Indian J. Pure Appl. Phys.*, Vol. 59, pp. 304–318, 2021.
[36] D.Yadav, B.Raj, and B.Raj, "Design and Simulation of Low Power Microcontroller for IoT Applications", *J. Sens. Lett.*, ASP USA, Vol. 18, pp. 401–409, 2020.
[37] S.Singh, and B.Raj, "A 2-D Analytical Surface Potential and Drain Current Modeling of Double-Gate Vertical t-Shaped Tunnel FET", *J. Comput. Electron.*, Springer, Vol. 19, pp. 1154–1163, 2020.
[38] J.Singh, and B.Raj, "An Accurate and Generic Window function for Nonlinear Memristor Model" *J. Comput. Electron.*, Springer, Vol. 18, No. 2, pp. 640–647, 2019.
[39] M.Kaur, N.Gupta, S.Kumar, B.Raj, and A. K.Singh, "Comparative RF and Crosstalk Analysis of Carbon Based Nano Interconnects", *IET Circ., Devices Syst.*, Vol. 15, No. 6, pp. 493–503, 2021.

[40] N.Kandasamy, F.Ahmad, D.Ajitha, B.Raj, and N.Telagam, "Quantum Dot Cellular Automata Based Scan Flip Flop and Boundary Scan Register", *IETE J. Res.*, Vol. 69, PP. 535-548, 2023.

[41] S. K.Sharma, B.Raj, and M.Khosla, "Enhanced Photosensitivity of Highly Spectrum Selective Cylindrical Gate In1-xGaxAs Nanowire MOSFET Photodetector", *Mod. Phys. Lett.-B*, Vol. 33, No. 12, p. 1950144, 2019.

[42] J.Singh, and B.Raj, "Design and Investigation of 7T2M NVSARM with Enhanced Stability and Temperature Impact on Store/Restore Energy", *IEEE Trans. Very Large Scale Integr. Syst.*, Vol. 27, No. 6, pp. 1322–1328, 2019.

[43] A. K.Bhardwaj, S.Gupta, B.Raj, and A.Singh,"Impact of Double Gate Geometry on the Performance of Carbon Nanotube Field Effect Transistor Structures for Low Power Digital Design", *J. Comput. Theor. Nanosci.*, ASP USA, Vol. 16, pp. 1813–1820, 2019.

[44] N.Jain, and B.Raj, "Thermal Stability Analysis and Performance Exploration of Asymmetrical Dual-k Underlap Spacer (ADKUS) SOI FinFET for Security and Privacy Applications", *Indian J. Pure Appl. Phys. (IJPAP)*, Vol. 57, pp. 352–360, 2019.

[45] A.Singh, M.Khosla, and B.Raj, "Design and Analysis of Dynamically Configurable Electrostatic Doped Carbon Nanotube Tunnel FET", *Microelectronics J.*, Elsevier, Vol. 85, pp. 17–24, 2019.

[46] N.Jain, and B.Raj, "Dual-k Spacer Region Variation at the Drain Side of Asymmetric SOI FinFET Structure: Performance Analysis towards the Analog/RF Design Applications", *J. Nanoelectron. Optoelectron.*, American Scientific Publishers, Vol. 14, pp. 349–359, 2019.

[47] J.Singh, S.Sharma, B.Raj, and M.Khosla, "Analysis of Barrier Layer Thickness on Performance of In1-xGaxAs Based Gate Stack Cylindrical Gate Nanowire MOSFET", *JNO*, ASP USA, Vol. 13, pp. 1473–1477, 2018.

[48] N.Jain, and B.Raj, "Analysis and Performance Exploration of High-k SOI FinFETs over the Conventional Low-k SOI FinFET toward Analog/RF Design", *J. Semicond. (JoS)*, IOP Science,Vol. 39, No. 12, pp. 124002-1-7, 2018.

[49] C.Goyal, J. S.Ubhi, and B.Raj, "A Reliable Leakage Reduction Technique for Approximate Full Adder with Reduced Ground Bounce Noise", *J. Math. Probl. Eng.*, Hindawi, Vol. 2018, Article ID 3501041, PP.1-16, 2018.

[50] Anuradha, J. Singh, B.Raj, and M.Khosla, "Design and Performance Analysis of Nano-Scale Memristor-Based Nonvolatile SRAM", *Journal of Sens. Lett.*, American Scientific Publishers USA, Vol. 16, pp. 798–805, 2018.

[51] G.Wadhwa, and B.Raj, "Parametric Variation Analysis of Charge-PlasmaBased Dielectric Modulated JLTFET for Biosensor Application", *IEEE Sensor Journal*, Vol. 18, No. 15, PP. 6070–6077, 2018.

[52] J.Singh, and B.Raj, "Comparative Analysis of Memristor Models for Memories Design", *JoS*, IoP, Vol. 39, No. 7, PP. 074006-1-12, 2018.

[53] D.Yadav, S. S.Chouhan, S. K.Vishvakarma, and B.Raj, "Application Specific Microcontroller Design for IoT Based WSN", *Sens. Lett.*, ASP, Vol. 16, pp. 374–385, 2018.

[54] G. Singh, R. K. Sarin, and B. Raj, "Fault-Tolerant Design and Analysis of Quantum-Dot Cellular Automata Based Circuits", *IEEE/IET Circuits, Devices & Systems*, Vol. 12, pp.638–664, 2018.

[55] J. Singh, and B. Raj, "Modeling of Mean Barrier Height Levying Various Image Forces of Metal Insulator Metal Structure to Enhance the Performance of Conductive Filament Based Memristor Model", *IEEE Nanotechnology*, Vol. 17, No. 2, pp. 268–275, 2018.

[56] A. Jain, S. Sharma, and B. Raj, "Analysis of Triple Metal Surrounding Gate (TM-SG) III-V Nanowire MOSFET for Photosensing Application", *OptoElectronics Journal*, Elsevier, Vol. 26, No. 2, PP. 141–148, 2018.

[57] A. Jain, S. Sharma, and B. Raj, "Design and Analysis of High Sensitivity Photosensor Using Cylindrical Surrounding Gate MOSFET for Low Power Sensor Applications", *Engineering Science and Technology, an International Journal (JESTECH)*, Elsevier's, Vol. 19, No. 4, pp. 1864–1870, 2016.

[58] A. Singh, M. Khosla, and B. Raj, "Analysis of Electrostatic Doped Schottky Barrier Carbon Nanotube FET for Low Power Applications,"*J. Mater. Sci. Mater. Electron.*, Springer,Vol. 28, PP. 1762–1768, 2017.

[59] G. Saiphani Kumar, A. Singh, B. Raj, "Design and Analysis of Gate All around CNTFET Based SRAM Cell Design", *J. Comput. Electron.*, Springer, Vol. 17, No.1, PP. 138–145, 2018.

[60] G. P. Singh, B. S. Sohi, and B. Raj, "Material Properties Analysis of Graphene Base Transistor (GBT) for VLSI Analog Circuits", *Indian J. Pure Appl. Phys. (IJPAP)*, Vol. 55, pp. 896–902, 2017.

[61] S. Kumar and B. Raj, "Estimation of Stability and Performance Metric for Inward Access Transistor Based 6T SRAM Cell Design Using n-type/p-type DMDG-GDOV TFET", *IEEE VLSI Circuits and Systems Letter*, Vol. 3, No. 2, PP. 25–39, 2017.

[62] S. Sharma, A. Kumar, M. Pattanaik, and B. Raj, "Forward Body Biased Multimode Multi-Threshold CMOS Technique for Ground Bounce Noise Reduction in Static CMOS Adders", *International Journal of Information and Electronics Engineering (IJIEE)*, Vol. 3, No. 3, PP. 567–572, 2013.

[63] H. Singh, P. Kumar, and B. Raj, "Performance Analysis of Majority Gate SET Based 1-bit Full Adder", *International Journal of Computer and Communication Engineering (IJCCE)*, IACSIT Press Singapore, ISSN: 20103743, Vol. 2, No. 4, PP.1-8, 2013.

[64] A. K. Bhardwaj, S. Gupta, and B. Raj, "Investigation of Parameters for Schottky Barrier (SB) Height for Schottky Barrier Based Carbon Nanotube Field Effect Transistor Device", *J. Nanoelectron. Optoelectron.*, ASP, Vol. 15, PP. 783–791, 2020.

[65] P. Bansal, and B. Raj, "Memristor: A Versatile Nonlinear Model for Dopant Drift and Boundary Issues", *JCTN*, American Scientific Publishers, Vol. 14, No. 5, PP. 2319–2325, 2017.

[66] N. Jain, and B. Raj, "An Analog and Digital Design Perspective Comprehensive Approach on Fin-FET (Fin-Field Effect Transistor) Technology - A Review", *Rev. Adv. Sci. Eng. (RASE)*, ASP, Vol. 5, PP. 1–14, 2016.

[67] S. Sharma, B. Raj, and M. Khosla, "Subthreshold Performance of In1-xGaxAs Based Dual Metal with Gate Stack Cylindrical/Surrounding Gate Nanowire MOSFET for Low Power Analog Applications", *J. Nanoelectron. Optoelectron.*, American Scientific Publishers, USA, Vol. 12, PP. 171– 176, 2017.

[68] B. Raj, A. K. Saxena, and S. Dasgupta, "Analytical Modeling for the Estimation of Leakage Current and Subthreshold Swing Factor of Nanoscale Double Gate FinFET Device" *Microelectron. Int.*, UK, Vol. 26, PP. 53–63, 2009.

[69] S. S. Soniya, G. Wadhwa, and B. Raj, "An Analytical Modeling for Dual Source Vertical Tunnel Field Effect Transistor", *Int. J. Recent Technol. Eng., (IJRTE)*, Vol. 8, No. 2, 2019.

[70] S. Singh, and B. Raj, "Design and Analysis of Hetrojunction Vertical T-Shaped Tunnel Field Effect Transistor", *J. Electron. Mater.*, Springer, Vol. 48, No. 10, PP. 6253–6260, 2019.

[71] C. Goyal, J. S. Ubhi, and B. Raj, "A Low Leakage CNTFET Based Inexact Full Adder for Low Power Image Processing Applications", *Int. J. Circuit Theory Appl.*, Wiley, Vol. 47, No. 9, PP. 1446–1458, 2019.

[72] B. Raj, A. K. Saxena, and S. Dasgupta, "A Compact Drain Current and Threshold Voltage Quantum Mechanical Analytical Modeling for FinFETs", *J. Nanoelectron. Optoelectron. (JNO)*, USA, Vol. 3, No. 2, PP. 163–170, 2008.

[73] G. Wadhwa, and B. Raj, "An Analytical Modeling of Charge Plasma Based Tunnel Field Effect Transistor with Impacts of Gate Underlap Region", *Superlattices Microstruct.*, Elsevier, Vol. 142, PP. 106512-22, 2020.

[74] S. Singh, and B. Raj, "Modeling and Simulation Analysis of SiGe Heterojunction Double GateVertical t-Shaped Tunnel FET", *Superlattices Microstruct.*, Elsevier,Vol. 142, PP. 106496-104, 2020.

[75] A. Singh, D. K. Saini, D. Agarwal, S. Aggarwal, M. Khosla, and B. Raj, "Modeling and Simulation of Carbon Nanotube Field Effect Transistor and Its Circuit Application", *J. Semicond. (JoS)*, IOP Science, Vol. 37, PP. 074001–074006, 2016.

[76] N. Jain, and B. Raj, "Device and Circuit Co-Design Perspective Comprehensive Approach on FinFET Technology - A Review", *J. Electron Devices*, Vol. 23, No. 1, PP. 1890–1901, 2016.

[77] S. Kumar and B. Raj, "Analysis of ION and Ambipolar Current for DualMaterial Gate-Drain Overlapped DG-TFET", *J. Nanoelectron. Optoelectron.*, American Scientific Publishers, USA, Vol. 11, PP. 323–333, 2016.

[78] N. Anjum, T. Bali, and B. Raj, "Design and Simulation of Handwritten Multiscript Character Recognition", *Int. J. Adv. Res. Comput. Commun. Eng.*, Vol. 2, No. 7, PP. 2544–2549, 2013.

[79] S. Sharma, B. Raj, and M. Khosla, "A Gaussian Approach for Analytical Subthreshold Current Model of Cylindrical Nanowire FET with Quantum Mechanical Effects", *Microelectronics J.*, Elsevier, Vol. 53, PP. 65–72, 2016.

[80] K. Singh, and B. Raj, "Performance and Analysis of Temperature Dependent Multi-Walled Carbon Nanotubes as Global Interconnects at Different Technology Nodes", *J. Comput. Electron.*, Springer, Vol. 14, No. 2, PP. 469–476, 2015.

[81] S. Kumar and B. Raj, "Compact Channel Potential Analytical Modeling of DG-TFET Based on Evanescent-Mode Approach", *J. Comput. Electron.*, Springer, Vol. 14, No. 2, PP. 820–827, 2015.

[82] K. Singh, and B. Raj, "Temperature Dependent Modeling and Performance Evaluation of Multi-Walled CNT and Single-Walled CNT as Global Interconnects", *J. Electron. Mater.*, Springer,Vol. 44, No. 12, PP. 4825–4835, 2015.

[83] V. K. Sharma, M. Pattanaik, and B. Raj, "INDEP Approach for Leakage Reduction in Nanoscale CMOS Circuits", *Int. J. Electron.*, Taylor & Francis, Vol. 102, No. 2, PP. 200–215, 2014.

[84] K. Singh, and B. Raj, "Influence of Temperature on MWCNT Bundle, SWCNT Bundle and Copper Interconnects for Nanoscaled Technology Nodes", *J. Mater. Sci. Mater. Electron.*, Springer, Vol. 26, No. 8, PP. 6134–6142, 2015.

[85] N. Anjum, T. Bali, and B. Raj, "Design and Simulation of Handwritten Gurumukhi and Devanagri Numerical Recognition", *Int. J. Comput. Appl.*, Published by Foundation of Computer Science, New York, USA,Vol. 73, No. 12, PP. 16–21, 2013.

[86] S. Khandelwal, V. Gupta, B. Raj, and R. D. Gupta, "Process Variability Aware Low Leakage Reliable Nano Scale DG-FinFET SRAM Cell Design Technique", *J. Nanoelectron. Optoelectron.*, Vol. 10, No. 6, PP. 810–817, 2015.

[87] V. K. Sharma, M. Pattanaik, and B. Raj, "ONOFIC Approach: Low Power High Speed Nanoscale VLSI Circuits Design", *Int. J. Electron.*, Taylor & Francis, Vol. 101, No. 1, PP. 61–73, 2014.

[88] S. Khandelwal, B. Raj, and R. D.Gupta, "FinFET Based 6T SRAM Cell Design: Analysis of Performance Metric, Process Variation and Temperature Effect", *J. Comput. Theor. Nanosci.*, ASP, USA, Vol. 12, PP. 2500–2506, 2015.

[89] S. Singh, Y. Shekhar, R. Jagdeep, S. Anurag, and B. Raj, "Impact of HfO2 in Graded Channel Dual Insulator Double Gate MOSFET", *J. Comput. Theor. Nanosci.*, American Scientific Publishers, Vol. 12, No. 6, PP. 950–953, 2015.

[90] V. K. Sharma, M. Pattanaik, and B. Raj, "PVT Variations Aware Low Leakage INDEP Approach for Nanoscale CMOS Circuits", *Microelectron. Reliab.*, Elsevier, Vol. 54, PP. 90–99, 2014.

[91] B. Raj, A. K. Saxena and S. Dasgupta, "Quantum Mechanical Analytical Modeling of Nanoscale DG FinFET: Evaluation of Potential, Threshold Voltage and Source/Drain Resistance", *Elsevier's J. Mater. Sci. Semicond. Process.*, Elsevier, Vol. 16, No. 4, PP. 1131–1137, 2013.

[92] M. Gopal, S. S. D. Prasad, and B. Raj, "8T SRAM Cell Design for Dynamic and Leakage Power Reduction", *Int. J. Comput. Appl.*, Published by Foundation of Computer Science, New York, USA,Vol. 71, No. 9, PP. 43–48, 2013.

[93] M. Pattanaik, B. Raj, S. Sharma, and A. Kumar, "Diode Based Trimode Multi-Threshold CMOS Technique for Ground Bounce Noise Reduction in Static CMOS Adders", *Adv. Mat. Res., Trans. Tech. Publications*, Switzerland, Vol. 548, PP. 885–889, 2012.

[94] B. Raj, A. K. Saxena, and S. Dasgupta, "Nanoscale FinFET Based SRAM Cell Design: Analysis of Performance Metric, Process Variation, Underlapped FinFET and Temperature Effect", *IEEE Circuits Syst. Mag.*, Vol. 11, No. 2, PP. 38–50, 2011.

[95] V. K. Sharma, M. Pattanaik, and B. Raj, "Leakage Current ONOFIC Approach for Deep Submicron VLSI Circuit Design", *Int. J. Electr. Comput. Electron. Commun. Eng.*, World Academy of Sciences, Engineering and Technology, Vol. 7, No. 4, PP. 239–244, 2013.

[96] T. Chawla, M. Khosla, and B. Raj, "Design and Simulation of Triple Metal Double-Gate Germanium on Insulator Vertical Tunnel Field Effect Transistor", *Microelectronics J.*, Elsevier, Vol. 114, PP. 105125-32, 2021.

[97] P. Kaur, S. S. Gill, and B. Raj, "Comparative Analysis of OFETs Materials and Devices for Sensor Applications", *J. Silicon*, Springer,Vol. 14, PP. 4463–4471, 2022.

[98] S. K. Sharma, P. Kumar, B. Raj, and Balwinder Raj, "In1-xGaxAs Double Metal Gate-Stacking Cylindrical Nanowire MOSFET for Highly Sensitive Photo Detector", *J. Silicon*, Springer, Vol. 14, PP. 3535–3541,2022.

[99] B. Raj, A. K. Saxena, and S. Dasgupta, "Analytical Modeling of Quasi Planar Nanoscale Double Gate FinFET with Source/Drain Resistance and Field Dependent Carrier Mobility: A Quantum Mechanical Study", *J. Comput. (JCP)*, Academy Publisher, FINLAND, Vol. 4, No. 9, PP.1-8, 2009.

[100] S. Bhushan, S. Khandelwal, and B. Raj, "Analyzing Different Mode FinFET Based Memory Cell at Different Power Supply for Leakage Reduction", *Seventh International Conference on Bio-Inspired Computing: Theories and Application*, (BIC-TA 2012), organized IIITM Gwalor, PP. 89–100, 14-16 Dec 2012.

[101] J. Singh, and B. Raj, "Temperature Dependent Analytical Modeling and Simulations of Nanoscale Memristor", *Engineering Science and Technology, an International Journal (JESTECH)*, Elsevier's, Vol. 21, PP. 862–868, 2018.

[102] S. Singh, S. Bala, B. Raj, and Balwinder Raj, "Improved Sensitivity of Dielectric Modulated Junctionless Transistor for Nanoscale Biosensor Design", *Sens. Lett.*, ASP, Vol. 18, PP. 328–333, 2020.

[103] V. Kumar, S. K. Vishvakarma, and B. Raj, "Design and Performance Analysis of ASIC for IoT Applications", *Sens. Lett.*, ASP,Vol. 18, PP. 31–38, 2020.

[104] A. Jaiswal, R. K.Sarin, B. Raj, and S. Sukhija, "A Novel Circular Slotted Microstrip-fed Patch Antenna with Three Triangle Shape Defected Ground Structure for Multiband Applications", *Adv. Electromagn. (AEM)*, Vol. 7, No. 3, PP. 56–63, 2018.

[105] G. Wadhwa, and BRaj, "Label Free Detection of Biomolecules Using Charge Plasma-Based Gate Underlap Dielectric Modulated Junctionless TFET", *J. Electron. Mater. (JEMS)*, Springer, Vol. 47, No. 8, PP. 4683– 4693, 2018.

[106] G. Singh, R. K. Sarin, and B. Raj, "Design and Performance Analysis of a New Efficient Coplanar Quantum-Dot Cellular Automata Adder", *Indian J. Pure Appl. Phys. (IJPAP)*, Vol. 55, PP. 97–103, 2017.

[107] A. Singh, M. Khosla, and B. Raj, "Design and Analysis of Electrostatic Doped Schottky Barrier CNTFET Based Low Power SRAM", *Int. J. Electron. Commun.*, (AEÜ), Elsevier,Vol. 80, PP. 67–72, 2017.

[108] P. Kaur, V. Pandey, and B. Raj, "Comparative Study of Efficient Design, Control and Monitoring of Solar Power Using IoT", *Sens. Lett.*, ASP, Vol. 18, PP. 419–426, 2020.

[109] G. Wadhwa, P. Kamboj, J. Singh, B. Raj, "Design and Investigation of Junctionless DGTFET for Biological Molecule Recognition", *Trans. Electr. Electron. Mater.*, Springer, Vol. 22, PP. 282–289, 2020.

[110] T. Chawla, M. Khosla, and B. Raj, "Optimization of Double-Gate Dual Material GeOI-Vertical TFET for VLSI Circuit Design", *IEEE VLSI Circuits and Systems Letter*, Vol. 6, No. 2, PP. 13–25, 2020.

[111] S. K. Verma, S. Singh, G. Wadhwa and B. Raj, "Detection of Biomolecules Using Charge-Plasma Based Gate Underlap Dielectric Modulated Dopingless TFET", *Trans. Electr. Electron. Mater. (TEEM)*, Springer, Vol. 21, PP. 528–535, 2020.

[112] N. Jain, and B. Raj, "Impact of Underlap Spacer Region Variation on Electrostatic and Analog/RF Performance of Symmetrical High-k SOI FinFET at 20 nm Channel Length", *J. Semicond. (JoS)*, IOP Science, Vol. 38, No. 12, PP. 122002-10, 2017.

10

Performance Evaluation of Prototype Microstrip Patch Antenna Fabrication Using Microwave Dielectric Ceramic Nanocomposite Materials for X-Band Applications

Srilali Siragam

Department of Electronics & Communication Engineering, Swarnandhra College of Engineering and Technology, Seetharamapuram, Narsapur, A.P., India

Abstract

This chapter presents the production and characterization of a sol-gel composite nanoparticle composed of titanium dioxide (TiO_2) and zinc aluminate ($ZnAl_2O_4$) as a prototype microstrip patch antenna. Initially, the molecular composition was used with the chemical formula $xTiO_2(1\text{-}x)\ ZnAl_2O_4$ and defined as $0.08TiO_2$ for $x = 8\%$. X-ray diffraction (XRD) was used to investigate the generated composite nanoparticles, which indicated the dominating peaks of $ZnAl_2O_4$ and TiO_2. The crystallite size of the nanoparticles was calculated to be 16.9 nm. Fourier transform infrared spectroscopy (FTIR), field-emission scanning electron microscopy (FESEM), and energy dispersive spectroscopy (EDS) were also used to evaluate the functional groups, shape, and elemental content of the composite nanoparticles. These nanoparticles were also used to fabricate a prototype patch antenna in order to test its performance. Finally, a fabricated patch antenna was suitable for the X-band applications with the operating frequency 8.49 GHz, return loss (RL) -19.97 dB and VSWR<2.

***Keywords*:** $ZnAl_2O_4$, microstrip patch antenna, VSWR, return loss, X-band

Email: srilalisep7@gmail.com

Balwinder Raj, Suman Lata Tripathi, Tarun Chaudhary, K. Srinivasa Rao and Mandeep Singh (eds.) Integrated Devices for Artificial Intelligence and VLSI, (247–270) © 2024 Scrivener Publishing LLC

10.1 Introduction

Because of their unique electrical properties, ceramic dielectric materials have been used to reduce component size and cost, which has revolutionized the wireless microwave communications market over the last few decades. Because the wavelength is inversely proportional to permittivity, the usage of high dielectric permittivity materials, in particular, results in size reduction. Materials, on the other hand, have become less appealing as interest in frequency swings from the industrial and scientific bands to the millimeter-wave zone. A low dielectric permittivity combined with a high-quality factor (Q) is required because the high Q reduces the dielectric loss while the low dielectric permittivity allows for rapid electrical signal conversion at ultrahigh frequencies.

Microwave ceramic materials are distinguished from conventional materials by their miniaturization, high integration, high reliability, high Q-factor, dielectric permittivity (ε_r), low dissipation loss and zero temperature coefficient of resonant frequency (τ_f) [1–3]. Duplexers, resonators, and substrates made of microwave-based materials have been widely used in communication systems [4–6].

Zinc aluminate ($ZnAl_2O_4$) is a kind of spinel that is also known as gahnite. The general formula was XY_2O_4, where X represented bivalent cations like barium, calcium. On the other hand, Y represents the octahedral face of an FCC structure with an Fd3m space group [7]. This Y site can be occupied by aluminium, ferrous, and other elements [8]. $ZnAl_2O_4$ is a broadband gap semiconductor with a polycrystalline optical band gap of 3.8 eV, implying that it is transparent to light at wavelengths higher than 320 nm [9]. Other researchers have documented various methods for producing $ZnAl_2O_4$, the simplest and least expensive of which is the sol-gel method [10, 11]. Each approach has unique properties that can be used during the experiment. However, there may be some difficulties in solid-state processes, which have several constraints such as lack of homogeneity, stoichiometric imbalance, a high sintering temperature, and a poor surface area, which can all affect the outcome of the $ZnAl_2O_4$. In comparison to previous approaches, the sol-gel methodology has proven to be a highly beneficial technology for the synthesis of $ZnAl_2O_4$ crystalline structure. This is owing to its capacity to produce pure and ultra-fine particles at low temperatures [12].

Surendran *et al.* were the first ones to investigate microwave dielectric properties of $ZnAl_2O_4$, which included permittivity of 8.5, a Q-factor of 56300 GHz, and a temperature coefficient of -79 ppm/°C. They claimed that by adding enough rutile TiO_2 to the $0.83ZnAl_2O_4$-$0.17TiO_2$ ceramic, they could achieve a good combination of dielectric parameters such as dielectric permittivity and Q-factor, and τ_f corresponding to 12.67, 1×10^5 GHz, and ~0.74 ppm/°C [13]. Similarly, Lei *et al.* found a comparable result in the $0.79ZnAl_2O_4$-$0.21TiO_2$ combination, with dielectric permittivity, Q-factor, and τ_f of 11.4, 71810 GHz, and 0.5 ppm per °C, respectively [14]. Finally, they suggested that the prepared materials are suitable for microwave applications.

Wang *et al.* investigated the dielectric properties of Mg_2SiO_4-Li_2TiO_3-LiF ceramics with $CaTiO_3$ addition synthesized using the usual solid-state method. After the $CaTiO_3$ was introduced, the relative density increased dramatically, resulting in a much-improved Q×*f*. As the sintering temperature increased, the dielectric permittivity and quality factor increased first and then declined. The ceramic material had a ε_r of 14.13, a Q-factor of 6,772, and a RL of -17.5 dB at 1.575 GHz. Finally, they stated that the fabricated patch antenna was best suited for GPS and co-fired ceramic applications with moderate temperatures [15].

Liu *et al.* demonstrated the fabrication of calcium/strontium nanoceramic composite materials based on Ga_2O_4. They noticed that as the sintering temperature increased, the morphology and microstructures density improved. $SrGa_2O_4$ dielectric permittivity and quality factors were 9.2 and 66,000 GHz, respectively, when sintered at 1275 °C. Furthermore, at the same temperature, $CaGa_2O_4$ material has ε_r of 10.6 and a Q-factor of 15,400. The $SrGa_2O_4$ ceramic patch antenna had -19.94 dB return loss and -1.38 dB efficiency. Finally, they notified us that $SrGa_2O_4$ ceramics are suited for 5G technology operating at mid and low-frequency bands [16].

Chinnaguruswamy *et al.* used sputtering technique to construct a multiband microstrip patch antenna. The suggested antenna was designed to function with minimum power from the transmitter and receivers. Furthermore, they stated that the fabricated patch antenna thickness was 50 times less than the traditional one. The fabricated patch antenna operates at 2.4 gigahertz, with -20 dB RL and 10 dB gain [17].

Naidu *et al.* used the sol-gel process to synthesize double lanthanide n-type semiconductors with $MgFe_2O_4$ doping. They discovered an increase in dielectric permittivity and dielectric loss ranging from 1.68 to 3.18 and

1.12×10^{-2} to 2.6×10^{-1}, respectively. Furthermore, at an operating frequency of 14.9 GHz, the standing wave ratio and gain are 1.66 dB and 5.73 dB, respectively, with 69.82 antenna efficiency [18].

Srilali *et al.* announced the $ZnAl_2O_4TiO_2$ composite ceramic nanoparticles synthesis for prototype patch antenna development. They found 13.3 nm crystallites size, dielectric permittivity from 22.12 to 21.63, and dielectric loss from 0.056 to 0.041; -30.72 dB return loss and 830 MHz bandwidth at operating frequency 4.85 GHz by adjusting the TiO_2 concentration. They claimed the fabricated patch antenna was suitable for C-band applications [19].

Ma *et al.* examined the properties of $LiWvO_6$-K_2MoO_4 ceramics using the cold sintering technique. They found the dielectric permittivity of 7.8, the Q-factor of 4514, and the temperature coefficient of the resonant frequency of ~1.3 ppm per °C. Additionally, they claimed the fabricated patch antenna efficiency was 89.5 percent with a bandwidth of 170 MHz resonated at 6.58 GHz [20].

Ji *et al.* investigated the properties of $CaSnSiO_5$-K_2MoO_4 composite ceramics by a cold sintering process. They found the dielectric permittivity and quality factors are 9.2 and ~6240 GHz. Further, they claimed the fabricated patch antenna has a bandwidth of 144 MHz, 5.7 dBi gain, and 88.4 percent efficiency resonated at 5.2 GHz, appropriate for 5 G communications [21].

The sol-gel production of composite nanoparticles containing $0.08TiO_2$-$0.92ZnAl_2O_4$ is described in this chapter. The fundamental characterizations have been verified, and the dielectric characteristics were examined using an LCR meter. The performance of a microstrip patch antenna made with the generated nanoparticles was assessed. The methodology for the synthesis process is covered in Section 10.3, and Section 10.4 debates the results. Finally, Section 10.5 concludes the entire paper.

10.2 Materials and Methods

Titanium tetra isopropoxide (TTIP, Ottokemi), zinc acetate $(CH3COO)_2Zn2H_2O)$, aluminum nitrate nonahydrate $(Al_2(NO_3)_3{\cdot}9H_2O,)$, ethanol $(C_2H_5OH,)$, ethylene glycol and nitric acid (HNO_3) were obtained and employed without further purification for the sol-gel production of $ZnAl_2O_4TiO_2$ dielectric ceramic nanoparticles [30–36].

First, 2.5 ml of distilled water and 38 ml of C_2H_5OH were combined together. After that, 2.25 ml of TTIP solution was mixed and stirred for 4 hours at 85°C. It was then processed and calcined at 700°C to get the final fine powder. In another beaker, 3.75 g $Al_2(NO_3)_3{\cdot}9H_2O$ was mixed with 10 ml ethanol before being mixed with 0.083 ml ethylene glycol under steady stirring conditions [37–43]. Finally, 1.65 g $(CH_3COO)_2Zn2H_2O$ and 0.1 g TiO_2 powder (prepared) was added to the previously described solution and mixed for 1 hour at 75°C. During this process, 0.06 ml of nitric acid was added to the solution to make a homogenous solution, which was then stirred for 1 hour at 75°C until a clear solution formed. As a result, $(1\text{-}x)ZnAl_2O_4xTiO_2$ powder was produced, where x represents the TiO_2 concentration. The sample was dried in an oven at 180°C for ½ hr. The sample was calcined for 1 hour at 700°C before being ground [44–49]. Using the synthesized composite nanoparticles cast on the FTO substrate, we implemented a prototype of a microstrip patch antenna. Following that, both sides of the FTO were coated with silver paste. A coaxial SMA connector was used to complete the fabrication of the 25 mm×15 mm microstrip patch antenna [50–56].

10.3 Results and Discussion

The prepared composite material was tested via XRD, FTIR, FESEM and EDS. XRD was used to investigate the crystalline phases of the $ZnAl_2O_4TiO_2$ ceramic nanoparticles. Figure 10.1 shows the XRD pattern of a $ZnAl_2O_4TiO_2$ sample recorded from 2θ = 20–80°. The crystalline structure of $ZnAl_2O_4TiO_2$ formation may be seen in the mixed anatase and rutile phases of TiO_2, as well as a tiny peak associated with the wurtzite hexagonal phase of ZnO.

This finding was also found to be similar to previously published studies [19, 22]. The amount of TiO_2 utilized in the preparation of nanoceramic composite material has a substantial impact on crystallite diameter. The crystallinity of the composite ceramic can be enhanced by changing the TiO_2 concentration [57–63]. In this study, we chemically produced TiO_2 nanoparticles, which we then used in the sol-gel synthesis of $ZnAl_2O_4TiO_2$. The crystallite size was calculated to be 16.9 nm from the prominent peak (311).

Figure 10.2 illustrates the FTIR spectrum of the prepared ceramic nanoparticles in the 1000–400 cm^{-1} range. The ZnO bands are related with

FTIR transmittance peaks at wave numbers 434, 442, 452, 533, and 544cm^{-1} [23, 24]. These peaks provide information about the chemical bonding of Zn and O. The peak at 492 cm^{-1} corresponds to the stretching vibration of $ZnAl_2O_4$, while the other peaks at 646, 513, and 501 cm^{-1} occur in the usual

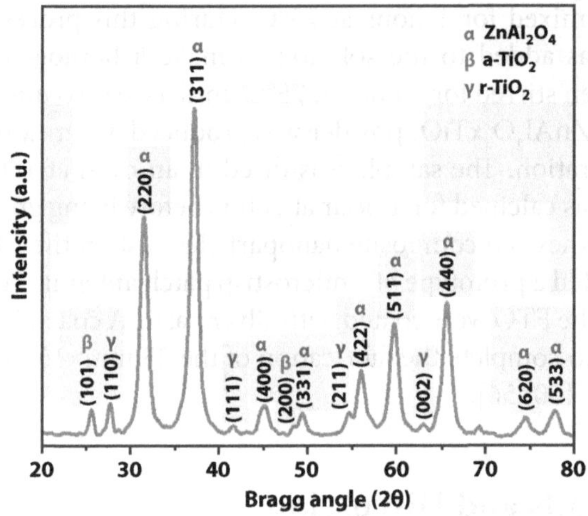

Figure 10.1 XRD pattern of $ZnAl_2O_4TiO_2$ nanoparticles.

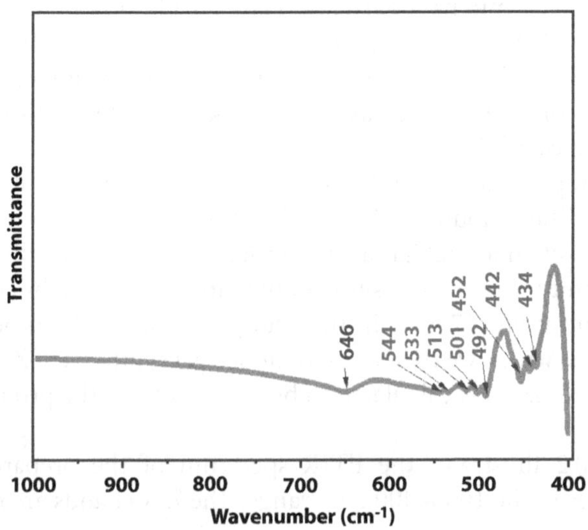

Figure 10.2 FTIR spectrum of $ZnAl_2O_4TiO_2$ nanoparticles.

range of 400–1000 cm^{-1}. This demonstrates that TiO_2 nanoparticles were formed in the prepared sample [25, 26].

Figure 10.3 (a) shows the use of FESEM to the morphology of $ZnAl_2O_4TiO_2$ ceramic nanoparticles. Spherical and agglomerated nanoparticles were identified. Figure 10.3 (b) depicts the histogram of the $ZnAl_2O_4TiO_2$ nanoparticles, and the average diameter of the produced ceramic nanoparticles was

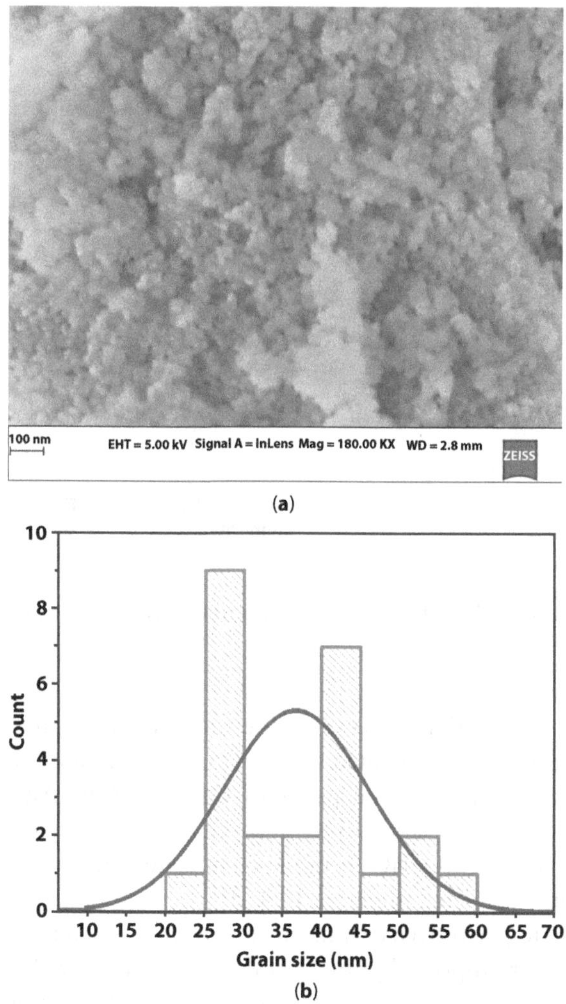

Figure 10.3 (a) FESEM image of 100 nm. (b) Histogram of $ZnAl_2O_4TiO_2$ nanoparticles.
(Continued)

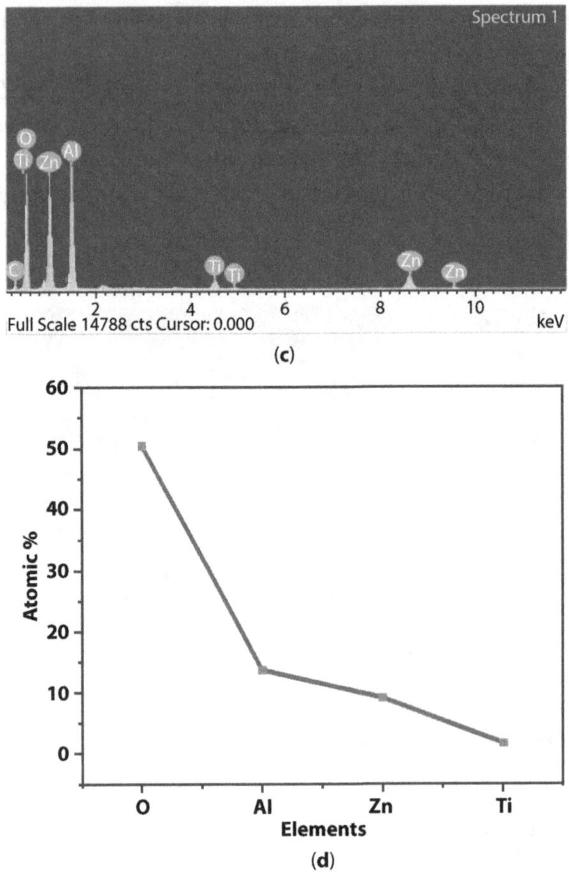

Figure 10.3 (Continued) (c) EDS spectrum. (d) Atomic percentage of the $ZnAl_2O_4TiO_2$ composite.

determined to be 36.67 nm [64–71]. The fundamental composition of the produced sample was investigated using EDS, as depicted in Figure 10.3 (c), which reveals the fundamental peaks of Oxygen, Zinc, Aluminium, and Titanium at 0.5, 1.0, 1.5, and 4.5 eV, respectively. Furthermore, the atomic percentage of the $ZnAl_2O_4TiO_2$ ceramic composite nanoparticles in relation to the elements is depicted in Figure 10.3 (d).

The dielectric permittivity value relates to the capacitance related with the charge carriers' dipole alignment and indicates how well a material can store electric energy when an electric field is applied. By applying the formula $\varepsilon_r = Cd/\varepsilon_0 A$, where, C, d, A and ε_0 are capacitance, distance, area and absolute permittivity of the sample. We were able to determine the dielectric permittivity after collecting the parallel capacitance values. Using an LCR meter with a frequency range of 100 hertz to 1 megahertz at room temperature, the dielectric properties of $TiO_2ZnAl_2O_4$ nanoparticles were investigated [72–77].

Figure 10.4 reveals the frequency-dependent fluctuation of dielectric permittivity. The reduced permittivity may be seen as the frequency increases. The dielectric permittivity was changed from 20.2 to 18.92 when the frequency range was raised from 100 kilo to 1 megahertz. Additionally, Roshini *et al.* informed a dielectric constant of 9.640 for the $TiO_2ZnAl_2O_4$ material. Likewise, Abdullah *et al.* determined the dielectric constant of SiO_2-$ZnAl_2O_4$ nanoparticles to be 8.5720. A dramatic reduction in dielectric permittivity was seen in the lower frequency range, which was determined to be constant in the higher frequency zone. The reduced polarization is responsible for this property of such materials [27].

Dielectric loss (tanδ) is a significant metric describing energy dissipation that must be investigated. Dielectric loss is sometimes viewed as microstructure faults, such as defects, collisions, porosity, natural orientation of the crystallite, and so on [28]. The dielectric loss of a $ZnAl_2O_4TiO_2$ nanocomposite ceramic sample as a function of frequency is depicted in Figure 10.5. The dielectric loss was reduced from 0.19 to 0.0185 as the frequency range was expanded from 100 kilohertz to 1 megahertz [78–83]. In general, as the frequency rises, the tanδ declines. The reason for this

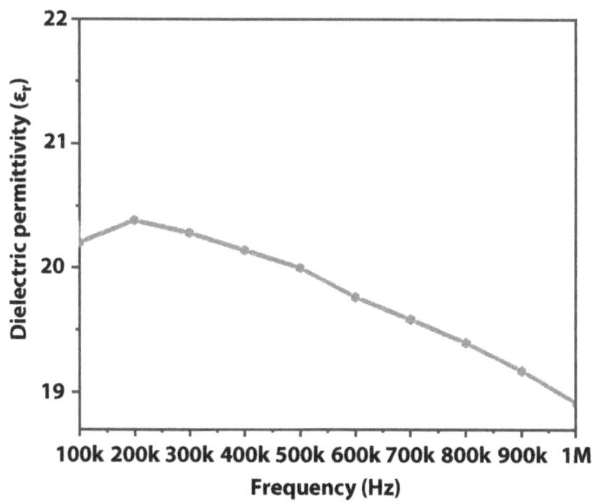

Figure 10.4 Dielectric permittivity w.r.t. frequency.

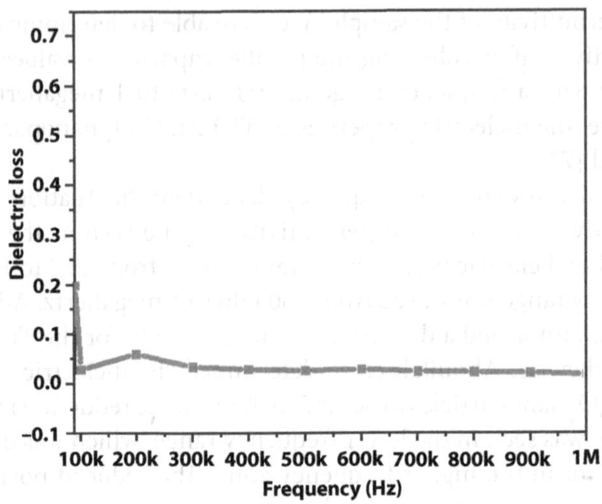

Figure 10.5 Dielectric loss w.r.t. frequency.

is that the hopping ions move more slowly than the applied electric field. The dielectric loss value increases in the lower frequency range and lowers further in the higher frequency range. In the initial instance, this is due to the high resistivity caused by the related effect of grain boundaries [84–91].

The return loss (RL) of the produced $ZnAl_2O_4TiO_2$ ceramic nanoparticles is depicted in Figure 10.6 at operating frequencies ranging from 6

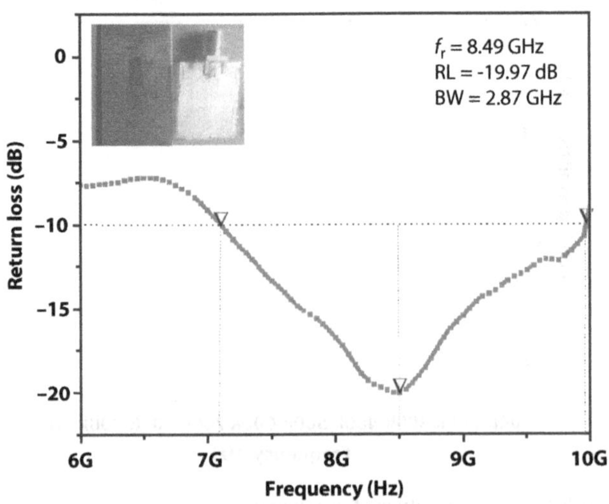

Figure 10.6 Return loss plot of $ZnAl_2O_4TiO_2$ prototype patch antenna.

to 10 GHz. At an operating frequency of 8.49 GHz and a bandwidth of 2.87 GHz, the return loss was measured to be -19.97 dB. Thirumanathan *et al.* informed the efficiency of a patch antenna composed of $CaCu_3Ti_4O_{12}$ dielectric material with an RL of -21.23 dB at an operating frequency of 2.37 GHz and 540 MHz bandwidth for wireless communication applications in a similar work [29]. Our $0.08TiO_2$-$0.92ZnAl_2O_4$ composite nanoparticle-based microstrip patch antenna had the lowest return loss. As a result, it was discovered to be suited for X-band communication. Figure 14.6's inset displays a digital image of a microstrip patch antenna with a dimension of 25×15 mm^2 [92–96].

In a perfect world, the maximum power transmitted from the transmitter to the receiving antenna is directly proportional to the voltage standing wave ratio (VSWR) [97–102]. Only when the input impedance and the transmitting antenna impedance are matched can this occur. The VSWR value can be determined using the formula VSWR = $V_{max}/V_{min} = (1+|\rho|)/(1-|\rho|) = (1+S_{11})/(1-S_{11})$. Figure 10.7 depicts the VSWR values about frequency, as the patch antenna's forwards and reflected powers were 99.07 percent and 0.92 percent, respectively. At an operational frequency of 8.49 GHz, the VSWR of the fabricated microstrip patch antenna was measured to be 1.29. In general, VSWR influences an antenna's reflection coefficient, and a higher ratio (V_{max}/V_{min}) designates a significant mismatch rather than perfect matching [103–108]. This mismatch or match is caused by the

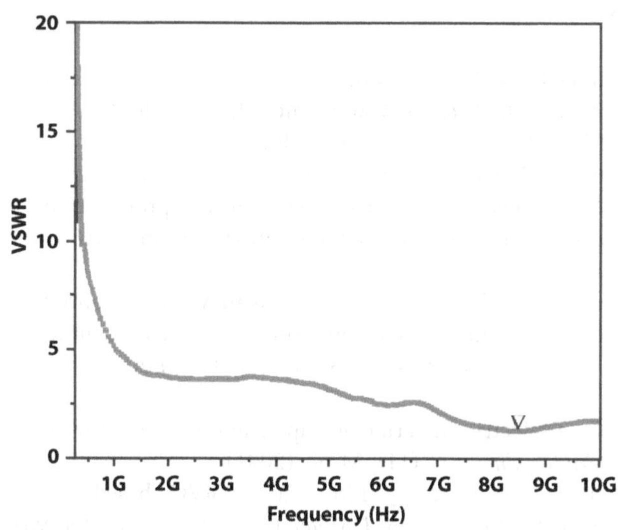

Figure 10.7 Voltage standing wave ratio of $ZnAl_2O_4TiO_2$ patch antenna.

minimum and maximum amplitudes of the standing wave. As shown in Figure 10.5, the VSWR value achieved is less than 2, suggesting a decreased return loss [109–116]. The microstrip patch antenna had a reflection coefficient of 0.13 [117–125]. Finally, the mismatch loss of the estimated antenna was 0.080 dB, while the reflected and forwarded powers of the microstrip patch antenna were 0.92 percent and 99.07 percent, respectively, using the formulas Reflected power (%) =100. $|\rho|^2$, Forward power (%) = 100. (1- $|\rho|^2$), and Mismatch loss = -10. Log (1- $|\rho|^2$) [126–132].

10.4 Conclusions

The prepared $ZnAl_2O_4TiO_2$ nanoparticles were examined with various measurements and fabricated the prototype patch antenna with the help of ceramic composition. The synthesized $ZnAl_2O_4TiO_2$ nanoparticles were crystalline, with an average diameter of 36.67 nm. The FTIR analysis revealed several vibration peaks connected to the functional groups. The existence of Ti, Al, Zn, and O was also confirmed by EDX spectroscopy at 4.5, 1.5, 1.0, and 0.5 eV, respectively. Furthermore, we developed a patch antenna with -19.97 dB return loss at 8.49 GHz operating frequency and a VSWR of < 2. A fabricated patch antenna based on $0.92TiO_2$-$0.08ZnAl_2O_4$ composite nanoparticles performs well in X-band applications.

References

1. D. Zhou, D. Guo, W. Li, L. Pang, X. Yao, D. Wang, I.M. Reaney, Novel temperature stable high-εr microwave dielectrics in the Bi_2O_3-TiO_2-V_2O_5 system, *J. Mater. Chem. C,* 23 (4) 5357-5362, (2016).
2. D. Zhou, J. Li, L. Pang, G. Chen, Z. Qi, D. Wang, I.M. Reaney, Crystal structure, infrared spectra, and microwave dielectric properties of temperature-stable zircontype (Y,Bi)VO_4 solid solution ceramics, *ACS Omega,* 1 (5), 963-970, (2016).
3. D. Zhou, L. Pang, D. Wang, Z. Q, I.M. Reaney, High quality factor, ultralow sintering temperature Li6B4O9 microwave dielectric ceramics with ultralow density for antenna substrates, *ASC Sustain. Chem. Eng.*, 6 (8), 11138-11143, (2018).
4. R.J. Cava, Dielectric materials for applications in microwave communications, *J. Mater. Chem.* 11 (11), 54-62, (2001).
5. D. Zhou, L. Pang, D. Wang, I.M. Reaney, $BiVO_4$ based high k microwave dielectric materials: a review, *J. Mater. Chem. C,* 6 (35), 9290-9313, (2018).

6. H.H. Guo, D. Zhou, W.F. Liu, L.X. Pang, D. Wang, J.Z. Su, Z.M. Qi, Microwave dielectric properties of temperature-stable zircon-type (Bi, Ce)VO_4 solid-solution ceramics, *J. Am. Ceram. Soc.*, 103 (1), 423-431, (2020).
7. R. Hill, J. Craig, G.V. Gibbs, Systematics of the spinel structure type, *Phys. Chem. Miner.*, 4, 317-339, (1979).
8. X. Duan, D. Yuan, Z. Sun, H. Sun, D. Xu, M. Lv, Synthesis and characterization of ZnAl2O4/SiO2 nanocomposites by sol–gel method, *J. Cryst. Growth*, 252, 4-8, (2003).
9. Y. Wu, J. Du, K.-L. Choy, L.L. Hench, J. Guo, Formation of interconnected microstructural ZnAl2O4 films prepared by sol–gel method, *Thin Solid Films*, 472, 150-156, (2005).
10. M.A. Valenzuela, J.P. Jacobs, P. Bosch, S. Reijne, B. Zapata, H.H. Brongersma, The influence of the preparation method on the surface structure of ZnAl2O4, *Appl. Catal., A*, 148, 315-324, (1997).
11. X. Wei, D. Chen, Synthesis and characterization of nanosized zinc aluminate spinel by sol-gel technique, *Mater. Lett.*, 60, 823-827, (2006).
12. S. Kurajica, E. Tkalcec, J. Sipusic, G. Matijasic, I. Brnardic, I. Simcic, Synthesis and characterization of nanocrystalline zinc aluminate spinel by sol-gel technique using modified alkoxide precursor, *J. Sol-Gel Sci. Technol.*, 46, 152-160, (2008).
13. K. P. Surendran, N. Santha, P. Mohanan, and M. T. Sebastian, "Temperature Stable Low Loss Ceramic Dielectrics in $(1\text{-}x)ZnAl_2O_4\text{-}xTiO_2$ System for Microwave Substrate Applications," *Eur. Phys. J. B*, 41, 301-306, (2004).
14. W. Lei, W. Z. Lu, J. H. Zhu, and X. H. Wang, "Microwave Dielectric Properties of $ZnAl_2O_4\text{-}TiO_2$ Spinel-Based Composites," *Mater. Lett.*, 61, 4066-4069, (2007).
15. Zhijian Wang, Feng Pan, Lanlan Liu, Qifeng Du, Ruitao Tang, Jun Ai, Hong Zhang and Ying Chen, Enhanced Microwave Dielectric Properties and Sintering Behaviors of $Mg_2SiO_4\text{-}Li_2TiO_3$-LiF Ceramics by Adding $CaTiO_3$ for LTCC and GPS Antenna Applications, *Crystals*, 12, 512, (2022).
16. Bing Liu, Ke Sha, Meng Fei Zhou, Kai Xin Song, Yu Hui Huang, Cheng Chao Hu, Novel low-εr MGa_2O_4 (M = Ca, Sr) microwave dielectric ceramics for 5 G antenna applications at the Sub-6 GHz band, 41(10), 5170-5175, (2021).
17. Balamurugan Chinnagurusamy, Marichamy Perumalsamy, Multiband microstrip patch antenna using copper nano radiating element for X band and C band applications, *Int. J. Numer. Modell. Electron. Networks Devices Fields*, (2021), doi:10.1002/jnm.2897.
18. V. Naidu, S. Ahamed Kandu, A. S. Kumar, M. Sivabharathy, R. Legadevi, C. Prakash & S. Pandian, Synthesis and Characterization of Novel Nanoceramic Magnesium Ferrite Material Doped with Samarium and Dysprosium for Designing-Microstrip Patch Antenna, Defect and Diffusion Forum, 332, 35-50, (2012).

19. Siragam, S., Dubey, R.S., Pappula, L. *et al.* Synthesis and investigation of dielectric ceramic nanoparticles for microstrip patch antenna applications. *Sci. Rep.*, 12, 3929, (2022). https://doi.org/10.1038/s41598-022-07899-6.
20. Mingtao Ma; Kaixin Song; Yuping Ji; Fayaz Hussain; Amir Khesro; Minmin Mao; Lingyun Xue; Ping Xu; Bing Liu;Z hilun Lu; Di Zhou; Dawei Wang; Shikuan Sun; 5G microstrip patch antenna and microwave dielectric properties of cold sintered $LiWVO_6$-K_2MoO_4 composite ceramics. *Ceram. Int.*, (2021), doi:10.1016/j.ceramint.2021.03.179.
21. Ji, Yuping; Song, Kaixin; Zhang, Shiyu; Lu, Zhilun; Wang, Ge; Li, Linhao; Zhou, Di; Wang, Dawei; Reaney, Ian M. Cold sintered, temperature-stable $CaSnSiO_5$-K_2MoO_4 composite microwave ceramics and its prototype microstrip patch antenna. *J. Eur. Ceram. Soc.*, (2020), S0955221920306956-doi:10.1016/j.jeurceramsoc.2020.08.053
22. Abazović, Nadica D.; Čomor, Mirjana I.; Dramićanin, Miroslav D.; Jovanović, Dragana J.; Ahrenkiel, S. Phillip; Nedeljković, Jovan M. (2006). Photoluminescence of Anatase and Rutile TiO_2 Particles. *J. Phys. Chem. B*, 110(50), 25366-25370, (2006).
23. Bhagaban Beheraa, Souvik Palb, Lizy Kanungob, Sunil Bhand, Sudhir Chandra, Synthesis and characterization of ZnO-$ZnAl_2O_4$ whiskers and their application in biosensors, *J. Nanosci. Lett.*, 5: 2, (2015).
24. Kalpana Handorea, Sanjay Bhavsara, Amit Hornea, Prakash Chhattisea, Kakasaheb Mohitebc, Jalinder Ambekard, Nishigandh Pandea & Vasant Chabukswara, Novel Green Route of Synthesis of ZnO Nanoparticles by Using Natural Biodegradable Polymer and Its Application as a Catalyst for Oxidation of Aldehydes, *J. Macromol. Sci. Part A Pure Appl. Chem.*, 51:12, 941-947, (2014).
25. Wang, Shi-Fa; Sun, Guang-Zhuang; Fang, Lei-Ming; Lei, Li; Xiang, Xia; Zu, Xiao-Tao. A comparative study of $ZnAl_2O_4$ nanoparticles synthesized from different aluminum salts for use as fluorescence materials. *Sci. Rep.*, 5, 12849. (2015), doi:10.1038/srep12849.
26. P. Murugakoothan, S. Ananth, P. Vivek, T. Arumanayagam, Natural Dye Extracts of Areca Catechu Nut as dye Sensitizer for Titanium dioxide Based Dye Sensitized Solar Cells, *J. Nano Electron. Phys.*, Vol. 6 No. 1, 01003(4pp), (2014).
27. Mohacek-Grosev, V., Vrankic, M., Maksimovic, A. & Mandic, V. Influence of titanium doping on the Raman spectra of nanocrystalline ZnAl2O4. *J. Alloy. Compd.*, 697, 90-95, (2017).
28. Zhang, L., Yan, J., Zhou, M., Yang, Y. & Liu, Y.-N. Fabrication and photocatalytic properties of spheres-in-spheres ZnO/$ZnAl_2O_4$ composite hollow microspheres. *Appl. Surf. Sci.*, 268, 237-245, (2013).
29. P. Thiruramanathana, A. Marikanib, S. Ravib, D. Madhavanc & G.S. Hikkud, Fabrication of miniaturized high bandwidth dielectric resonator on patch

(DRoP) antenna using high dielectric $CaCu_3Ti_4O_{12}$ nanoparticles, Vol. 747, pp. 1033-1042, (2018).

30. A. Bhardwaj, P. Kumar, B. Raj, and S. Anand, "Design and Performance Optimization of Doping-Less Vertical Nanowire TFET Using Gate Stack Technique," *JEMS*, Springer, Vol. 41, No. 7, pp. 4005–4013, 2022.
31. J. Singh, and B. Raj, "Tunnel Current Model of Asymmetric MIM Structure Levying Various Image Forces to Analyze the Characteristics of Filamentary Memristor", *Appl. Phys. A*, Springer, Vol. 125, No. 3, pp. 203.1 to 203.11, 2019.
32. C. Goyal, J. S. Ubhi, and B. Raj, "Low Leakage Zero Ground Noise Nanoscale Full Adder Using Source Biasing Technique", *J. Nanoelectron. Optoelectron.*, American Scientific Publishers,Vol. 14, pp. 360–370, 2019.
33. G.Singh, R. K.Sarin, and B.Raj, "A Novel Robust Exclusive-OR Function Implementation in QCA Nanotechnology with Energy Dissipation Analysis", *J. Comput. Electron.*, Springer, Vol. 15, No. 2, pp. 455–465, 2016.
34. T. Wadhera, D. Kakkar, G. Wadhwa, and B. Raj, "Recent Advances and Progress in Development of the Field Effect Transistor Biosensor: A Review", *J. Electron. Mater.*, Springer,Vol. 48, No. 12, pp. 7635–7646, 2019.
35. G. Wadhwa, P. Kamboj, and B. Raj, "Design Optimisation of Junctionless TFET Biosensor for High Sensitivity", *Adv. Nat. Sci.: Nanosci. Nanotechnol.*, Vol. 10, No. 7, p. 045001, 2019.
36. P. Bansal, and B. Raj, "Memristor Modeling and Analysis for Linear Dopant Drift Kinetics", *J. Nanoeng. Nanomanuf.*, American Scientific Publishers, Vol. 6, pp. 1–7, 2016.
37. A. Singh, M. Khosla, and B. Raj, "Circuit Compatible Model for Electrostatic Doped Schottky Barrier CNTFET," *J. Electron. Mater.*, Springer, Vol. 45, No. 12, pp. 4825–4835, 2016.
38. Ashima, D. Vaithiyanathan, and B. Raj, "Performance Analysis of Charge Plasma induced Graded Channel Si Nanotube", *JER*, EMSME Special Issue, Vol. 9, pp. 146–154, 2021.
39. A. S. Tomar, V. K. Magraiya, and B. Raj, "Scaling of Access and Data Transistor for High Performance DRAM Cell Design", *Quantum Matter*, Vol. 2, pp. 412–416, 2013.
40. N. Jain, and B. Raj, "Parasitic Capacitance and Resistance Model Development and Optimization of Raised Source/Drain SOI FinFET Structure for Analog Circuit Applications", *J. Nanoelectron. Optoelectronins*, ASP, USA, Vol. 13, pp. 531–539, 2018.
41. S. Singh, S. K. Vishvakarma, and B.Raj, "Analytical Modeling of Split-Gate Junction-Less Transistor for a Biosensor Application", *Sens. Bio-Sens.*, Elsevier, Vol. 18, pp. 31–36, 2018.

42. M. Gopal, and B. Raj, "Low Power 8T SRAM Cell Design for High Stability Video Applications", *ITSI Trans. Electr. Electron. Eng.*, Vol.1, No. 5, pp. 91–97, 2013.
43. B. Raj, J.Mitra, D. K. Bihani, V. Rangharajan, A. K. Saxena, and S. Dasgupta, "Analysis of Noise Margin, Power and Process Variation for 32 nm FinFET Based 6T SRAM Cell", *JCP*, Academy Publisher, FINLAND, Vol. 5, No. 6, PP. 1-6, 2010.
44. D. Sharma, R. Mehra, and B. Raj, "Comparative Analysis of Photovoltaic Technologies for High Efficiency Solar Cell Design", *Superlattices Microstruct.*, Elsevier, Vol. 153, pp. 106861, 2021.
45. P. Kaur, A. S. Buttar, and B. Raj, "A Comprehensive Analysis of Nanoscale Transistor Based Biosensor: A Review", *Indian J. Pure Appl. Phys.*, Vol. 59, pp. 304–318, 2021.
46. D. Yadav, B. Raj, and B. Raj, "Design and Simulation of Low Power Microcontroller for IoT Applications", *J. Sens. Lett.*, ASP USA, Vol. 18, pp. 401–409, 2020.
47. S. Singh, and B. Raj, "A 2-D Analytical Surface Potential and Drain Current Modeling of Double-Gate Vertical t-Shaped Tunnel FET", *J. Comput. Electron.*, Springer, Vol. 19, pp. 1154–1163, 2020.
48. J. Singh, and B. Raj, "An Accurate and Generic Window function for Non-linear Memristor Model" *J. Comput. Electron.*, Springer, Vol. 18, No. 2, pp. 640–647, 2019.
49. M. Kaur, N. Gupta, S. Kumar, B. Raj, and A. K. Singh, "Comparative RF and Crosstalk Analysis of Carbon Based Nano Interconnects", *IET Circuits Devices Syst.*, Vol. 15, No. 6, pp. 493–503, 2021.
50. N. Kandasamy, F. Ahmad, D. Ajitha, B. Raj, and N. Telagam, "Quantum Dot Cellular Automata Based Scan Flip Flop and Boundary Scan Register", *IETE J. Res.*, Vol.69, PP. 535-548, 2023.
51. S. K. Sharma, B. Raj, and M. Khosla, "Enhanced Photosensitivity of Highly Spectrum Selective Cylindrical Gate In1-xGaxAs Nanowire MOSFET Photodetector", *Mod. Phys. Lett. B-B*, Vol. 33, No. 12, p. 1950144, 2019.
52. J. Singh, and B. Raj, "Design and Investigation of 7T2M NVSARM with Enhanced Stability and Temperature Impact on Store/Restore Energy", *IEEE Transactions on Very Large Scale Integration Systems*, Vol. 27, No. 6, pp. 1322–1328, 2019.
53. A. K. Bhardwaj, S. Gupta, B. Raj, and A. Singh, "Impact of Double Gate Geometry on the Performance of Carbon Nanotube Field Effect Transistor Structures for Low Power Digital Design", *Comput. Theor. Nanosci.*, ASP USA, Vol. 16, pp. 1813–1820, 2019.
54. N. Jain, and B. Raj, "Thermal Stability Analysis and Performance Exploration of Asymmetrical Dual-k Underlap Spacer (ADKUS) SOI FinFET for Security and Privacy Applications", *IJPAP*, Vol. 57, pp. 352–360, 2019.

55. A. Singh, M. Khosla, and B. Raj, "Design and Analysis of Dynamically Configurable Electrostatic Doped Carbon Nanotube Tunnel FET", *Microelectron. J.*, Elsevier, Vol. 85, pp. 17–24, 2019.
56. N. Jain, and B. Raj, "Dual-k Spacer Region Variation at the Drain Side of Asymmetric SOI FinFET Structure: Performance Analysis towards the Analog/RF Design Applications", *J. Nanoelectron. Optoelectron.*, American Scientific Publishers, Vol. 14, pp. 349–359, 2019.
57. J. Singh, S. Sharma, B. Raj, and M. Khosla, "Analysis of Barrier Layer Thickness on Performance of In1-xGaxAs Based Gate Stack Cylindrical Gate Nanowire MOSFET", *JNO*, ASP USA, Vol. 13, pp. 1473–1477, 2018.
58. N. Jain, and B. Raj, "Analysis and Performance Exploration of High-k SOI FinFETs over the Conventional Low-k SOI FinFET toward Analog/RF Design", *JoS*, IOP Science, Vol. 39, No. 12, pp. 124002-1-7, 2018.
59. C. Goyal, J. S. Ubhi, and B. Raj, "A Reliable Leakage Reduction Technique for Approximate Full Adder with Reduced Ground Bounce Noise", *J. Math. Prob. Eng.*, Hindawi, Vol. 2018, Article ID 3501041, PP.1-16, 2018.
60. Anuradha, J. Singh, B. Raj, and M. Khosla, "Design and Performance Analysis of Nano-Scale Memristor-Based Nonvolatile SRAM", *J. Sens. Lett.*, American Scientific Publishers USA, Vol. 16, pp. 798–805, 2018.
61. G. Wadhwa, and B. Raj, "Parametric Variation Analysis of Charge-Plasma-Based Dielectric Modulated JLTFET for Biosensor Application", *IEEE Sens. J.*, Vol. 18, No. 15, PP. 6070 - 6077, 2018.
62. J. Singh, and B. Raj, "Comparative Analysis of Memristor Models for Memories Design", *JoS*, IoP, Vol. 39, No. 7, PP. 074006-1-12, 2018.
63. D. Yadav, S. S. Chouhan, S. K. Vishvakarma, and B. Raj, "Application Specific Microcontroller Design for IoT Based WSN", *Sens. Lett*, ASP, Vol. 16, pp. 374–385, 2018.
64. G. Singh, R. K. Sarin, and B. Raj, "Fault-Tolerant Design and Analysis of Quantum-Dot Cellular Automata Based Circuits", *IEEE/IET Circuits Devices Syst.*, Vol. 12, pp.638–664, 2018.
65. J. Singh, and B. Raj, "Modeling of Mean Barrier Height Levying Various Image Forces of Metal Insulator Metal Structure to Enhance the Performance of Conductive Filament Based Memristor Model", *IEEE Nanotechnol.*, Vol. 17, No. 2, pp. 268–275, 2018.
66. A. Jain, S. Sharma, and B. Raj, "Analysis of Triple Metal Surrounding Gate (TM-SG) III-V Nanowire MOSFET for Photosensing Application", *Opto. Electron. J.*, Elsevier, Vol. 26, No. 2, PP. 141–148, 2018.
67. A. Jain, S. Sharma, and B. Raj, "Design and Analysis of High Sensitivity Photosensor Using Cylindrical Surrounding Gate MOSFET for Low Power Sensor Applications", *Eng. Sci. Technol. Int. J.*, Elsevier's, Vol. 19, No. 4, pp. 1864–1870, 2016.

68. A. Singh, M. Khosla, and B. Raj, "Analysis of Electrostatic Doped Schottky Barrier Carbon Nanotube FET for Low Power Applications," *J. Mater. Sci.: Mater. Electron.*, Springer, Vol. 28, PP. 1762–1768, 2017.
69. G. Saiphani Kumar, A. Singh, B. Raj, "Design and Analysis of Gate All around CNTFET Based SRAM Cell Design", *J. Comput. Electron.*, Springer, Vol. 17, No.1, PP. 138–145, 2018.
70. G. P. Singh, B. S. Sohi, and B. Raj, "Material Properties Analysis of Graphene Base Transistor (GBT) for VLSI Analog Circuits", *IJPAP*, Vol. 55, pp. 896–902, 2017.
71. S. Kumar and B. Raj, "Estimation of Stability and Performance Metric for Inward Access Transistor Based 6T SRAM Cell Design Using n-type/p-type DMDG-GDOV TFET", *IEEE VLSI Circuits and Systems Letter*, Vol. 3, No. 2, PP. 25–39, 2017.
72. S. Sharma, A. Kumar, M. Pattanaik, and B. Raj, "Forward Body Biased Multimode Multi-Threshold CMOS Technique for Ground Bounce Noise Reduction in Static CMOS Adders", *Int. J. Inf. Electron. Eng.*, Vol. 3, No. 3, PP. 567–572, 2013.
73. H. Singh, P. Kumar, and B. Raj, "Performance Analysis of Majority Gate SET Based 1-bit Full Adder", *IJCCE*, IACSIT Press Singapore, ISSN: 20103743, Vol. 2, No. 4, PP.1-8, 2013.
74. A. K. Bhardwaj, S. Gupta, and B. Raj, "Investigation of Parameters for Schottky Barrier (SB) Height for Schottky Barrier Based Carbon Nanotube Field Effect Transistor Device", *J. Nanoelectron. Optoelectron.*, ASP, Vol. 15, PP. 783–791, 2020.
75. P. Bansal, and B. Raj, "Memristor: A Versatile Nonlinear Model for Dopant Drift and Boundary Issues", *JCTN*, American Scientific Publishers, Vol. 14, No. 5, PP. 2319–2325, 2017.
76. N. Jain, and B. Raj, "An Analog and Digital Design Perspective Comprehensive Approach on Fin-FET (Fin-Field Effect Transistor) Technology - A Review", *RASE*, ASP, Vol. 5, PP. 1–14, 2016.
77. S. Sharma, B. Raj, and M. Khosla, "Subthreshold Performance of In1-xGaxAs Based Dual Metal with Gate Stack Cylindrical/Surrounding Gate Nanowire MOSFET for Low Power Analog Applications", *J. Nanoelectron. Optoelectron.*, American Scientific Publishers, USA, Vol. 12, PP. 171– 176, 2017.
78. B. Raj, A. K. Saxena, and S. Dasgupta, "Analytical Modeling for the Estimation of Leakage Current and Subthreshold Swing Factor of Nanoscale Double Gate FinFET Device" *Microelectron. Int.*, UK, Vol. 26, PP. 53–63, 2009.
79. S. S. Soniya, G. Wadhwa, and B. Raj, "An Analytical Modeling for Dual Source Vertical Tunnel Field Effect Transistor", *IJRTE*, Vol. 8, No. 2, 2019.
80. S. Singh, and B. Raj, "Design and Analysis of Hetrojunction Vertical T-Shaped Tunnel Field Effect Transistor", *J. Electron. Mater.*, Springer, Vol. 48, No. 10, PP. 6253–6260, 2019.

81. C. Goyal, J. S. Ubhi, and B. Raj, "A Low Leakage CNTFET Based Inexact Full Adder for Low Power Image Processing Applications", *Int. J. Circuit Theory Appl.*, Wiley, Vol. 47, No. 9, PP. 1446–1458, 2019.
82. B. Raj, A. K. Saxena, and S. Dasgupta, "A Compact Drain Current and Threshold Voltage Quantum Mechanical Analytical Modeling for FinFETs", *J. Nanoelectron. Optoelectron.*, USA, Vol. 3, No. 2, PP. 163–170, 2008.
83. G. Wadhwa, and B. Raj, "An Analytical Modeling of Charge Plasma Based Tunnel Field Effect Transistor with Impacts of Gate Underlap Region", *Superlattices Microstruct.*, Elsevier, Vol. 142, PP. 106512-22, 2020.
84. S. Singh, and B. Raj, "Modeling and Simulation Analysis of SiGe Heterojunction Double GateVertical t-Shaped Tunnel FET", *Superlattices Microstruct.*, Elsevier, Vol. 142, PP. 106496-104, 2020.
85. A. Singh, D. K. Saini, D. Agarwal, S. Aggarwal, M. Khosla, and B. Raj, "Modeling and Simulation of Carbon Nanotube Field Effect Transistor and Its Circuit Application", *JoS*, IOP Science, Vol. 37, PP. 074001–074006, 2016.
86. N. Jain, and B. Raj, "Device and Circuit Co-Design Perspective Comprehensive Approach on FinFET Technology - A Review", *J. Electron Devices*, Vol. 23, No. 1, PP. 1890–1901, 2016.
87. S. Kumar and B. Raj, "Analysis of ION and Ambipolar Current for Dual-Material Gate-Drain Overlapped DG-TFET", *J. Nanoelectron. Optoelectron.*, American Scientific Publishers, USA, Vol. 11, PP. 323–333, 2016.
88. N. Anjum, T. Bali, and B. Raj, "Design and Simulation of Handwritten Multiscript Character Recognition", *IJARCCE*, Vol. 2, No. 7, PP. 2544–2549, 2013.
89. S. Sharma, B. Raj, and M. Khosla, "A Gaussian Approach for Analytical Subthreshold Current Model of Cylindrical Nanowire FET with Quantum Mechanical Effects", *Microelectron. J.*, Elsevier, Vol. 53, PP. 65–72, 2016.
90. K. Singh, and B. Raj, "Performance and Analysis of Temperature Dependent Multi-Walled Carbon Nanotubes as Global Interconnects at Different Technology Nodes", *J. Comput. Electron.*, Springer, Vol. 14, No. 2, PP. 469–476, 2015.
91. S. Kumar and B. Raj, "Compact Channel Potential Analytical Modeling of DG-TFET Based on Evanescent–Mode Approach", *J. Comput. Electron.*, Springer, Vol. 14, No. 2, PP. 820–827, 2015.
92. K. Singh, and B. Raj, "Temperature Dependent Modeling and Performance Evaluation of Multi-Walled CNT and Single-Walled CNT as Global Interconnects", *J. Electron. Mater.*, Springer, Vol. 44, No. 12, PP. 4825–4835, 2015.
93. V. K. Sharma, M. Pattanaik, and B. Raj, "INDEP Approach for Leakage Reduction in Nanoscale CMOS Circuits", *Int. J. Electron.*, Taylor & Francis, Vol. 102, No. 2, PP. 200–215, 2014.

94. K. Singh, and B. Raj, "Influence of Temperature on MWCNT Bundle, SWCNT Bundle and Copper Interconnects for Nanoscaled Technology Nodes", *J. Mater. Sci.: Mater. Electron.*, Springer, Vol. 26, No. 8, PP. 6134–6142, 2015.
95. N. Anjum, T. Bali, and B. Raj, "Design and Simulation of Handwritten Gurumukhi and Devanagri Numerical Recognition", *Int. J. Comput. Appl.*, Published by Foundation of Computer Science, New York, USA, Vol. 73, No. 12, PP. 16–21, 2013.
96. S. Khandelwal, V. Gupta, B. Raj, and R. D. Gupta, "Process Variability Aware Low Leakage Reliable Nano Scale DG-FinFET SRAM Cell Design Technique", *J. Nanoelectron. Optoelectron.*, Vol. 10, No. 6, PP. 810–817, 2015.
97. V. K. Sharma, M. Pattanaik, and B. Raj, "ONOFIC Approach: Low Power High Speed Nanoscale VLSI Circuits Design", *Int. J. Electron.*, Taylor & Francis, Vol. 101, No. 1, PP. 61–73, 2014.
98. S. Khandelwal, B. Raj, and R. D. Gupta, "FinFET Based 6T SRAM Cell Design: Analysis of Performance Metric, Process Variation and Temperature Effect", *J. Comput. Theor. Nanosci.*, ASP, USA, Vol. 12, PP. 2500–2506, 2015.
99. S. Singh, Y. Shekhar, R. Jagdeep, S. Anurag, and B. Raj, "Impact of HfO2 in Graded Channel Dual Insulator Double Gate MOSFET", *J. Comput. Theor. Nanosci.*, American Scientific Publishers, Vol. 12, No. 6, PP. 950–953, 2015.
100. V. K. Sharma, M. Pattanaik, and B. Raj, "PVT Variations Aware Low Leakage INDEP Approach for Nanoscale CMOS Circuits", *Microelectron. Reliab.*, Elsevier, Vol. 54, PP. 90–99, 2014.
101. S. Singh, and B. Raj, "Parametric Variation Analysis on Hetero-Junction Vertical t-Shape TFET for Suppressing Ambipolar Conduction", *Indian J. Pure Appl. Phys.*, Vol. 58, PP. 478–485, 2020.
102. S. S. Soniya, G. Wadhwa, and B. Raj, "Design and Analysis of Dual Source Vertical Tunnel Field Effect Transistor for High Performance", *Trans. Electr. Electron. Mater.*, Springer, Vol. 21, PP. 74–82, 2019.
103. M. Kaur, N. Gupta, S. Kumar, B. Raj, and A. K. Singh, "RF Performance Analysis of Intercalated Graphene Nanoribbon Based Global Level Interconnects", *J. Comput. Electron.*, Springer, Vol. 19, PP.1002–1013, 2020.
104. G. Wadhwa, B. Raj, "Design and Performance Analysis of Junctionless TFET Biosensor for High Sensitivity" *IEEE Nanotechnol.*, Vol. 18, PP. 567–574, 2019.
105. J. Singh, and B. Raj, "Enhanced Nonlinear Memristor Model Encapsulating Stochastic Dopant Drift", *JNO*, ASP, Vol.14, PP. 958–963, 2019.
106. B. Raj, A. K. Saxena and S. Dasgupta, "Quantum Mechanical Analytical Modeling of Nanoscale DG FinFET: Evaluation of Potential, Threshold Voltage and Source/Drain Resistance", *Elsevier's J. Mater. Sci. Semicond. Process.*, Elsevier, Vol. 16, No. 4, PP. 1131–1137, 2013.

107. M. Gopal, S. S. D. Prasad, and B. Raj, "8T SRAM Cell Design for Dynamic and Leakage Power Reduction", *Int. J. Comput. Appl.,* Published by Foundation of Computer Science, New York, USA, Vol. 71, No. 9, PP. 43–48, 2013.
108. M. Pattanaik, B. Raj, S.Sharma, and A. Kumar, "Diode Based Trimode MultiThreshold CMOS Technique for Ground Bounce Noise Reduction in Static CMOS Adders", *Adv. Mater. Res. Trans. Tech.*, Publications, Switzerland, Vol. 548, PP. 885–889, 2012.
109. B. Raj, A. K. Saxena, and S. Dasgupta, "Nanoscale FinFET Based SRAM Cell Design: Analysis of Performance Metric, Process Variation, Underlapped FinFET and Temperature Effect", *IEEE Circuits Syst. Mag.*, Vol. 11, No. 2, PP. 38–50, 2011.
110. V. K. Sharma, M. Pattanaik, and B. Raj, "Leakage Current ONOFIC Approach for Deep Submicron VLSI Circuit Design", *Int. J. Electr. Comput. Electron. Commun. Eng.,* World Academy of Sciences, Engineering and Technology, Vol. 7, No. 4, PP. 239–244, 2013.
111. T. Chawla, M. Khosla, and B. Raj, "Design and Simulation of Triple Metal Double-Gate Germanium on Insulator Vertical Tunnel Field Effect Transistor", *Microelectron. J.*, Elsevier, Vol. 114, PP. 105125-32, 2021.
112. P. Kaur, S. S. Gill, and B. Raj, "Comparative Analysis of OFETs Materials and Devices for Sensor Applications", *J. Silicon*, Springer, Vol. 14, PP. 4463–4471, 2022.
113. S. K. Sharma, P. Kumar, B. Raj, and Balwinder Raj, "In1-xGaxAs Double Metal Gate-Stacking Cylindrical Nanowire MOSFET for Highly Sensitive Photo Detector", *J. Silicon*, Springer, Vol. 14, PP. 3535–3541, 2022.
114. B. Raj, A. K. Saxena, and S. Dasgupta, "Analytical Modeling of Quasi Planar Nanoscale Double Gate FinFET with Source/Drain Resistance and Field Dependent Carrier Mobility: A Quantum Mechanical Study", *JCP*, Academy Publisher, FINLAND, Vol. 4, No. 9, PP.1-8, 2009.
115. S. Bhushan, S. Khandelwal, and B. Raj, "Analyzing Different Mode FinFET Based Memory Cell at Different Power Supply for Leakage Reduction", *Seventh International Conference on Bio-Inspired Computing: Theories and Application*, (BIC-TA 2012), organized IIITM Gwalor, PP. 89–100, 14-16 Dec 2012.
116. J. Singh, and B. Raj, "Temperature Dependent Analytical Modeling and Simulations of Nanoscale Memristor", *Eng. Sci. Technol. Int. J.*, Elsevier's, Vol. 21, PP. 862–868, 2018.
117. S. Singh, S. Bala, B. Raj, and Balwinder Raj, "Improved Sensitivity of Dielectric Modulated Junctionless Transistor for Nanoscale Biosensor Design", *Sens. Lett.*, ASP, Vol. 18, PP. 328–333, 2020.
118. V. Kumar, S. K. Vishvakarma, and B. Raj, "Design and Performance Analysis of ASIC for IoT Applications", *Sens. Lett.*, ASP, Vol. 18, PP. 31–38, 2020.

119. A. Jaiswal, R. K. Sarin, B. Raj, and S. Sukhija, "A Novel Circular Slotted Microstrip-fed Patch Antenna with Three Triangle Shape Defected Ground Structure for Multiband Applications", *Advanced Electromagnetic (AEM)*, Vol. 7, No. 3, PP. 56–63, 2018.
120. G. Wadhwa, and BRaj, "Label Free Detection of Biomolecules Using Charge Plasma-Based Gate Underlap Dielectric Modulated Junctionless TFET", *JEMS*, Springer, Vol. 47, No. 8, PP. 4683– 4693, 2018.
121. G. Singh, R. K. Sarin, and B. Raj, "Design and Performance Analysis of a New Efficient Coplanar Quantum-Dot Cellular Automata Adder", *Indian J. Pure Appl. Phys.*, Vol. 55, PP. 97–103, 2017.
122. A. Singh, M. Khosla, and B. Raj, "Design and Analysis of Electrostatic Doped Schottky Barrier CNTFET Based Low Power SRAM", *AEU Int. J. Electron. Commun.*, Elsevier, Vol. 80, PP. 67–72, 2017.
123. P. Kaur, V. Pandey, and B. Raj, "Comparative Study of Efficient Design, Control and Monitoring of Solar Power Using IoT", *Sens. Lett.*, ASP, Vol. 18, PP. 419–426, 2020.
124. G. Wadhwa, P. Kamboj, J. Singh, B. Raj, "Design and Investigation of Junctionless DGTFET for Biological Molecule Recognition", *Trans. Electr. Electron. Mater.*, Springer, Vol. 22, PP. 282–289, 2020.
125. T. Chawla, M. Khosla, and B. Raj, "Optimization of Double-Gate Dual Material GeOI-Vertical TFET for VLSI Circuit Design", *IEEE VLSI Circuits and Systems Letter*, Vol. 6, No. 2, PP. 13–25, 2020.
126. S. K.Verma, S. Singh, G. Wadhwa and B. Raj, "Detection of Biomolecules Using Charge-Plasma Based Gate Underlap Dielectric Modulated Dopingless TFET", *Trans. Electr. Electron. Mater. (TEEM)*, Springer, Vol. 21, PP. 528–535, 2020.
127. N. Jain, and B. Raj, "Impact of Underlap Spacer Region Variation on Electrostatic and Analog/RF Performance of Symmetrical High-k SOI FinFET at 20 nm Channel Length", *JoS*, IOP Science, Vol. 38, No. 12, PP. 122002-10, 2017.
128. S. Singh, and B. Raj, "Analytical Modeling and Simulation Analysis of T-Shaped III-V Heterojunction Vertical T-FET", *Superlattices Microstruct.*, Elsevier, Vol. 147, PP. 106717-25, 2020.
129. G. Singh, R. K. Sarin, and B. Raj, "Design and Analysis of Area Efficient QCA Based Reversible Logic Gates", *Microprocess. Microsyst.*, Elsevier, Vol. 52, PP. 59–68, 2017.
130. A. Singh, M. Khosla, and B. Raj, "Compact Model for Ballistic Single Wall CNTFET under Quantum Capacitance Limit", *JoS*, IOP Science, Vol. 37, PP. 104001–104008, 2016.

131. S. Singh, M. Khosla, G. Wadhwa, and B. Raj, "Design and Analysis of DoubleGate Junctionless Vertical TFET for Gas Sensing Applications", *Appl. Phys. A*, Springer, Vol. 127, No. 16, PP.883-890, 2021.
132. I. Singh, B. Raj, M. Khosla, and B. K. Kaushik, "Potential MRAM Technologies for Low Power SoCs", *SPIN World Scientific Publisher*, SCIE; Vol. 10, No. 04, PP. 2050027-35, 2020.

11

Build and Deploy a Smart Speaker with Biometric Authentication and Advanced Voice Interaction Capabilities

Gur Sharan Kant* **and Kavi Bhushan**

SCRIET CCS University, Meerut, Uttar Pradesh, India

Abstract
Advances in the semiconductor era have increased the overall performance and capabilities of chipsets while reducing size and cost. Additionally, advances in AI frameworks and libraries are enabling more AI with less assistance on IoT devices at the edge. Sensors are essential in today's environment, providing seamless data for the development of new smart applications. An example is a smart home situation with many connected devices. In such a place, there is easy and fast access to web services and personal statistics, including calendar, notes, email, reminders, bank and more users connect 0.33 Birthday Celebration talents or skills from the Amazon store to their smart speaker. In addition, smart home devices such as smart security cameras, video doorbells, smart stores, smart carbon monoxide video indicators and smart door locks have come very close to smart homes. With smart speakers that add special requirements. Considering that the smart machine interacts with the products and devices used by the user account, every day people who own a physical device can use it. They have the most advanced technology available for smart speakers via voice commands. In this case, poverty, home security and other new aspects of the consumer suffer. Recently, Tensor Cam brought AI cameras, and Toshiba's Symbio and Facebook's Portal are smart speakers that support digital cameras and AI capabilities. Although they support cameras, they currently do not have an authentication process other than aa003e33992aa1e-42449a037e2560bf2. This article discusses how smart speakers face cybersecurity risks due to the lack of authentication mechanisms, and discusses improvements to the modern 49a2d564f1275e1c4e633abc331547db, digital camera-enabled, microphone array- based Alexa smart speaker prototype.

**Corresponding author*: gskant9319@gmail.com

Balwinder Raj, Suman Lata Tripathi, Tarun Chaudhary, K. Srinivasa Rao and Mandeep Singh (eds.) Integrated Devices for Artificial Intelligence and VLSI, (271–294) © 2024 Scrivener Publishing LLC

***Keywords*:** Alexa voice service, snowboy hotword recognition, smart speaker design, microphone arrangement, ReSpeaker, speech algorithms, open CV, smart speaker authentication

11.1 Introduction

Recent advances in technology (especially the Internet of Things and artificial intelligence) have a huge impact on our daily lives [20, 32, 34]. In a smart home scenario, multiple smart devices connect and work together to achieve a common goal [21, 23, 28]. Smart speakers are one of the smart devices adopted by many users and become an essential part of the smart home. Artificial intelligence helps create new smart speakers that can understand voice-based commands and control the integrated system of the smart home. While voice-based commands provide a simple system to tackle complex processes, it also offers a security feature, placing control of the system in the translation hands of all users who have access to the smart speaker and can issue voice commands. There is an urgent need to implement biometric authentication mechanisms for smart speakers to increase the security of integration without affecting the rich user experience. Due to the lack of reliability of existing voice authentication systems, the best solution is to make the technology more familiar. When someone claims to be the registered user of the smart speaker, it is important to prove that the user is who they claim to be [1, 2].

This may include providing information that the user knows (PIN code or password) to the smart speaker authentication system, or using information that the user owns (physical token) or uses identity (biometric data). Biometric authentication is most convenient because authentication is part of the user and allows hands-free smart speaker authentication. Voice authentication relies on the user's unique voice to analyze the user's voice to identify the person. Voice recognition is ideal for hands-free use of standalone devices such as smartphones, smart speakers and car stereos because it is inexpensive to integrate, known and easy for most users, contactless (contactless) and cleaner. However, its disadvantages are that it is not as accurate as other biometrics [11], requires additional survival checks, and background noise affects speech coherence [15].

Biometric authentication solutions such as Knomi [10] provide a set of biometric matching and life detection algorithms that use face and voice for authentication. Similarly, Hands Free by Sensory [4] uses a biometric fusion algorithm that combines computer vision, noise processing and machine learning to provide face and voice recognition and close-to-equipment identification. SDKs for such multimode authentication systems are designed to develop applications for smartphones and tablets rather than smart speakers due to their low hardware specifications [24, 30, 31]. Recently released Tensor Cam's AI camera, Toshiba's Symbio, and Facebook's Portal are smart talking cameras with AI features [22, 29, 33]. Although they support the camera, they are no more reliable than making phone calls. The modern Alexa smart speakers discussed in this article are designed to use hardware (Raspberry Pi, Re Speaker v2, Raspberry Pi camera, regular speaker). By adding a camera module and introducing facial recognition algorithms, a biometric-based authentication system was developed for the Alexa smart speaker. This face recognition algorithm is trained with a deep neural network to recognize and recognize faces for authentication. It can also recognize and recognize faces when people look at them, so Alexa only wakes up when it recognizes a familiar face. A microphone array with a built-in chip with DSP-based audio algorithms was chosen to capture, process and output audio from Alexa to ensure seamless interaction between users and Alexa. Our proof-of-concept model demonstrates a rich user experience when interacting with a smart speaker by providing an additional layer of recognition and enabling good communication with the device [14].

11.2 Cybersecurity Risk as Smart Speakers Don't Have an Authentication Process

Users wake up the Alexa AI voice assistant by searching for the wake word "Alexa", start interacting with the Alexa smart speaker, and then start speaking normally. Currently, some Alexa devices support a voice profile [5] to provide personalized interaction with support functions. To do this, users must train Alexa with their own voice and then associate the learning voice with the user's Alexa account. But the feature is sound only based on user authentication, not authentication. First, the current voice biometrics is limited to a few Alexa-powered features and cannot be used as a method of voice biometric authentication for all smart speakers.

a. **Attacks by Curious Children:** There is always a risk of a child making audio purchases from the smart speaker without knowing the user's account link.

b. **Naughty Neighbor Stories:** If a neighbor wants to get caught in the crime. It can send commands to smart speakers using ultrasonic frequencies that humans cannot hear but smart speakers can detect.
c. **"This parrot tried to buy food by talking to Alexa"** [12]: A parrot delivered strawberries, light bulbs, water bottles, etc. to the online shopping cart owner bought things successfully. Using needles can avoid these tasks, but it's also possible for parrots to learn and reuse needles.
d. **Speech Problems While Watching TV:** Even just watching TV or listening to the radio can wake you up and affect the smart speaker.
e. **Physical Access:** Anyone near the smart speaker can wake it up, interact with it, and access the user's calendar, reminders, etc. can gather information from related applications
f. **Biometric Overlay Attack** [13]: An attacker can access voice commands that work by playing pre-recorded clips of the victim's voice or emulating the victim's voice [17]. 4 444
g. **Bad Orders:** A person can give bad orders that are interpreted as meaningless by the human ear, but are interpreted as orders given by a wise speaker. Such commands can be added to online videos or TV commercials to turn off the device [8].

Because smart speakers are always listening, they can enable such security with devices that can generate negative messages [35]. TV news voice makes Amazon Echo order for dollhouse [16]. One way to solve these problems is to send a list, email, bank, etc. is to provide voice biometrics-based authentication for third-party applications that we are important to, such as Connect to Alexa.

This can be done by integrating third-party voice biometric APIs such as ArmorVox [7]. In this way, raw audio data captured by smart speakers can be provided to third-party APIs, which may cause privacy and data security issues in the future. This system also does not provide a way to authenticate all smart speakers, which puts the system at risk. Unfortunately,

Most of today's smart speakers do not have an authentication method, so it is often ineffective in solving the above problems. The model developed and described in this article interacts with the Alexa API by providing voice data recorded by an array of microphones. In addition to calling the wake word Alexa, it can also wake up with biometric (face recognition) based technology. In this session, the importance of biometric-based smart speaker authentication was discussed, and in the next session, the development of biometric-enabled smart speaker models will be discussed.

11.3 Related Work

Auth [13] recommends continuous authentication of voice assistants to avoid the threat of smart speakers. VAuth is an integrated system that includes devices such as glasses, headphones and neckbands. The system detects vibrations in the user's body and directs them to commands received by the Assistant's voice microphone. While VAuth is 97% accurate, simply paying, managing and moving the link to the user is not the solution to identify the smart speaker. Daon's IdentityX [6] is a multi-brand, multi-vendor identity services platform that provides additional biometric authentication using smartphones only, while financial services use it via Alexa. This process involves the use of a second device (a smartphone) and not all smart speakers are certified, making Alexa vulnerable to nearby devices when no one is using it (as explained in Chapter 1). When the user sends important commands to the smart speaker, the smart speaker will send music to determine if the user is in the room. According to the company, the Echo Safe system can recognize the user's presence with 93.13 percent accuracy when executing important commands. Echo Safe is the only solution to the voice command attack, not another flaw [37–45].

11.4 Overview of Biometric Authentication and the Voice Algorithm-Based Smart Speaker

The primary goal of this project is to provide a face biometric-based authentication method for all smart speakers [46–53]. For this purpose, the camera module is integrated into the smart speaker module (see Figure 11.1). Smart speakers don't always have an authentication process that opens the door for any nearby speaker to access the user's personal information. Since the model covered in this post has the camera module and is also equipped with Alexa wake-up script as facial recognition, it provides an additional method of recognition. As shown in Figure 11.2, registered users must first look at the camera to confirm the system, then invite Alexa to speak and start a normal conversation with Alexa. Section 4.3 describes the algorithm used to wake up the system when it recognizes a face pattern. The second goal is to capture and send high-quality, high-quality voice feedback to

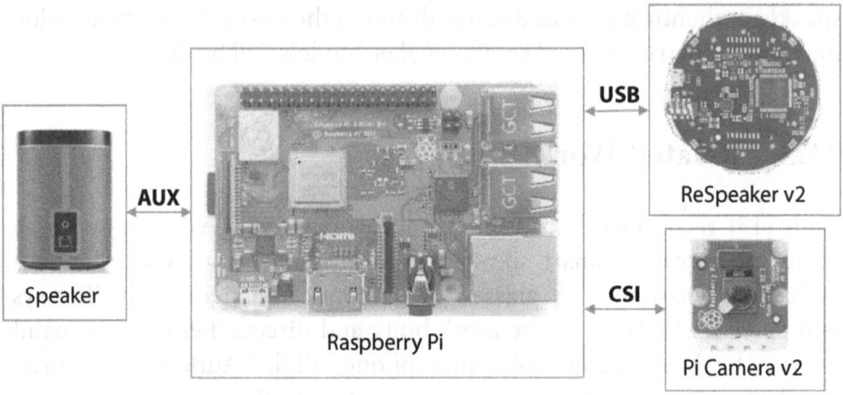

https://www.researchgate.net/figure/Hardware-prototype-of-smart-speaker- system_fig1_338984616

Figure 11.1 Smart speaker system hardware prototype.

Alexa for a full-duplex interaction between the user and Alexa. Instead of a microphone, the Repeater v2 microphone array is used for this because it separates speech from noise. Additionally, the microphone array has a built-in high-performance processor with advanced DSP (digital signal processing) based audio algorithms, allowing users to interact with smart speakers from a distance of approximately 5 meters or more while walking around the room. Chapter 4.2 discusses the role of this microphone array and the benefits of using it to capture, process and send the audio output to Alexa. The third goal is to improve the user experience by ensuring that the smart speaker does not open an error when no motion is spoken and Alexa wake words are detected with voice output. For this purpose, the third-party recommendation engine is integrated with the Alexa Voice Service C++ SDK as described in Section 4.5, as described in Section 4.4.

Hardware Composition of Smart Speaker Prototype

Today's speaker is made using a commercial product with an advanced microphone (ReSpeaker v2) with built-in DSP, a camera module (Raspberry Pi camera) and a speaker regularly connected to a single computer (see Figure 11.2).

a) A Choice Board: Beagle Bone Black, Orange Pi 3, Latte Panda 2G/32G and Banana PI M4 are the SBCs we need.

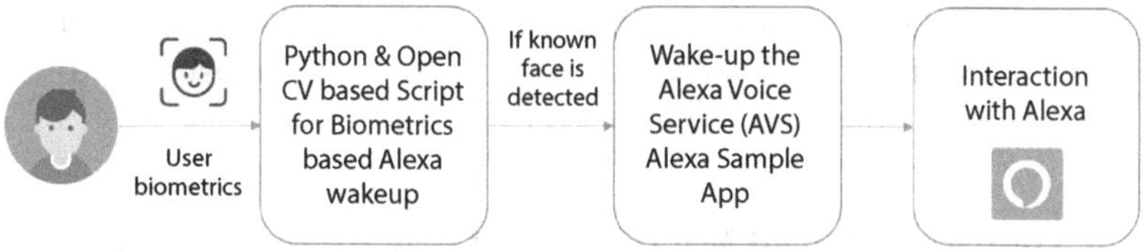

https://www.researchgate.net/figure/High-level-system-diagram-of-Alexa-smart-speaker-prototype-with-biometricsbased-wakeup_fig6_338984616

Figure 11.2 High-level diagram of Alexa smart speaker model with biometric-based stand-up feature.

b) The Raspberry Pi 3 b+ model was chosen for its design, price/performance, low power consumption, compatibility with products, community development tips, advice and support (Pi 4 not yet released). As shown in Figure 11.2, this Raspberry Pi uses a Linux SBC Python script written using an external library. The script used here is responsible for waking the Alexa model app when a face is detected from the live view [54–61].

c) Camera unit selection: Raspberry Pi Camera V2 is the first choice for computer vision using time, due to its 1080p 30fps video encoding and 5MP quality image capture. Because the camera is directly connected to the GPU via the CSI connector (as shown in Figure 11.2), there is little load on the Pi CPU and can be used for other tasks. Most low-cost webcams do not have built-in coding like the Pi camera. As a result, the webcam uses more CPU resources and lowers the overall performance of the system.

d) Microphone array selection: The microphone is an essential part of the smart speaker. Since we need to preprocess the sound with speech algorithms, we focus on microphone arrays with built-in advanced DSP algorithms. ReSpeaker v2, MatrixCreator, PS3 eye, Conexant 4-microphone development kit, Mini DSP UMA- 8, Microsemi Acu Edge ZLK38AVS are the microphone arrays we need. ReSpeaker v2 has great success in searching hot words from long distances. It was augmented and tested in a quiet room, a white noise room, and a room with background music [19]. PS3 Eye has better ReSpeaker v2, but ReSpeaker v2 was chosen for this project because Raspberry Pi camera with CSI interface provides better support for Open CV environment than PS3 Eye camera. The second reason is that ReSpeaker has 12 RGB LED pixel rings that can be used for visual feedback in addition to the speaker [62–66].

Voice Algorithm-Based Microphone Array to Realize Potential for Voice Interaction

The firmware of the XVF-3000 chip (now in ReSpeaker v2 hardware) creates a 6-channel microphone output to the Linux system via USB. Zero channel features audio processing using advanced DSP algorithms. Channels one to four contain raw data from microphones corresponding to each channel number. Channel 5 provides the raw data, which is a combination of all the raw data from ReSpeaker v2's four microphones. Figure 11.3 shows the role of this microphone array. Here, the audio signal from channel 0 is used to generate confirmation messages and also as voice feedback for Alexa. The advantages of using ReSpeaker v2 with Alexa are listed below.

a. Remote audio capture: Stand up and interact with smart speakers by capturing and processing raw microphone data from a distance of five meters or more.
b. USB Audio Class 1.0 (UAC 1.0): USB Audio is used to send music from Raspberry Pi to ReSpeaker v2's built-in digital-to-analog converter (DAC). Class 1.0 can transmit high resolution data up to 24-bit/96 kHz. This bypasses the Raspberry Pi's internal audio and allows the USB DAC to play Alexa's voice in higher quality.
c. Twelve Programmable RGB LED Pixel Rings: ReSpeaker v2's RGB LED pixel ring is used to indicate that the speaker

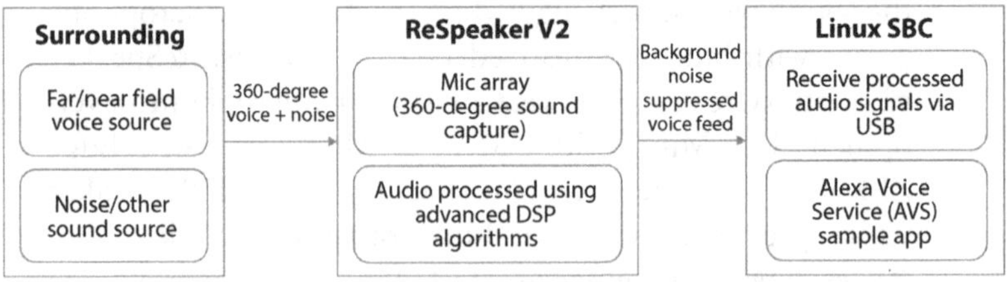

https://www.researchgate.net/figure/Flow-diagram-to-illustrate-microphone-arrays-role-Capture- and-provide-speech-algorithm_fig2_33898461

Figure 11.3 Flow chart illustrating the role of the microphone array: capture and delivery of the voice processed by the speech algorithm to the Alexa Voice Service.

signal (source) is coming. The pixel ring library is used to address the LED pixels through the USB interface and change the color and brightness according to the requirements of the main function.

d. **Digital signal processing algorithms on the ReSpeaker V2:**

   I. Beamforming: All MEMS microphones are omnidirectional. This means that they respond in the same way to sound from surrounding microphones. Arrangements or line layouts can be created by arranging several microphones in an array. In this way, we can capture and follow the position of the smart speaker user's voice in the room. As the smart speaker user interacts with the smart speaker and moves around the room, the angle of the microphone beam adjusts to match the sound. This makes it possible to speak directly in the direction of the user and suppresses noise or reverberation from other directions.

   II. Noise canceling: In acoustic beamforming, active microphone noise cancellation and control is achieved by aligning the microphones in the microphone array. If the direction of the noise relative to the microphone array is known, an acoustic beam former can be designed to pass the signal through the respective noise source and filter the music from other instructions. This microphone array noise reduction method is best for situations where one person's voice needs to be heard while several people are speaking. Noise reduction removes static noise (high pitched sounds) and background noise.

   III. Echo Cancellation: In any room, people's voices are reflected from hard surfaces in the room, such as windows or TV screens. Echo cancellation removes these reflections and clears the speech signal.

   IV. Acoustic Echo Cancellation: Sometimes when dealing with electronic devices

Biometric Authentication with Alexa Wakeup

As appear in Figure 11.4. While running the face map, the face is detected by the dsd frame captured by the Pi digital camera, and the face-embedded

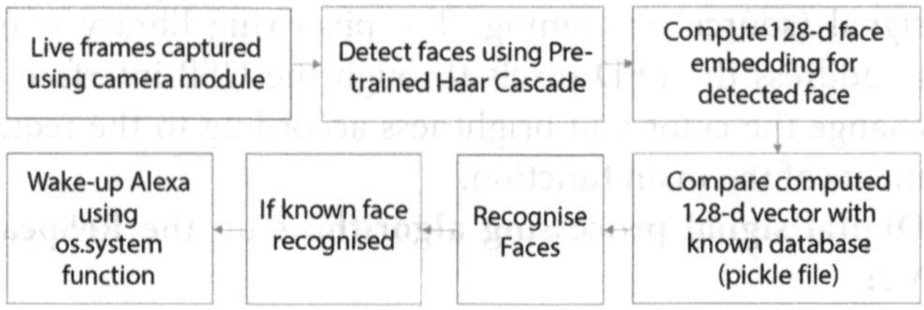

https://www.researchgate.net/figure/Flow-diagram-of-face- recognition-script_fig7_338984616

Figure 11.4 Flow chart of face recognition script.

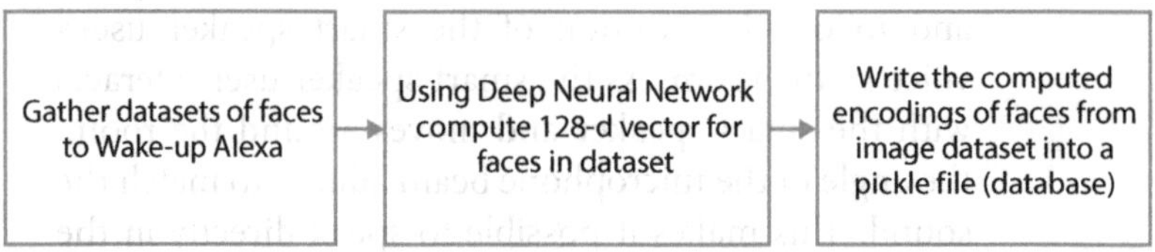

https://www.researchgate.net/figure/Flow-diagram-for- preparing-database-of-faces_fig3_338984616

Figure 11.5 Flow chart for preparing the face database.

128-dimensional face visibility is calculated from the deep metric network. The calculated 128- dimensional face embeds are then compared with the recognized data of the calculated face encodings of the registered face to complement the face from the live frames. The script wakes up Alexa when facial recognition is recognized, while the ReSpeaker's RGB LED pixel ring turns green, providing visual feedback to the person. Before running the face analysis script [67–72]. The subscript shown in Figure 11.5 must be run to encode the 128 dimensional vectors on the faces in the dataset (the directory containing the faces.jpg file) and store the encodings in a .pickle file that will later be used as a database (while making all the important text of the face) identify the face from the live frames and compete try to do. Because the Pi has limited computing power, memory, and GPU, its valuable resources must be freed up for different applications to run. Therefore, many powerful algorithms capable of transmitting data above 10 FPS, such as Eigen faces and LBP (Local Binary Patterns), are now deprecated [73–77].

Alexa Voice Service C++ SDK

At the same time, when procuring smart devices and using scripts, the Pi needs to be registered as a system on the Amazon Developer Console and create a protection profile [84–87]. Follow the step-by-step instructions in [9] to set up the cloud and set the permission process for the config.json created when creating the Alexa AVS sample app from its SDK. Based entirely on the AVS Gadget SDK, this C++ library allows us to integrate Alexa into our smart speaker model [78–83]. The interaction of the smart speaker with the AVS is done using the Alexa model application developed for the Raspberry Pi via the venerable SDK. Python model of Alexa Voice Service application [18] was tested using Raspberry Pi and ReSpeaker v2 before using Alexa AVS C++ SDK.

Following consequences have been observed.

a. After talking to Alexa for a while, Alexa's voice became muffled and muffled.
   The only solution is after restarting the Pi
b. After Alexa raises the word, there is a minimal delay (about 0.5 seconds) before t he voice is sent to the Alexa cloud.

As mentioned in [9] and Figure 11.6, many things join the C++ AVS SDK via the audio stream. Initially, the process uses signal processing algorithms on the input and output audio channels to create clear sound. If raw data from ReSpeaker's 4 microphones is input as input, the third audio signal processor mixes and gives voice action to the next part in the

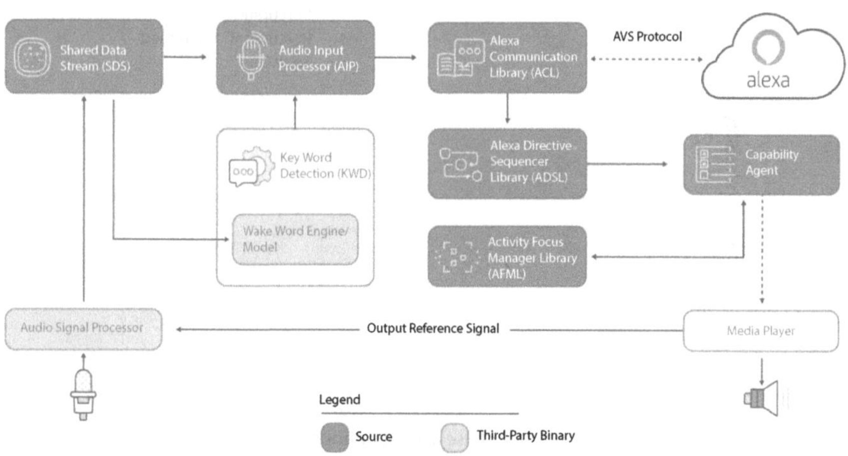

https://www.researchgate.net/figure/Data-flow-between-components-of-AVS-Device-C-SDK- 9_fig4_338984616

Figure 11.6 Data flow of AVS Device C++ SDK components [9].

model [83–88]. But here we saw a single circuit completed with the help of DSP in ReSpeaker v2's XVF-3000 chip. The performance of the closed section of this model is shown in [9] and Figure 11.6. KITT.ai's Snowboy and Sensory [3] are two third-party motion sensors; Alexa raised the word from input to enable mobile interaction. Both engines were tested with all Alexa smart speakers installed as ReSpeaker v2. Snowboy Wake Work Engine was chosen as a plugin for building Alexa AVS mode apps as it uses less than 8% of the Raspberry Pi CPU and is more efficient with search terms [89–93].

Snowboy Upgrades his Vocabulary Diary to Know that Alexa is Raising the Word

The Snowboy machine ensures that the sound is no longer misunderstood when the sound is unknown [94–100]. Wake-up phrase search engine accuracy is measured by plotting the false positive rate per hour (number of false positives) against the ignore detection rate (percentage of false positives) of audio not recognized by the engine. Figure 11.7 shows the ROC curve of the 4-character search engine [101–106]. Here the Snowboy wake word engine has the lowest search cost and higher accuracy than any other engine. The reason for the integration of voice logic with voice-based AI assistance and intelligent voice [107–115].

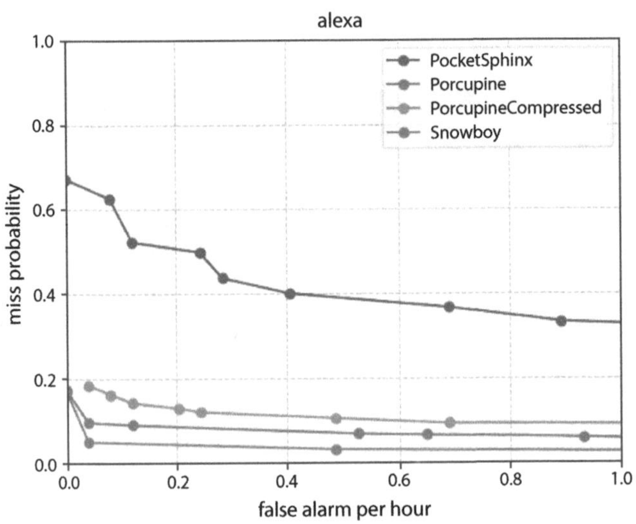

https://www.researchgate.net/figure/ROC-curves-for-popular-wake-word-engines- 2_fig5_338984616

Figure 11.7 Study of miss probability.

a. Privacy: The microphone does not always need to be monitored on
b. Cost: impractical and expensive when statistics are fully transferred to the cloud
c. Power consumption: Voice assistants in smartphones, wearables, and smart audio systems have the longest latency [121–125]

11.5 Conclusion and Discussion

It is estimated that by 2023, more than 50% of the Irish population will be native speakers [36] and by 2021 the world's native English speaker ownership will exceed tablets [36]. Likewise, the smart technology that supports the camera will soon replace ordinary noise and eventually become an important part of our daily lives [116–120]. This article highlights the cybersecurity risks that smart speaker users face due to the lack of authentication processes and discusses the improvements made to Alexa smart speakers today that support cameras and microphone arrays to address these risks. Add to that all the biometrics-based gadget wake-up and microphone array-based interaction [121–126]. Since this smart speaker model is a Linux-based system with a camera, it can achieve special control functions at the request of the user, providing complete audio and video performance. We plan to expand our current work to more applications that require voice commands, such as smart companies (online meetings) and different voice assistants [25–27].

Acknowledgements

This publication has emanated from research supported with the aid of a research grant from Science Foundation Ireland (SFI) underneath Grant Number SFI/16/RC/3918 (Confirm), and SFI/12/RC/2289 P2 (Insight) co-funded through the European Regional Development Fund.

References

1. A guide to the security of voice-activated smart speakers an istr special report, https://www.symantec.com/content/dam/symantec/docs/security-center/white-papers/istr-security-voice-activated-smart-speakers-en.pdf
2. Picovoice benchmark, https://github.com/Picovoice/wakeword-benchmark

3. Home — sensory (2014), https://www.sensory.com/
4. Trulysecure — sensory (2014), https://www.sensory.com/products/technologies/ trulysecure/
5. Amazon.com help: About alexa voice profiles (2019), https://www.amazon.com/ gp/help/customer/display.html?nodeId=202199440
6. Daon — multi-factor mobile biometric authentication — identityx platform - daon(07 2019), https://www.daon.com/
7. Welcome - armorvox (2019), https://cloud.armorvox.com/
8. Alanwar, A., Balaji, B., Tian, Y., Yang, S., Srivastava, M.: Echosafe. Proceedings of the 1st ACM Workshop on the Internet of Safe Things - SafeThings'17 (2017)
9. Alexa: alexa/avs-device-sdk, https://github.com/alexa/avs-device-sdk
10. Biometrics, A.: Knomi™ - mobile biometric authentication framework - aware biometrics (2017), https://www.aware.com/knomi-mobile-biometric-authentication/
11. Biometrics, A.: Voice authentication technology - aware biometrics software (2018), https://www.aware.com/voice-authentication/
12. Charlton, A.: This parrot keeps trying to buy food by speaking to alexa (12 2018). https://www.gearbrain.com/parrot-uses-amazon-alexa-262363 3611.html
13. Feng, H., Fawaz, K., Shin, K.G.: Continuous authentication for voice assistants. *Proceedings of the 23rd Annual International Conference on Mobile Computing and Networking - MobiCom* '17 (2017), https://dl.acm.org/citation.cfm?doid= 3117811.3117823
14. Gebhart, A.: Fooling amazon and google's voice recognition isn't hard (11 2017), https://www.cnet.com/news/fooling-amazon-and-googles-voice-recognition-isnt-hard/
15. Kenarsari, A.: Yet another wake-word detection engine (04 2018), https://medium.com/@alirezakenarsarianhari/yet-another-wake-word-detection-engine-a2486d36d8d4
16. Liptak, A.: Amazon's alexa started ordering people dollhouses after hearing its name on tv (01 2017), https://www.theverge.com/2017/1/7/14200210/amazon-alexa-tech-news-anchor-order-dollhouse
17. Panjwani, S., Prakash, A.: Crowdsourcing attacks on biometric systems, https://www.usenix.org/system/files/conference/soups2014/soups14-paper-panjwani.pdf
18. Respeaker: Alexa, https://github.com/respeaker/Alexa
19. Rouchon, C.: Benchmarking microphone arrays: Respeaker, conexant, microsemi acuedge, matrix creator, minidsp. . . (08 2017), https://medium.com/snips-ai/benchmarking-microphone-arrays-respeaker-conexant-microsemi-acuedge-matrix-creator-minidsp-950de8876fda

20. Sudharsan, B., Breslin, J.G., Ali, M.I.: Adaptive strategy to improve the quality of a. communication for iot edge devices. In: *2020 IEEE 6th World Forum on Internet of Things (WF-IoT)*. pp. 1–6. IEEE (2020).
21. Sudharsan, B., Breslin, J.G., Ali, M.I.: Edge2train: A framework to train machine learning models (svms) on resource-constrained iot edge devices. In: *Proceedings of the 10th International Conference on the Internet of Things*. pp. 1–8 (2020).
22. Zhao, W.; Chellappa, R.; Phillips, P.J.; Rosenfeld, A. Face recognition: A literature survey. *ACM Comput. Surv. (CSUR)*, **2003**, i. *35*, 399–458. [CrossRef]
23. Maltoni, D.; Maio, D.; Jain, A.K.; Prabhakar, S. *Handbook of Fingerprint Recognition*; Springer Science & Business Media: Berlin/Heidelberg, Germany, 2009.
24. Daugman, J. New methods in iris recognition. *IEEE Trans. Syst. Man Cybern. Part B (Cybernetics)*, **2007**, *37*, 1167–1175. [CrossRef]
25. Lu, G.; Zhang, D.; Wang, K. Palmprint recognition using eigenpalms features. *Pattern Recognit. Lett.*, **2003**, *24*, 1463–1467. [CrossRef]
26. Chen, H.; Bhanu, B. Contour matching for 3D ear recognition. In *Proceedings of the 2005 Seventh IEEE Workshops on Applications of Computer Vision (WACV/MOTION'05)*, Breckenridge, CO, USA, 5–7 January 2005; Volume 1, pp. 123–128.
27. Ganapathi, I.I.; Ali, S.S.; Vu, N.S.; Prakash, S.; Werghi, N. A Survey of 3D Ear Recognition Techniques. *ACM Comput. Surv. (CSUR)*, **2022**. [CrossRef]
28. Nejati, H.; Zhang, L.; Sim, T.; Martinez-Marroquin, E.; Dong, G. Wonder ears: Identification of Identical Twins from Ear Images. In *Proceedings of the 21st International Conference on Pattern Recognition (ICPR2012)*, Tsukuba, Japan, 11–15 November 2012; pp. 1201–1204.
29. Sharkas, M. Ear recognition with ensemble classifiers; A deep learning approach. *Multimed. Tools Appl.*, **2022**, 1–27. [CrossRef]
30. Zhu, Q.; Mu, Z. PointNet++ and Three Layers of Features Fusion for Occlusion Three- Dimensional Ear Recognition Based on One Sample per Person. *Symmetry*, **2020**, *12*, 78. [CrossRef]
31. Ramos-Cooper, S.; Gomez-Nieto, E.; Camara-Chavez, G. VGGFace-Ear: An Extended Dataset for Unconstrained Ear Recognition. *Sensors*, **2022**, *22*, 1752. [CrossRef] [PubMed]
32. Lei, Y.; Qian, J.; Pan, D.; Xu, T. Research on Small Sample Dynamic Human Ear Recognition Based on Deep Learning. *Sensors*, **2022**, *22*, 1718. [CrossRef]
33. Lei, J.; Zhou, J.; Abdel-Mottaleb, M. Gender classification using automa-ti-cally detected and aligned 3D ear range data. In *Proceedings of the 2013 International Conference on Biometrics (ICB)*, Madrid, Spain, 4–7 June 2013; pp. 1–7. [CrossRef]
34. Chen, H.; Bhanu, B.; Wang, R. Performance evaluation and prediction for 3D ear recognition. In *International Conference on Audio- and Video-Based Biometric Person Authentication*; Springer: Berlin/Heidelberg, Germany, 2005; 288, pp. 748–757.

35. Chen, H.; Bhanu, B. Efficient recognition of highly similar 3D objects in range images. *IEEE Trans. Pattern Anal. Mach. Intell.*, **2008**, *31*, 172–179. [CrossRef]
36. Chen, H.; Bhanu, B. Human ear recognition in 3D. *IEEE Trans. Pattern Anal. Mach. Intell.*, **2007**, *29*, 718–737. [CrossRef]
37. Chen, L.; Mu, Z.; Nan, B.; Zhang, Y.; Yang, R. TDSIFT: A new descriptor for 2D and 3D ear recognition. In *Proceedings of the Eighth International Conference on Graphic and Image Processing (ICGIP 2016), International Society for Optics and Photonics*, Tokyo, Japan, 29–31 October 2016; Volume 10225, p. 102250C.
38. Ganapathi, I.I.; Prakash, S.; Dave, I.R.; Joshi, P.; Ali, S.S.; Shrivastava, A.M. Ear recognition in 3D using 2D curvilinear features. *IET Biom.*, **2018**, *7*, 519–529. [CrossRef]
39. Ganapathi, I.I.; Prakash, S.; Ali, S.S. Secure Multimodal Access with 2D and 3D Ears. In *Machine Learning for Intelligent Multimedia Analytics*; Springer: Singapore, 2021; pp. 1–20.
40. Ganapathi, I.I.; Ali, S.S.; Prakash, S. Geometric statistics-based descriptor for 3D ear recognition. *Vis. Comput.*, **2020**, *36*, 161–173. [CrossRef]
41. Ganapathi, I.I.; Ali, S.S.; Prakash, S. Multi-resolution Local Descriptor for 3D Ear Recognition. In *Proceedings of the BIOSIG 2019—18th International Conference of the Biometrics Special Interest Group*, Gesellschaft für Informatik eV, Darmstadt, Germany, 18–20 September 2019.
42. Islam, S.M.; Davies, R.; Bennamoun, M.; Owens, R.A.; Mian, A.S. Multibiometric human recognition using 3D ear and face features. *Pattern Recognit.*, **2013**, *46*, 613–627. [CrossRef]
43. Islam, S.M.; Davies, R.; Bennamoun, M.; Mian, A.S. Efficient detection and recognition of 3D ears. *Int. J. Comput. Vis.*, **2011**, *95*, 52–73. [CrossRef]
44. Prakash, S.; Gupta, P. Human recognition using 3D ear images. *Neurocomputing*, **2014**, *140*, 317–325. [CrossRef]
45. Sun, X.; Wang, G.; Wang, L.; Sun, H.; Wei, X. 3D ear recognition using local salience and principal manifold. *Graph. Model.*, **2014**, *76*, 402–412. [CrossRef]
46. Sun, X.P.; Li, S.H.; Han, F.; Wei, X.P. 3D ear shape matching using joint α-entropy. *J. Comput. Sci. Technol.*, **2015**, *30*, 565–577. [CrossRef]
47. Maity, S.; Abdel-Mottaleb, M. 3D ear segmentation and classification through indexing. *IEEE Trans. Inf. Forensics Secur.*, **2014**, *10*, 423–435. [CrossRef]
48. Zeng, H.; Dong, J.Y.; Mu, Z.C.; Guo, Y. Ear recognition based on 3D keypoint matching. In *Proceedings of the IEEE 10th International Conference on Signal Processing Proceedings*, Beijing, China, 24–28 October 2010; pp. 1694–1697.
49. Zeng, H.; Zhang, R.; Mu, Z.; Wang, X. Local feature descriptor based rapid 3D ear recognition. In *Proceedings of the 33rd Chinese Control Conference*, Nanjing, China, 28–30 July 2014; pp. 4942–4945.
50. Zhang, Y.; Mu, Z.; Yuan, L.; Zeng, H.; Chen, L. 3D ear normalization and recognition based on local surface variation. *Appl. Sci.*, **2017**, *7*, 104. [CrossRef]

51. Cadavid, S.; Abdel-Mottaleb, M. Human identification based on 3D ear models. In *Proceedings of the 2007 First IEEE International Conference on Biometrics: Theory, Applications, and Systems*, Crystal City, DC, USA, 27–29 September 2007; pp. 1–6.
52. Cadavid, S.; Abdel-Mottaleb, M. 3-D ear modeling and recognition from video sequences using shape from shading. *IEEE Trans. Inf. Forensics Secur.*, **2008**, *3*, 709–718. [CrossRef]
53. Mahoor, M.H.; Cadavid, S.; Abdel-Mottaleb, M. Multi-modal ear and face modeling and recognition. In *Proceedings of the 2009 16th IEEE International Conference on Image Processing (ICIP)*, Cairo, Egypt, 7–10 November 2009; pp. 4137–4140.
54. Passalis, G.; Kakadiaris, I.A.; Theoharis, T.; Toderici, G.; Papaioannou, T. Towards fast 3D ear recognition for real-life biometric applications. In *Proceedings of the 2007 IEEE Conference on Advanced Video and Signal Based Surveillance*, Genoa, Italy, 5–7 September 2007; pp. 39–44.
55. Theoharis, T.; Passalis, G.; Toderici, G.; Kakadiaris, I.A. Unified 3D face and ear recognition using wavelets on geometry images. *Pattern Recognit.*, **2008**, *41*, 796–804. [CrossRef]
56. De Tré, G.; De Mol, R.; Vandermeulen, D.; Claes, P.; Hermans, J.; Nielandt, J. Human centric recognition of 3D ear models. *Int. J. Comput. Intell. Syst.*, **2016**, *9*, 296–310. [CrossRef]
57. De Tré, G.; Vandermeulen, D.; Hermans, J.; Claeys, P.; Nielandt, J.; Bronselaer, A. Bipolar comparison of 3D ear models. In *International Conference on Information Processing and Management of Uncertainty in Knowledge-Based Systems*; Springer: Cham, Switzerland, 2014; pp. 160–169.
58. Mursalin, M.; Islam, S.M.S. Deep Learning for 3D Ear Detection: A Complete Pipeline from Data Generation to Segmentation. *IEEE Access*, **2021**, *9*, 164976–164985. [CrossRef]
59. Qi, C.R.; Yi, L.; Su, H.; Guibas, L.J. PointNet++: Deep Hierarchical Feature Learning on Point Sets in a Metric Space. In *Proceedings of the Advances in Neural Information Processing systems*, Long Beach, CA, USA, 4–9 December 2017; pp. 5099–5108.
60. Kingma, D.P.; Ba, J. Adam: A method for Stochastic Optimization. *arXiv*, **2014**, arXiv:1412.6980.
61. Mursalin, M.; Islam, S.M.S. EpNet: A Deep Neural Network for Ear Detection in 3D Point Clouds. In *International Conference on Advanced Concepts for Intelligent Vision Systems*; Springer: Cham, Switzerland, 2020; pp. 15–26.
62. Mursalin, M.; Islam, S.M.S.; Gilani, S.Z. 3D Morphable Ear Model: A Complete Pipeline from Ear Segmentation to Statistical Modeling. In *Proceedings of the 2021 Digital Image Computing: Techniques and Applications (DICTA)*, Gold Coast, Australia, 29 November–1 December 2021; pp. 1–6.
63. Amberg, B.; Romdhani, S.; Vetter, T. Optimal Step Nonrigid ICP Algorithms for Surface Registration. In *Proceedings of the 2007 IEEE Conference on*

Computer Vision and Pattern Recognition, Minneapolis, MN, USA, 17–22 June 2007; pp. 1–8. [CrossRef]

64. Gilani, S.Z.; Mian, A.; Shafait, F.; Reid, I. Dense 3D face correspondence. *IEEE Trans. Pattern Anal. Mach. Intell.*, **2017**, *40*, 1584–1598. [CrossRef] [PubMed]
65. Yan, P.; Bowyer, K.W. Biometric recognition using 3D ear shape. *IEEE Trans. Pattern Anal. Mach. Intell.*, **2007**, *29*, 1297–1308. [CrossRef] [PubMed]
66. Zhou, J.; Cadavid, S.; Abdel-Mottaleb, M. An efficient 3-D ear recognition system employing local and holistic features. *IEEE Trans. Inf. Forensics Secur.*, **2012**, *7*, 978–991. [CrossRef]
67. Anjana Bhardwaj, Pradeep Kumar, B. Raj, Sunny Anand, "Design and Performance Optimization of doping-less Vertical Nanowire TFET using Gate Stack Technique", *J. Electron. Mater. (JEMS)*, Springer, Vol. 41, Issuse 7, PP. 4005-4013, 2022.
68. Jeetendra Singh, B Raj, "Tunnel Current Model of Asymmetric MIM Structure Levying Various Image Forces to Analyze the Characteristics of Filamentary Memristor", *Appl. Phys. A*, Springer, Vol. 125, No.3, pp. 203.1 to 203.11 Feb 2019.
69. Candy Goyal, Jagpal Singh Ubhi and B Raj, Low Leakage Zero Ground Noise Nanoscale Full Adder using Source Biasing Technique, *J. Nanoelectron. Optoelectron.*, American Scientific Publishers, Vol. 14, PP. 360–370, March 2019.
70. Gurmohan Singh, R. K. Sarin and B Raj, "A Novel Robust Exclusive-OR Function Implementation in QCA Nanotechnology with Energy Dissipation Analysis", *J. Comput. Electron.*, Springer, Volume 15, Issue 2, PP. 455-465, June 2016.
71. Tanu Wadhera, Deepti Kakkar, Girish wadhwa, B Raj, "Recent Advances and Progress in Development of the Field Effect Transistor Biosensor: A Review" *J. Electron. Mater.*, Springer, Volume 48, Issue 12, pp 7635–7646, December 2019.
72. Girish Wadhwa, Priyanka Kamboj, Balwinder Raj" Design optimisation of junctionless TFET biosensor for high sensitivity" *Adv. Nat. Sci.: Nanosci. Nanotechnol.*, Vol. 10, PP. 045001 (7), (2019).
73. Priya Bansal, B Raj, "Memristor modeling and analysis for linear dopant drift kinetics," *J. Nanoeng. Nanomanufactur.*, American Scientific Publishers, Vol. 6, pp. 1–7, 2016.
74. Amandeep Singh, Mamta Khosla, B Raj, "Circuit Compatible Model for Electrostatic Doped Schottky Barrier CNTFET," *J. Electron. Mater.*, Springer, Vol. 45, Issue 12, pp 4825-4835, 2016.
75. Ashima, D Vaithiyanathan, B Raj," Performance Analysis of Charge Plasma induced Graded Channel Si nanotube", *J. Eng. Res. (JER)*, EMSME Special Issue, Vol.9, pp. 146-154 Aug 2021.

76. Abhishek Singh Tomar, Vijay Kumar Magraiya, and B Raj, Scaling of Access and Data Transistor for High Performance DRAM Cell Design, *Quantum Matter*, Vol. 2, PP. 412-416 Oct. 2013.
77. Neeraj Jain, B Raj, "Parasitic Capacitance and Resistance Model Development and Optimization of Raised Source/Drain SOI FinFET Structure for Analog Circuit Applications, *J. Nanoelectron. Optoelectron.*, ASP, USA, Vol. 13, PP. 531–539, Apl 2018.
78. Shradhya Singh, S. K. Vishvakarma and B Raj, "Analytical Modeling of SplitGate Junction-Less Transistor for a Biosensor Application", *Sens. Biosens.*, Elsevier, Vol. 18, Pages 31-36, April 2018.
79. Maisagalla Gopal, and Balwinder Raj, Low Power 8T SRAM Cell Design for High Stability Video Applications, *ITSI Trans. Electr. Electron. Eng.*, Vol. 1(5), pp. 91-97, 2013.
80. Balwinder Raj, Jatin Mitra, Deepak Kumar Bihani, Rangharajan V, A.K. Saxena and S. Dasgupta, "Analysis of Noise Margin, Power and Process Variation for 32 nm FinFET Based 6T SRAM Cell" *J. Comput. (JCP)*, Academy Publisher, FINLAND, vol. 5, issue 6, PP.1-8, 2010.
81. Divya Sharma, Rajesh Mehra and B Raj, "Comparative Analysis of Photovoltaic Technologies for High Efficiency Solar Cell Design", *Superlattices Microstruct.*, Elsevier, Volume 153, PP. 106861, May 2021.
82. Pawandeep Kaur, Avtar Singh Buttar, B Raj, "A comprehensive Analysis of Nanoscale Transistor based Biosensor: A Review, *Indian J. Pure Appl. Phys.*, Vol. 59, pp. 304-318, April 2021.
83. Divya Yadav, Balwant Raj and B Raj, "Design and Simulation of Low Power Microcontroller for IoT Applications", *J. Sens. Lett.*, ASP, Vol. 18, PP. 401–409, May 2020.
84. Shailendra Singh, B Raj,"A 2-D Analytical Surface Potential and Drain current Modeling of Double-Gate Vertical t-shaped Tunnel FET", *J. Comput. Electron.*, Springer, Vol. 19, PP.1154–1163, Apl 2020.
85. Jeetendra Singh, B Raj, "An Accurate and Generic Window function for Nonlinear Memristor Model" *J. Comput. Electron.*, Springer, Volume 18 Issue 2, PP. 640-647 June 2019.
86. Manjit Kaur; Neena Gupta; Sanjeev Kumar; B Raj; Arun Kumar Singh, "Comparative RF and Crosstalk Analysis of Carbon Based Nano Interconnects" *IET Circuits Devices Syst.*, Vol. 15(6), PP. 493-503, Feb 2021.
87. Nehru Kandasamy, Firdous Ahmad, D. Ajitha, B Raj, and Nagarjuna Telagam "Quantum Dot Cellular Automata based Scan Flip Flop and Boundary Scan Register, *IETE J. Res.*, Volume 66, PP. 535-548, 2020.
88. Sharma, S. K.., Raj, B., Khosla, M., "Enhanced Photosensivity of Highly Spectrum Selective Cylindrical Gate In1-xGaxAs Nanowire MOSFET Photodetector, *Mod. Phys. Lett. B*, Vol. 33, No. 12, PP. 1950144 (2019).
89. Jeetendra Singh, B Raj, "Design and Investigation of 7T2M NVSARM with Enhanced Stability and Temperature Impact on Store/Restore Energy", *IEEE Trans. Very Large Scale Integr. Syst.*, Vol. 27, Issue 6, PP. 1322-1328 June 2019.

90. Anil Kumar Bhardwaj, Sumeet Gupta, B Raj, Amandeep Singh, "Impact of Double Gate Geometry on the Performance of Carbon Nanotube Field Effect Transistor Structures for Low Power Digital Design", *Comput. Theor. Nanosci.*, ASP, Vol. 16, PP. 1813–1820, 2019.
91. Neeraj Jain, B Raj, "Thermal Stability Analysis and Performance Exploration of Asymmetrical Dual-k underlap Spacer (ADKUS) SOI FinFET for Security and Privacy Applications", *Indian J. Pure Appl. Phys. (IJPAP)*, Vol. 57, pp. 352-360, May 2019.
92. Amandeep Singh, Mamta Khosla, B Raj, "Design and Analysis of Dynamically Configurable Electrostatic Doped Carbon Nanotube Tunnel FET", *Microelectronics J.*, Elsevier, Volume 85, Pages 17-24, March 2019.
93. Neeraj Jain and Balwinder Raj, "Dual-k Spacer Region Variation at the Drain Side of Asymmetric SOI FinFET Structure: Performance Analysis towards the Analog/RF Design Applications", communicated in "*J. Nanoelectron. Optoelectron.*", American Scientific Publishers, Vol. 14, PP. 349–359, March 2019.
94. Jeetendra Singh, Sanjeev Sharma, B Raj, Mamta Khosla Analysis of barrier layer thickness on performance of In1-xGaxAs based Gate Stack Cylindrical Gate Nanowire MOSFET" *JNO*, ASP, Vol. 13, PP 1473-1477, Oct 2018.
95. Neeraj Jain, B Raj, "Analysis and Performance Exploration of High-k SOI FinFETs Over the Conventional Low-k SOI FinFET toward Analog/RF Design", *Journal of Semiconductors (JoS)*, IOP Science, Vol.39, Issue 12, PP- 124002-1-7, Dec 2018.
96. Candy Goyal, Jagpal Singh Ubhi and B Raj, A reliable leakage reduction technique for approximate full adder with reduced ground bounce noise, *J. Math. Prob. Eng.*, Hindawi, Volume 2018, Article ID 3501041, 16 pages, 15 Oct 2018.
97. Anuradha, Jeetendra Singh, B Raj and Mamta Khosla, "Design and Performance Analysis of Nano-scale Memristor-based Nonvolatile SRAM" *J. Sens. Lett.*", American Scientific Publishers, Vol. 16, PP 798–805, Oct 2018.
98. Girish Wadhwa, B Raj, "Parametric Variation Analysis of Charge-Plasma-based Dielectric Modulated JLTFET for Biosensor Application" *IEEE Sens. J.*, Vol. 18, no. 15, PP. 6070–6077, 2018.
99. Jeetendra Singh, B Raj, "Comparative Analysis of Memristor Models for Memories Design", *JoS*, IoP, Vol. 39, No. 7, PP. 074006-1-12, July 2018.
100. Divya Yadav, Shailesh Singh Chouhan, Santosh Kumar Vishvakarma and B Raj, "Application Specific Microcontroller Design for IoT based WSN", *Sens. Lett.*, ASP, Vol. 16, PP. 374–385, May 2018.
101. Gurmohan Singh, R. K. Sarin and B Raj, "Fault-Tolerant Design and Analysis of Quantum-Dot Cellular Automata Based Circuits", *IEEE/IET Circuits, Devices Syst.*, Vol. 12, PP. 638–64, 2018.
102. Jeetendra Singh, B Raj, "Modeling of Mean Barrier Height Levying Various Image Forces of Metal Insulator Metal Structure to Enhance the Performance

of Conductive Filament Based Memristor Model", *IEEE Nanatechnol.*, Vol. 17, No. 2, PP. 268-267, March 2018.

103. Aakash Jain, Sanjeev Sharma, B Raj, "Analysis of Triple Metal Surrounding Gate (TM-SG) III-V Nanowire MOSFET for Photosensing Application", *Optoelectronics Journal*, Elsevier, Volume 26, Issue 2, Pages 141-148, May 2018.
104. Aakash Jain, Sanjeev Sharma, B Raj, "Design and Analysis of High Sensitivity Photosensor Using Cylindrical Surrounding Gate MOSFET For Low Power Sensor Applications", *Eng. Sci. Technol. Int. J.*, Elsevier's, Volume 19, Issue 4, Pages 1864–1870, December 2016.
105. Amandeep Singh, Mamta Khosla, B Raj, "Analysis of Electrostatic Doped Schottky Barrier Carbon Nanotube FET for Low Power Applications," *J. Mater. Sci.: Mater. Electron.*, Springer, Vol.28, PP. 1762–1768, 2017.
106. G. Saiphani Kumar, Amandeep Singh, B Raj, "Design and Analysis of Gate All Around CNTFET based SRAM cell Design", *J. Comput. Electron.*, Springer, Volume 17, Issue 1, pp 138–145, March 2018.
107. Gurinder pal Singh, B. S. Sohi, Balwinder Raj, "Material Properties Analysis of Graphene Base Transistor (GBT) for VLSI Analog Circuits, *Indian J. Pure Appl. Phys. (IJPAP)*, Vol. 55, pp. 896-902, Dec-2017.
108. Amandeep Singh, Mamta Khosla, B Raj, "Comparative Analysis of Carbon Nanotube Field Effect Transistor and Nanowire Transistor for Low Power Circuit Design," *J. Nanoelectron. Optoelectron.*, American Scientific Publishers, USA, Vol. 11, PP. 388-393, June 2016.
109. Sunil Kumar and B Raj, "Estimation of Stability and Performance metric for Inward Access Transistor based 6T SRAM Cell Design using n-type/p-type DMDG-GDOV TFET" *IEEE VLSI Circuits and Systems Letter*, Volume 3, Issue 2, PP 25-39, June2017
110. Shashikant Sharma, Anjan Kumar, Manisha Pattanaik and B Raj, "Forward Body Biased Multimode Multi-Threshold CMOS Technique for Ground Bounce Noise Reduction in Static CMOS Adders" *Int. J. Inf. Electron. Eng.*, pp. 567-572, Vol. 3, issue 3, 2013.
111. Hamendra Singh, Pankaj Kumar, Balwinder Raj, "Performance Analysis of Majority Gate SET Based 1-bit Full Adder", *International Journal of Computer and Communication Engineering (IJCCE)*, IACSIT Press Singapore, ISSN: 2010-3743, Volume 2, No. 4, PP.1-8, 2013.
112. Anil Kumar Bhardwaj, Sumeet Gupta and B Raj, "Investigation of Parameters for Schottky Barrier (SB) Height for Schottky Barrier Based Carbon Nanotube Field Effect Transistor Device", *J. Nanoelectron. Optoelectron.*, ASP, Vol. 15, PP. 783–791, July 2020.
113. Priya Bansal, B Raj, "Memristor: A versatile nonlinear model for dopant drift and boundary issues," *JCTN*, American Scientific Publishers, Vol. 14 (5), PP. 2319-2325, May 2017.
114. Neeraj Jain, B Raj, "An Analog and Digital Design Perspective Comprehensive Approach on Fin-FET (Fin-Field Effect transistor) Technology - A Review",

Reviews in Advanced Sciences and Engineering (RASE), ASP, Vol. 5, PP. 1-14, 2016.

115. Sanjeev Sharma, B Raj, Mamta Khosla, "Subthreshold Performance of In1-xGaxAs based Dual Metal with Gate Stack Cylindrical/Surrounding Gate Nanowire MOSFET for Low Power Analog Applications, *J. Nanoelectron. Optoelectron.*, American Scientific Publishers, USA, Vol.12, PP. 171–176, 2017.
116. Balwinder Raj, A. K. Saxena and S. Dasgupta, "Analytical Modeling for the Estimation of Leakage Current and Subthreshold Swing Factor of Nanoscale Double Gate FinFET Device" *Microelectron. Int.*, UK, vol. 26, pp. 53-63, 2009.
117. Soniya, Shailendra Singh, Girish Wadhwa and Balwinder Raj, "An Analytical Modeling for Dual Source Vertical Tunnel Field Effect Transistor" submitted to '*Int. J. Recent Technol. Eng. (IJRTE)*' at Volume-8 Issue-2, PP. 603-608, July 2019.
118. Shailendra Singh, B Raj, "Design and analysis of hetrojunction Vertical T-shaped Tunnel Field Effect Transistor", *J. Electron. Mater.*, Springer, Volume 48, Issue 10, pp 6253–6260, October 2019.
119. Candy Goyal, Jagpal Singh Ubhi and B Raj, "A Low Leakage CNTFET based Inexact Full Adder for Low Power Image Processing Applications", *Int. J. Circuit Theory Appl.*, Wiley, Volume 47, Issue 9, Pages 1446-1458, September 2019.
120. B Raj, A. K. Saxena and S. Dasgupta, "A Compact Drain Current and Threshold Voltage Quantum Mechanical Analytical Modeling for FinFETs" *J. Nanoelectron. Optoelectron.*, USA, vol. 3, no. 2, pp. 163-170, 2008.
121. Girish Wadhwa, B Raj, "An Analytical Modeling of Charge Plasma based Tunnel Field Effect Transistor with Impacts of Gate underlap Region" *Superlattices Microstruct.*, Elsevier, Vol. 142, PP. 106512, June 2020.
122. Shailendra Singh, B Raj,"Modeling and Simulation analysis of SiGe hetrojunction Double GateVertical t-shaped Tunnel FET", *Superlattices Microstruct.*, Elsevier Volume 142, PP. 106496, June 2020.
123. Amandeep Singh, Dinesh Kumar Saini, Dinesh Agarwal, Sajal Aggarwal, Mamta Khosla, B Raj, "Modeling and Simulation of Carbon Nanotube Field Effect Transistor and its Circuit Application," *J. Semiconduct. (JoS)*, IOP Science, Vol.37, PP. 074001-6, July 2016.
124. Neeraj Jain, Balwinder Raj, "Device and Circuit Co-Design Perspective Comprehensive Approach on FinFET Technology - A Review, *J. Electron Devices*, Vol. 23, Num. 1, pp 1890-1901, 2016.

125. Sunil Kumar and B Raj, "Analysis of I_{ON} and Ambipolar Current for DualMaterial Gate-drain Overlapped DG-TFET," *J. Nanoelectron. Optoelectron.*, American Scientific Publishers, USA, Vol.11, PP. 323-333, June 2016.
126. Naveed Anjum, Tarun Bali and B Raj, Design and Simulation of Handwritten Multiscript Character Recognition, *IJARCCE*, Vol.2 (7), PP. 2544-2549, July 2013.

12

Boron-Based Nanomaterials for Intelligent Drug Delivery Using Computer-Aided Tools

Jupinder Kaur[1]*, Ravinder Kumar[2] and Rajan Vohra[3]

[1]*Vishwakarma University, Pune, India*
[2]*Guru Nanak Dev University, Amritsar, India*
[3]*Lovely Professional University, Phagwara, Punjab, India*

Abstract

Recently, nanostructure-based drug delivery systems have become the center of attraction for researchers. In the last decade, various studies have revealed that fullerenes can be effectively utilized as carriers to transport various drugs to the affected cells. Thus, the potential application of borospherene (B_{40}) as a carrier for delivery of anti-cancer drug, 5-Fluorouracil, to prostate cancer cells is investigated using the density functional theory. To achieve targeted drug delivery, anisamide molecule is attached as ligand to the nano-carrier. It is deduced that the drug prefers to interact with the boron fullerene via its fluorine and oxygen atoms. The interaction of the drug and ligand at both the 6- and 7-membered rings of the borospherene is considered. The complex with ligand attached at the heptagonal ring and drug at the hexagonal ring is found to be the most stable with adsorption energy of -0.76eV. In order to get more insight, electronic properties viz. Density of states and Eigen states are also calculated for the most stable complex. This complex is found to be stable even in the aqueous medium. A pH stimulated drug release phenomenon is also proposed based on the knowledge that the cancer cells have lesser pH in comparison to the healthy cells. Thus, this study reveals that borospherene can be efficiently utilized as a potential nano-carrier for delivery of 5-Fluorouracil drug to prostate cancer cells.

Keywords: Borospherene (B_{40}), drug delivery, 5-fluorouracil, prostate cancer cells, DFT

**Corresponding author*: jupinder.kaur1993@gmail.com

Balwinder Raj, Suman Lata Tripathi, Tarun Chaudhary, K. Srinivasa Rao and Mandeep Singh (eds.) Integrated Devices for Artificial Intelligence and VLSI, (295–328) © 2024 Scrivener Publishing LLC

12.1 Introduction

Nanomaterials have been successfully utilized in numerous applications as they demonstrate excellent structural and electronic properties [1–6]. One of the most important applications of nanomaterials is their use as carriers for drug delivery to affected cells without any negative impact on the healthy cells [7–14]. Recently, fullerenes have been exploited as drug carriers for a number of drugs as they have remarkable structural properties [15–21]. Buckminsterfullerene is a popular material that has been investigated to deliver drugs like warfarin, ibuprofen, favipiravir, etc. [22–26] as it has excellent structural properties and is readily available. Although a lot of research has been done related to utilization of carbon fullerenes as nano-carriers in drug delivery systems, yet boron fullerenes still remain unexplored. So, in the present research work, the possibility of utilizing borospherene (B_{40}) as carrier in drug delivery has been exploited. B_{40} was synthesized by Zhai *et al.* in 2014 [27]. It is a very stable molecule comprising of hexagonal and heptagonal rings on its surface. Under non-equilibrium conditions, it demonstrates negative differential behavior [28]. It can be incorporated as an excellent sensor to detect the presence of many gasses like CO, CO_2, H_2, NO, NO_2, NH_3, H_2S, SO_2, CH_4, etc. due to the existence of both acidic and basic sites on its surface [29–36]. The encapsulation of Li, K and their oxides modifies the optical properties of borospherene [37, 38] while that of Fe, Ni, Co and Mg alters the spintronic properties [39]. The stability of borospherene remains intact on placement of Li, K, Na, Ba and Tl inside the cage [40]. Doping and encapsulation with other elements can effectively alter the properties of B_{40} [41–43]. As borospherene is the only boron fullerene which has been successfully synthesized and has exceptional structural properties, it has been chosen in the present study to exploit its feasibility as nano-carrier in drug delivery systems.

5-fluorouracil (or Adrucil) is widely utilized for treatment of cancer [44, 45]. However, oral consumption of this drug has very severe side effects like loss of appetite, hair loss, inflammation of mouth, low blood cell count, inflammation of skin and harmful impact on the neighboring healthy cells [46–49]. Although efforts have been made to find suitable materials to deliver this drug to the target cancer cells [50–54], further studies are necessary to find ways to minimize the side effects. Thus, we have chosen this drug, to investigate its interaction with the B_{40} fullerene utilizing density functional theory calculations.

12.2 Computational Details

DFT (density functional theory) calculations [55–58] in combination with NEGF (Non-equilibrium Green's Function) [59] were carried out to get insight about the capability of borospherene (B_{40}) in delivering anti-cancer drug; 5F-Uracil (FU) to prostate cancer cells [60–62]. Generalized gradient approximation (GGA) of Perdew, Burke and Ernzerhof (PBE) is applied along-with double zeta-polarized basis set (DZP-BS) [63] in VNL (Virtual NanoLab) software which is a graphical interface of ATK (Atomistix ToolKit) [64]. All the structures were optimized till the forces converged to less than 0.05eV/Å.

In order to explore the possibility of utilizing B_{40} in drug delivery, targeted drug delivery is the most important phenomenon as targeted delivery of drug aids in drug delivery with minimum side effects and very less toxicity [65]. Targeted drug delivery can be achieved by two methods namely, active targeting and passive targeting [66]. The active targeting includes modification of the surface of the nano-carrier to target the cancer cells by attaching targeting agents popularly known as ligands. The passive targeting is achieved by reconnoitering the improved permeability and retention (EPR) effect of the cancer cells, as these cells have leaky and wide blood vessels due to angiogenesis. The pore size in these leaky blood vessels is very large ranging from 100 to 780 nm [67–69] and any nanoparticle having size less than this can easily penetrate through and reach the cancer cells easily [70, 71]. Both the active and passive targeting can be explored together for targeted drug delivery. To utilize the phenomenon of active targeting, a ligand is attached to the B_{40} fullerene [65]. The ligand that is chosen in this research work is anisamide (also called 2-methoxy benzamide) because the sigma receptors present in the prostate cells can easily detect anisamide (AN) and hence it is an excellent ligand to be utilized for targeted drug delivery to prostate cancer cells [72, 73].

As B_{40} fullerene has hexagonal holes at the upper and lower surfaces and heptagonal rings at the sides [27], so the adsorption of anisamide is considered on both these rings. The interaction of anisamide with B_{40} was studied by calculating the adsorption energies of both the optimized B_{40}+AN complexes utilizing the following equation: -

$$E_{ad} = E(B_{40} + AN) - E(AN) - E(FU) \tag{12.1}$$

Where E (B_{40} + AN), E (AN) and E (FU) are the energies of B_{40}+Anisamide, Anisamide and 5F-uracil respectively.

Next, the FU molecule is adsorbed on the 6- and 7-membered rings of both the B_{40}+AN complexes one by one. Several interactions between the FU and B_{40}+AN complexes were considered and the interaction of oxygen and fluorine atoms of FU were found to be the most favorable active sites for interaction with the B_{40} fullerene. The adsorption energies of these optimized structures were determined as follows:-

$$E_{ad} = E(B_{40} + AN + FU) - E(B_{40} + AN) - E(FU) \quad (12.2)$$

Here E (B_{40} + AN + FU) denotes the energy of B_{40}+Anisamide+5F-uracil complex and E (B_{40} + AN) and E (FU) represent total energies of B_{40}+Anisamide and 5F-uracil respectively [76].

12.3 Results and Discussion

The geometry of the B_{40} fullerene is identical to the experimentally obtained structure by Zhai *et al.* having D_{2d} symmetry (Figure 12.1(a)) [27]. The borospherene molecule comprises of four 7- and two 6-membered rings oriented at the sides and top and bottom respectively. The bond lengths at equilibrium for B-B bonds in the 6-membered rings are 1.63Å and 1.76Å respectively while for the 7-membered rings the B-B bond lengths are 1.62Å, 1.68Å, 1.59Å and 1.73Å respectively. Two BBB angles are noticed in the 6-membered rings having values, 114.26° and 119.94° and

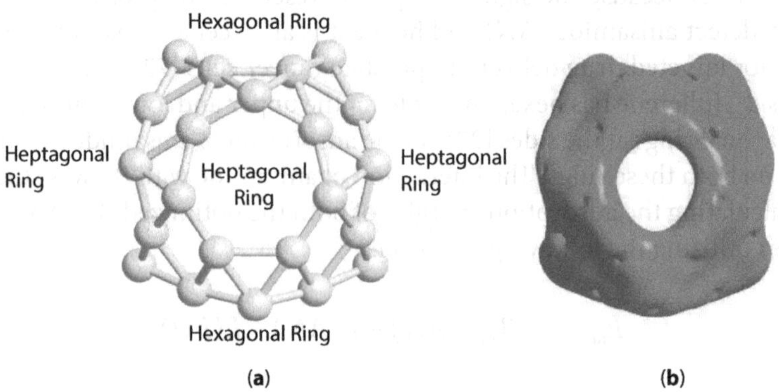

Figure 12.1 (a) Optimized structure of borospherene. (b) Electron density of borospherene.

three BBB angles are visualized in the 7-membered rings viz., 126.48°, 128.78° and 122.37°. To garner knowledge about the electronic properties of B_{40} the electron density of the borospherene is calculated as shown in Figure 12.1(b). It is clear from the figure that B_{40} has a dense electron cloud thus making it favorable for interaction with other molecules. Next, the density of states (DOS) for the borospherene molecule is determined as shown in Figure 12.2 (a). The DOS is helpful in predicting the quantum states that are available for transmission in the proximity of E_F (Fermi level) [77–81]. The borospherene depicts a very sharp HOMO peak in comparison to the LUMO. To further analyze the behavior of the HOMO and LUMO, the Eigen states of both these orbitals are contemplated as shown in Figures 12.2(b–c). The LUMO orbital showcases more delocalization in comparison to HOMO, thus implying the dominance of LUMO in transmission.

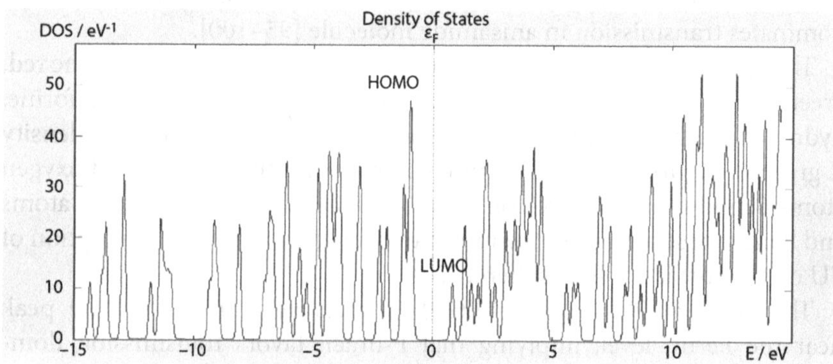

(a)

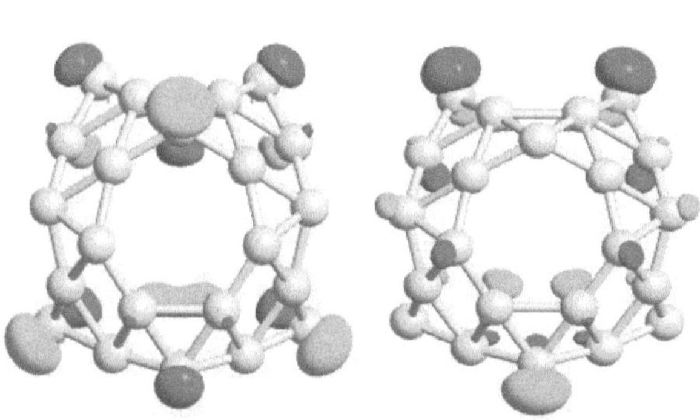

(b) (c)

Figure 12.2 (a) DOS (The dotted line at 0eV represents the Fermi level), (b) HOMO and (c) LUMO of borospherene.

The optimized structure of the anisamide molecule which is used as a ligand is shown in Figure 12.3(a). The white, gray, red and blue color atoms represent hydrogen, carbon, oxygen and nitrogen atoms respectively. Figure 12.3(b) denotes the electron density of anisamide [82–86]. The pink color around the oxygen atoms signifies that the oxygen atoms are the most appropriate active sites for interaction of Anisamide with the Borospherene as they have larger density of electrons. To get more information about the electronic properties of anisamide, the DOS is calculated as shown in Figure 12.4(a). From the figure, it is seen that both the HOMO and LUMO peaks are identical for Anisamide [87–94]. Also, more number of peaks is noticed above the Fermi level, this implies that LUMO-dominant transmission occurs. To further explore the contribution of both the HOMO and LUMO, the Eigen states are determined as given in Figures 12.4(b–c). The HOMO orbital shows negligible delocalization around one oxygen atom while the LUMO orbital shows significant delocalization around both the oxygen atoms and four carbon atoms. This indicates that the LUMO orbital dominates transmission in anisamide molecule [95–100].

The optimized structure of 5F-uracil is shown in Figure 12.5(a). The red, green, white, gray and blue colored atoms represent the oxygen, fluorine, hydrogen, carbon and nitrogen atoms respectively. The electron density is given in Figure 12.5(b). The pink color around the fluorine and oxygen atoms denotes that the electron cloud is concentrated around these atoms and hence these atoms are the most favorable active sites for adsorption of FU on the surface of Borospherene.

The DOS of FU (Figure 12.6(a)) depicts very sharp HOMO peak near the Fermi level, implying that F-uracil favors transmission dominated by HOMO [101–106]. The Eigen states of HOMO and LUMO

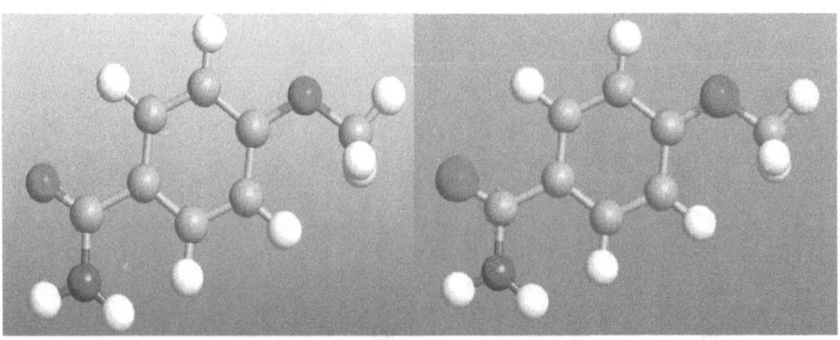

(a) (b)

Figure 12.3 (a) Optimized structure and (b) electron density of anisamide.

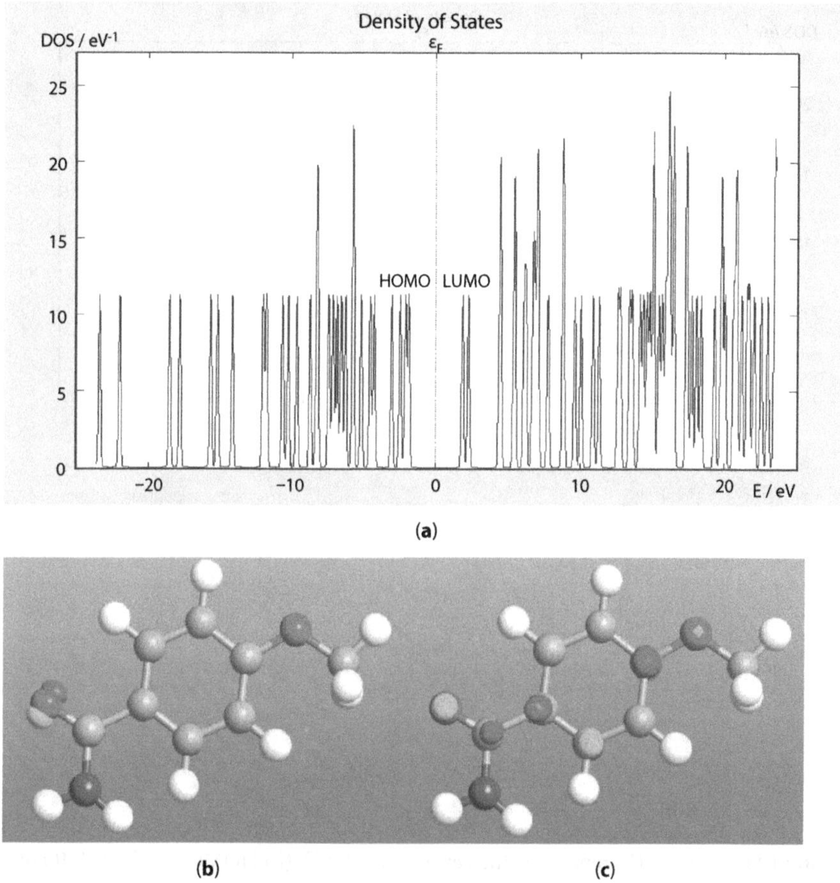

Figure 12.4 (a) DOS (The dotted line represents the Fermi level (0eV)), (b) HOMO and (c) LUMO of optimized structure of anisamide.

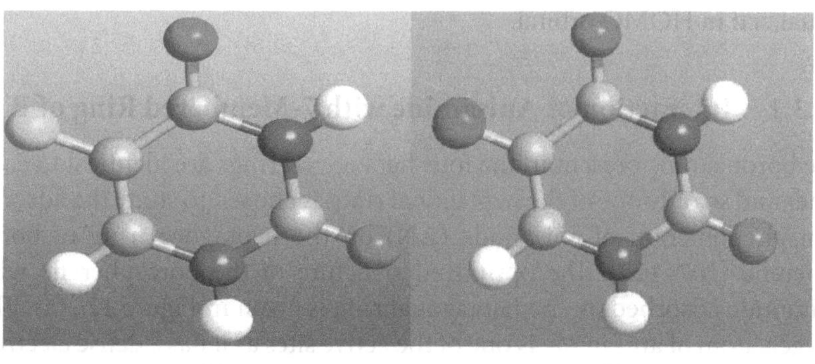

Figure 12.5 (a) Optimized structure and (b) electron density of 5F-uracil.

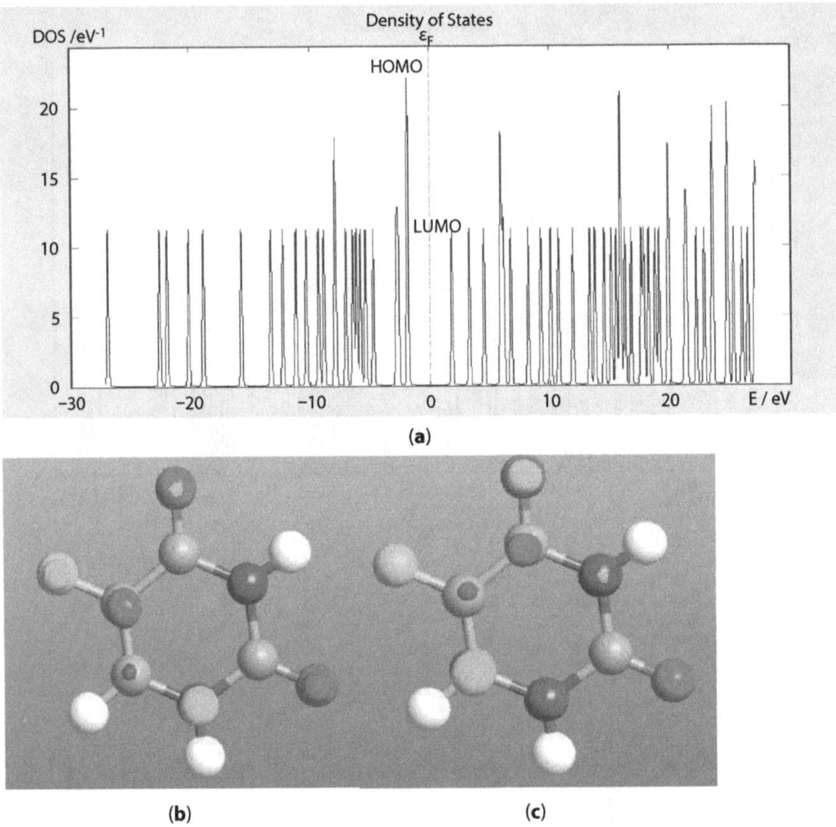

Figure 12.6 (a) DOS (E_F represents the Fermi level (0eV)), (b) HOMO and (c) LUMO of optimized FU.

(Figures 12.6(b–c)) also support the DOS results as more delocalization is visualized in HOMO orbital.

12.3.1 Interaction of Anisamide with 7-Membered Ring of B_{40}

The boron atoms present in the four heptagonal rings are identical to each other and so only one of the heptagonal rings is chosen to study the adsorption phenomenon of anisamide (AN) on the heptagonal ring of borospherene [107–112]. The optimized structure of the borospherene with anisamide adsorbed on the heptagonal ring is given in Figure 12.7(a). The oxygen atom of anisamide is one of the active sites as it has a dense electron cloud which interacts with the boron fullerene (Figure 12.3(b)). The interaction distance between the oxygen atom of AN and boron atom of B_{40} is

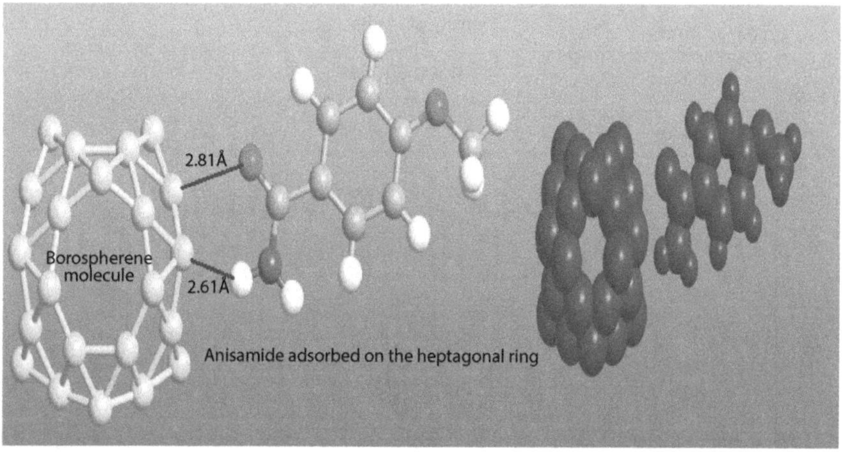

(a) (b)

Figure 12.7 (a) Optimized structure and (b) electron density of anisamide adsorbed on the heptagonal ring of B_{40}. The black lines represent the interaction distances.

2.81Å and that between hydrogen atom of AN and boron of borospherene is 2.61Å. The adsorption energy for this complex is -0.61eV and very little charge transfer (0.016e) occur. There is no structural change in both the molecules thus implying that AN is physisorbed on the 7-membered ring of B_{40}. Further the electron density of the B_{40}+7AN (7 denotes the adsorption on heptagonal ring) is calculated as shown in Figure 12.7(b). Due to physisorption no orbital overlapping occurs between the two molecules.

To get more information about the electronic properties DOS is determined for the B_{40}+7AN complex (Figure 12.8(a)). A very sharp HOMO peak is visualized near the Fermi level (0eV). On comparing the DOS of B_{40} (Figure 12.2(a)) with this B_{40}+AN complex, it is noticed that the magnitude of DOS increases for the fullerene-AN complex. In addition the number of peaks drastically intensifies for the B_{40}+7AN complex, thus denoting that more number of quantum states are available for transmission in this complex [113–118].

Next, the molecular energy spectrum for the complex given in Figure 12.7(a) is evaluated and the HOMO, LUMO and HLG are obtained which are given in Table 12.1. The HLG obtained is 1.56eV. The Eigen states of the HOMO and LUMO are comprehended to evaluate the contribution of these orbitals in transmission (Figures 12.8(b–c)). The delocalization is more pronounced for the LUMO orbital with major contribution of the pi bonds. This signifies that LUMO orbital plays the dominant role in transmission in this complex [119–123].

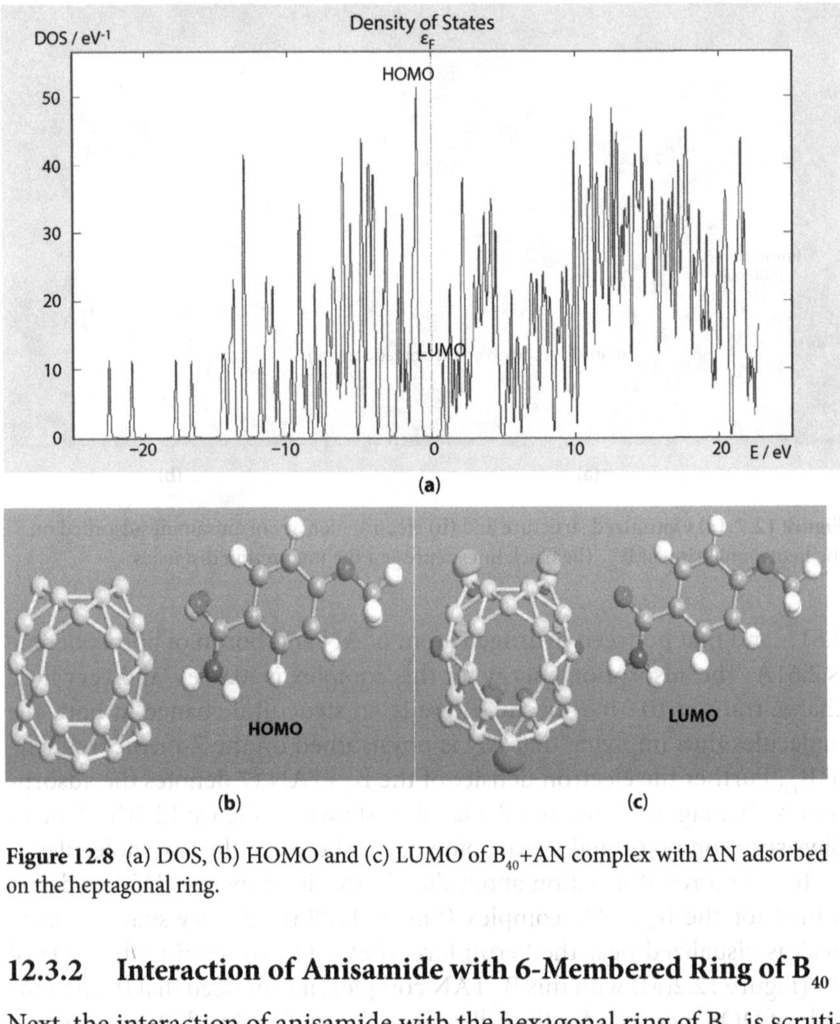

Figure 12.8 (a) DOS, (b) HOMO and (c) LUMO of B_{40}+AN complex with AN adsorbed on the heptagonal ring.

12.3.2 Interaction of Anisamide with 6-Membered Ring of B_{40}

Next, the interaction of anisamide with the hexagonal ring of B_{40} is scrutinized. As all the boron atoms in each of the hexagonal rings have identical bond lengths and bond angles, only one hexagonal ring is selected to study the interaction of AN. The optimized structure of this complex is shown in Figure 12.9(a). The interaction distance is 2.56Å and 2.72Å between the B-H and B-O atoms respectively. The adsorption energy is -0.63eV. No structural alterations in the molecules, large interaction distances and small charge transfer (0.026e) implies that AN is physisorbed on the hexagonal ring of borospherene.

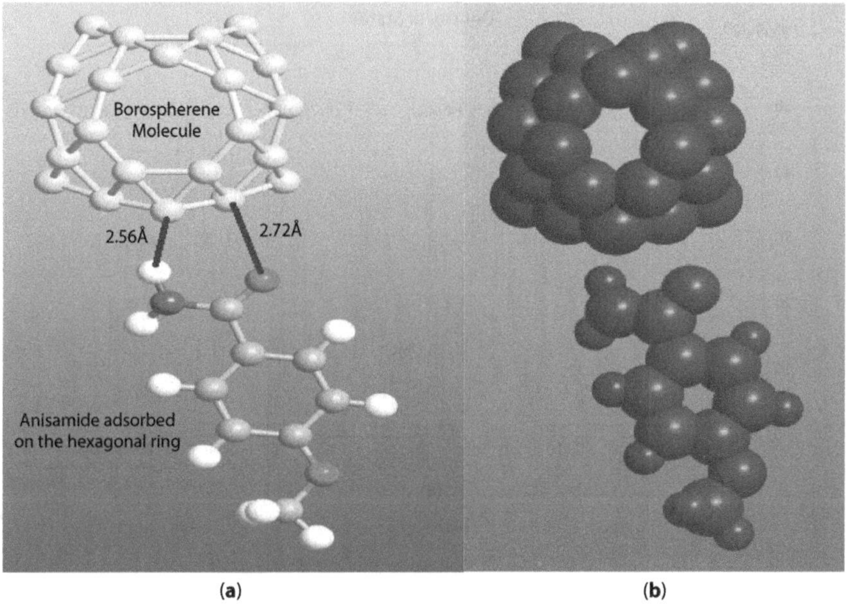

Figure 12.9 (a) Optimized structure and (b) electron density of B_{40}+AN complex with adsorbed on the hexagonal ring. The black lines represent the interaction distances.

Next, the electron density is calculated (Figure 12.9(b)) to determine the orbital overlapping between the molecules. No overlapping is visualized as the AN molecule is physisorbed on the 6-membered ring of B_{40}. DOS is determined to scrutinize the electronic properties of this complex (Figure 12.10(a)). Sharp HOMO peak is noticed near the Fermi level. The shape and peaks of the DOS are similar for B_{40}+AN complexes with AN adsorbed on both the hex- and heptagonal rings. Therefore, it is deduced that anisamide adsorption on both the 6-and 7-membered rings of borospherene has similar impact on the electronic properties of the borospherene [124–127].

Further the molecular energy spectra is calculated and the values of HOMO, LUMO and HLG are tabulated in Table 12.1. The HLG is 1.62eV which is slightly higher than the B_{40}+7AN complex (1.58eV). The Eigen states of the HOMO and LUMO are analyzed next (Figures 12.10(b–c)). From the figure, clearly more delocalization is portrayed by the LUMO orbital. Thus, it denotes that LUMO orbital is significant in transmission in this complex [128–133].

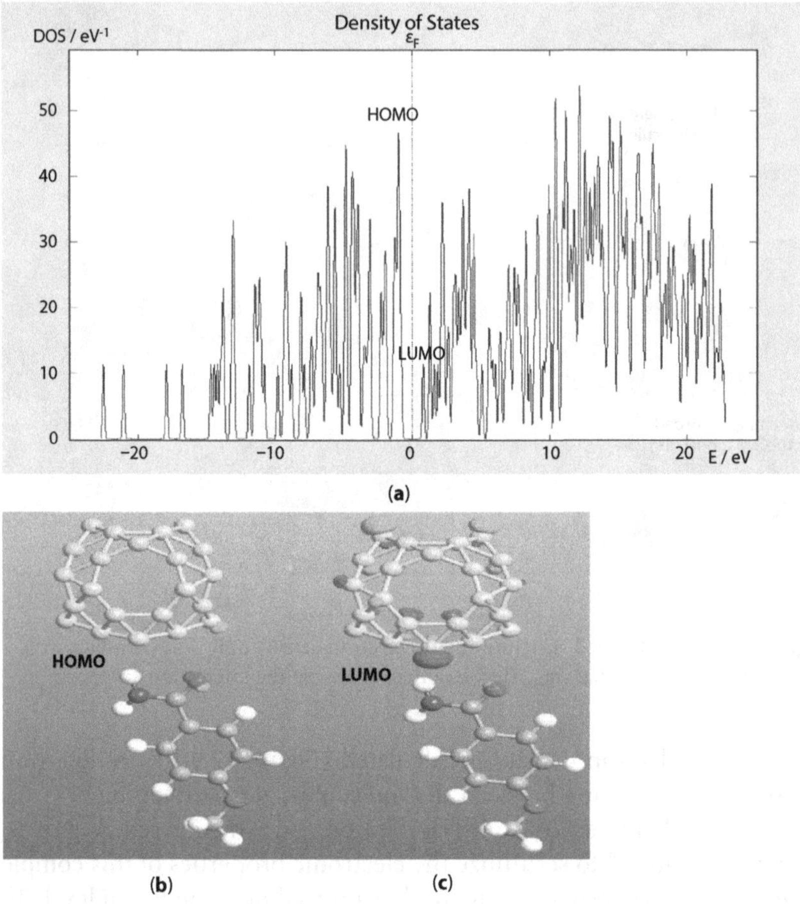

Figure 12.10 (a) DOS, (b) HOMO and (c) LUMO of B_{40}+AN complex with AN adsorbed on the hexagonal ring.

12.3.3 Interaction of 5F-Uracil with the Heptagonal Ring of B_{40}+7AN Complex (AN on Heptagonal Ring)

Firstly, the interaction of FU with the heptagonal ring of B_{40}+7AN is considered. The optimized structure for the B_{40}+7AN+7FU complex (where 7 represents the adsorption on the heptagonal ring) is given in Figure 12.11(a). The electron density obtained for FU (Figure 12.5(b)), clearly portray that the oxygen and fluorine atoms of FU have dense electron cloud distributed around them and hence are the most favorable active sites for adsorption. The interaction distance is 2.97Å and 3.15Å for F-B and O-B respectively. The adsorption energy is -0.72eV. From the figure it is visible that FU is

Table 12.1 Adsorption energies, HOMO, LUMO and HLG of all the structures.

S. no.	Structure	E_{ad} (eV)	HOMO (eV)	LUMO (eV)	HLG (eV)
1.	B_{40}	-	-0.83	0.83	1.66
2.	FU	-	-1.81	1.81	3.62
3.	AN	-	-1.87	1.87	3.74
4.	B_{40}+7AN	-0.61	-0.78	0.78	1.56
5.	B_{40}+6AN	-0.63	-0.81	0.81	1.62
6.	B_{40}+7AN+7FU	-0.72	-0.79	0.79	1.58
7.	B_{40}+7AN+6FU	-0.76	-0.81	0.81	1.62
8.	B_{40}+6AN+7FU	-0.71	-0.81	0.81	1.62
9.	B_{40}+6AN+6FU	-0.68	-0.84	0.84	1.68

physisorbed on the B_{40}+7AN complex as the interaction distances are very large, there are no structural deformations and little charge transfer (0.081e).

The electron density (Figure 12.11(b)) of this complex portrays that no orbital overlapping exists between the three molecules. The electronic properties are further explored by investigating the DOS (Figure 12.12(a)). Comparing the DOS with the DOS for B_{40}+7AN (Figure 12.8(a)), it is deduced that the shape of the graphs is identical with slight variation in the magnitude of the peaks [134–139]. Therefore the adsorption of FU on the heptagonal ring of B_{40}+7AN complex does not affect the electronic properties much, thus further confirming the physisorption of FU on the complex.

The value of HOMO, LUMO and HLG obtained from the molecular spectra analysis is given in Table 12.I. The HLG is 1.58eV which is slightly lower than the B_{40}+7AN complex (1.56eV). This further assures that FU is physisorbed on the complex. To harness information about the molecular orbitals, the Eigen states for HOMO and LUMO are evaluated (Figures 12.12(b–c)). The dominance of LUMO orbital in transmission is noticed as it is more delocalized with pi bonds actively participating in transmission.

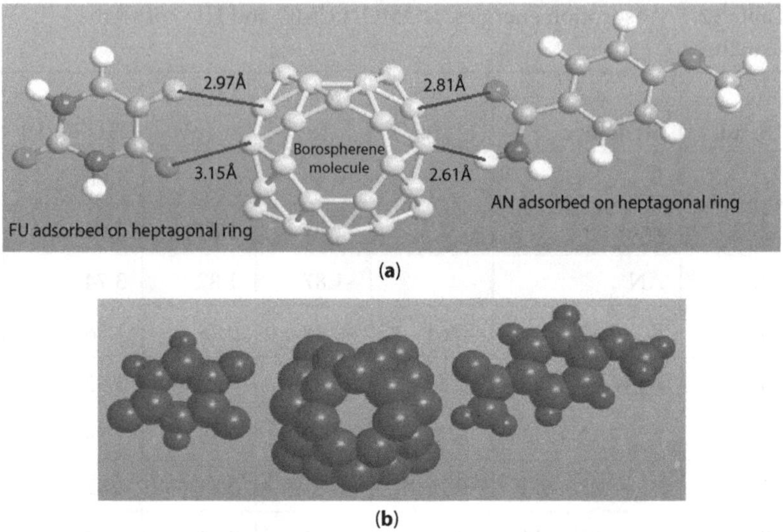

Figure 12.11 (a) Optimized structure of B_{40}+7AN+7FU complex. The black lines represent the interaction distances, (b) Electron density of B_{40}+7AN+7FU complex.

12.3.4 Interaction of 5F-Uracil with the Hexagonal Ring of B_{40}+7AN Complex (AN on Heptagonal Ring)

The optimized structure for the B_{40}+7AN+6FU complex (where the numbers 6 and 7 represent the adsorption on hexagonal and heptagonal rings respectively) is shown in Figure 12.13(a). FU is physisorbed on the hexagonal ring of B_{40}+7AN complex with adsorption energy -0.76eV, small charge transfer (0.042e) and interaction distances of 2.73Å and 2.77Å for O-B and F-B respectively. The electron density (Figure 12.13(b)) shows no overlapping of orbitals for the three molecules in the complex.

The electronic properties are investigated by calculating the DOS and molecular energy spectra. The DOS (Figure 12.14(a)) shows similar trend as the B_{40}+7AN complex but the peaks have reduced magnitude. The HLG is 1.62eV. Higher value of HLG compared to the B_{40}+7AN complex (1.56eV) is the reason behind the reduced magnitude of DOS peaks. Thus it is inferred that B_{40}+7AN+6FU complex has lesser transmission. The Eigen states of the HOMO and LUMO (Figures 12.14(b–c)) depict significant role of LUMO in transmission due to more delocalized structure.

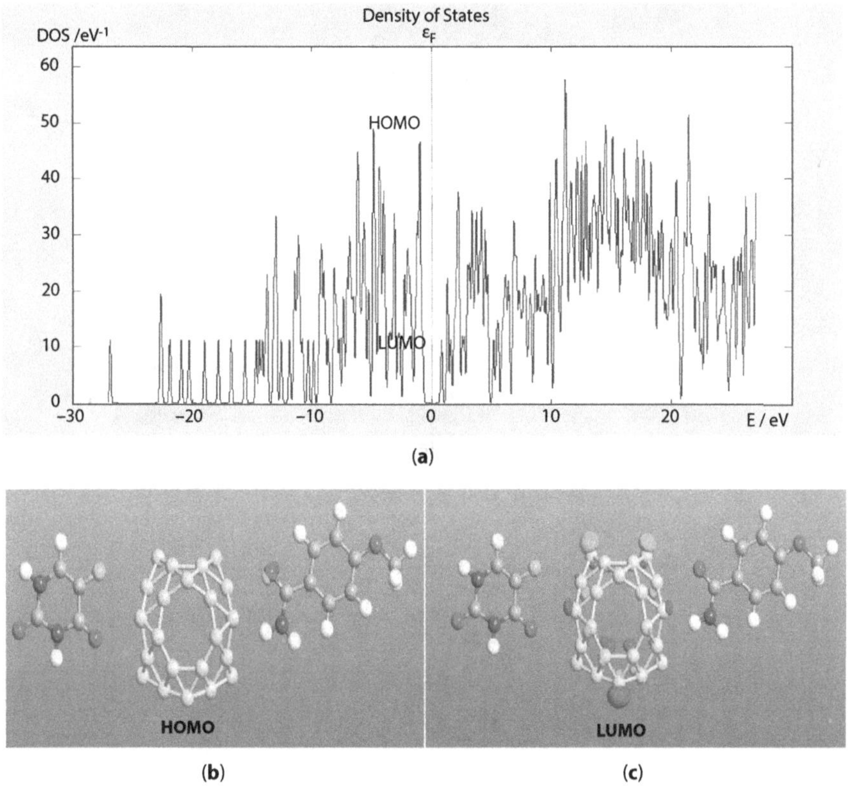

Figure 12.12 (a) DOS, (b) HOMO and (c) LUMO of B_{40}+7AN+7FU complex.

12.3.5 Interaction of 5F-Uracil with the Heptagonal Ring of B_{40}+6AN Complex (AN on Hexagonal Ring)

The optimized structure of B_{40}+6AN+7FU is given in Figure 12.15(a). The adsorption energy of the complex is -0.71eV and the interaction distances are 2.99Å and 3.01Å for F-B and O-B respectively. Thus FU is physisorbed on the heptagonal ring of B_{40}+6AN complex. The electron density (Figure 12.15(b)) also supports this deduction as no overlapping orbitals are visualized in the three molecules in the complex.

The electronic properties of this complex are scrutinized further by determining the density of states and molecular energy spectra. DOS (Figure 12.16(a)) is analogous to the DOS for B_{40}+7AN+6FU (Figure 12.14(a)) with only minor variation in the magnitude of the peaks.

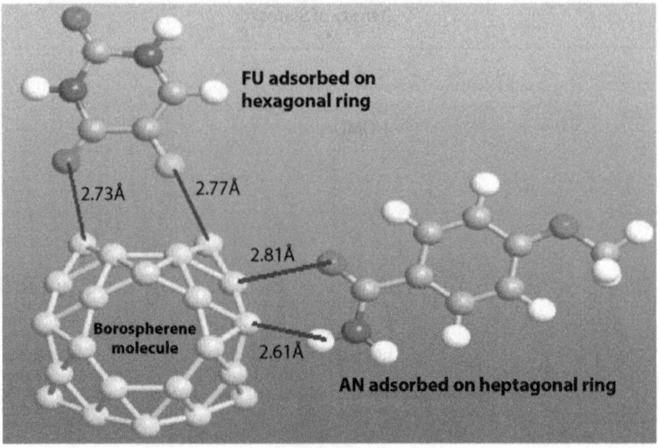

(a)

(b)

Figure 12.13 (a) Optimized structure of B_{40}+7AN+6FU complex (The black lines represent the interaction distances). (b) Electron density for optimized structure of B_{40}+7AN+6FU complex.

The HLG is 1.62eV which is same as B_{40}+7AN+6FU. This is the reason that these two complexes have identical DOS plots. The Eigen states of the LUMO (Figure 12.12(c)) are more delocalized, therefore the contribution of this orbital is more significant in the transmission than the HOMO.

12.3.6 Interaction of 5F-Uracil with the Hexagonal Ring of B_{40}+6AN Complex (AN on Hexagonal Ring)

The optimized structure of B_{40}+6AN+6FU complex (Figure 12.17(a)) has adsorption energy of -0.68eV and interaction distances of 2.76Å and 2.81Å for O-B and F-B respectively. No structural changes, large interaction distances and no orbital overlapping (Figure 12.17(b)) implies the

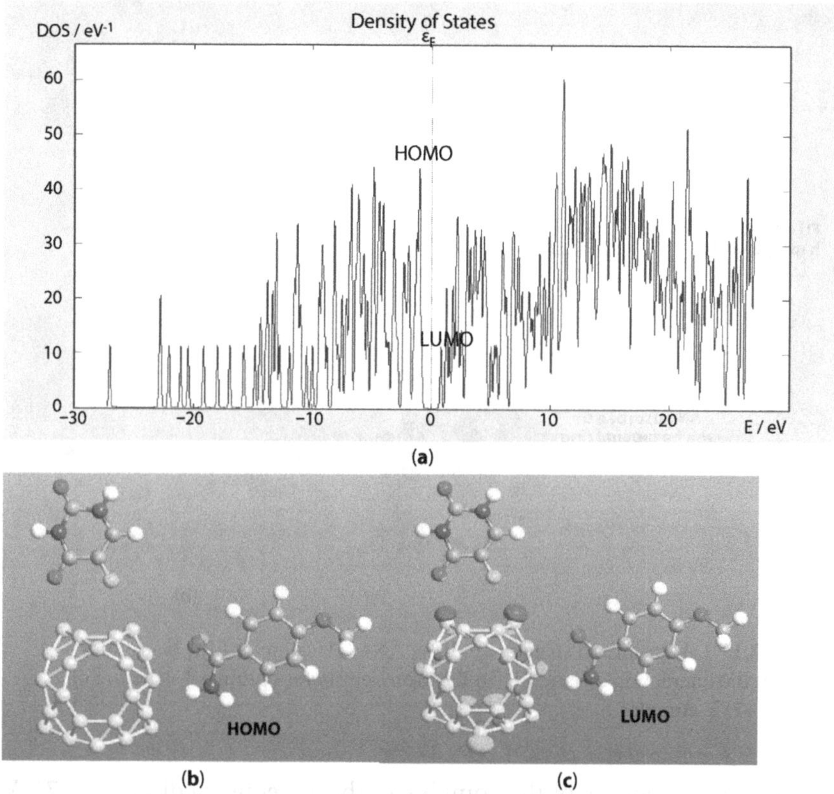

Figure 12.14 (a) DOS, (b) HOMO and (c) LUMO for the B_{40}+7AN+6FU complex.

physisorption of FU on the B_{40}+6AN complex. The DOS (Figure 12.18(a)) obtained is similar to the ones obtained for other structures with slightly reduced magnitude of peaks. The HLG is 1.68eV which is the highest amongst all the B_{40}+AN+ FU complexes. Due to this, this complex has lowest magnitude of DOS peaks. The more delocalized LUMO orbital plays significant role in transmission due to dominant behavior of pi bonds (Figure (12.18(c)).

12.3.7 Stability in Aqueous Solution

To investigate the stability in aqueous solution, B_{40}+7AN+6FU complex is chosen as it has the most stable configuration due to largest value of adsorption energy (-0.76eV). A water molecule is adsorbed on the surface of the complex and the structure is optimized (Figure 12.19).

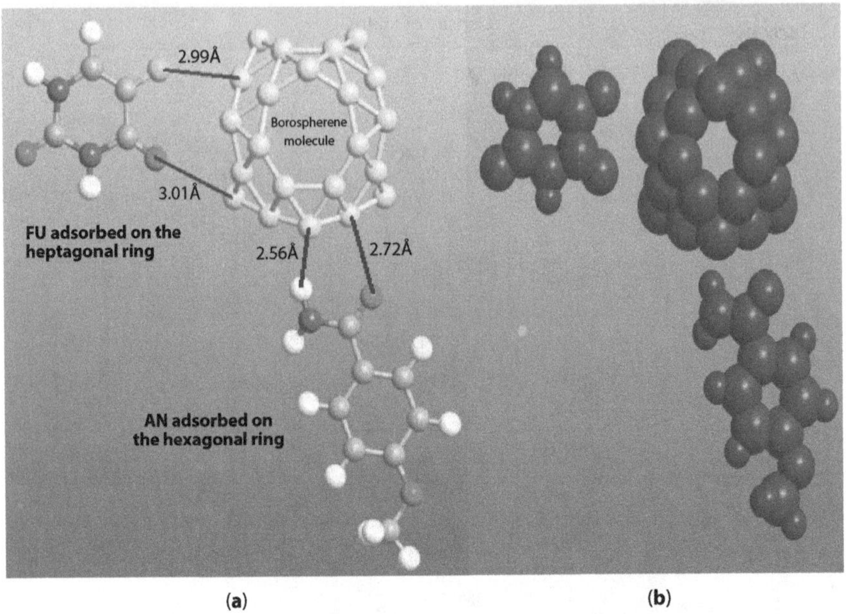

Figure 12.15 (a) Optimized structure of B_{40}+6AN+7FU complex (The black lines represent the interaction distances). (b) Electron density for optimized structure of B_{40}+6AN+7FU complex.

The adsorption energy of the complex in the aqueous medium is -0.71eV which is quiet close to the original value (-0.76eV). Thus it is deduced that the B_{40}+7AN+6FU complex is stable in aqueous solution and can easily travel in the bloodstream to the targeted cells.

12.3.8 Drug Release

The most significant step in the process of drug delivery is the release of drug from the nano-carrier to the target cells. A number of exogenous (like temperature stimulated, magnetic field stimulated, ultrasound stimulated etc.) and endogenous (pH, enzyme, redox etc.) stimuli responsive systems have been explored for drug release [74]. Here, we exploit the endogenous stimuli viz., pH stimulated drug release as the pH of the cancer cells (less than 6) is lower than the normal pH of the blood (7.35–7.45) [75]. In this context the effect of pH is scrutinized on the most stable complex i.e. B_{40}+7AN+6FU as it has the highest adsorption energy (-0.76eV). The release of drug, FU from the nano-carrier is investigated using the same DFT calculations. The H^+ species has the tendency to attach to the O and

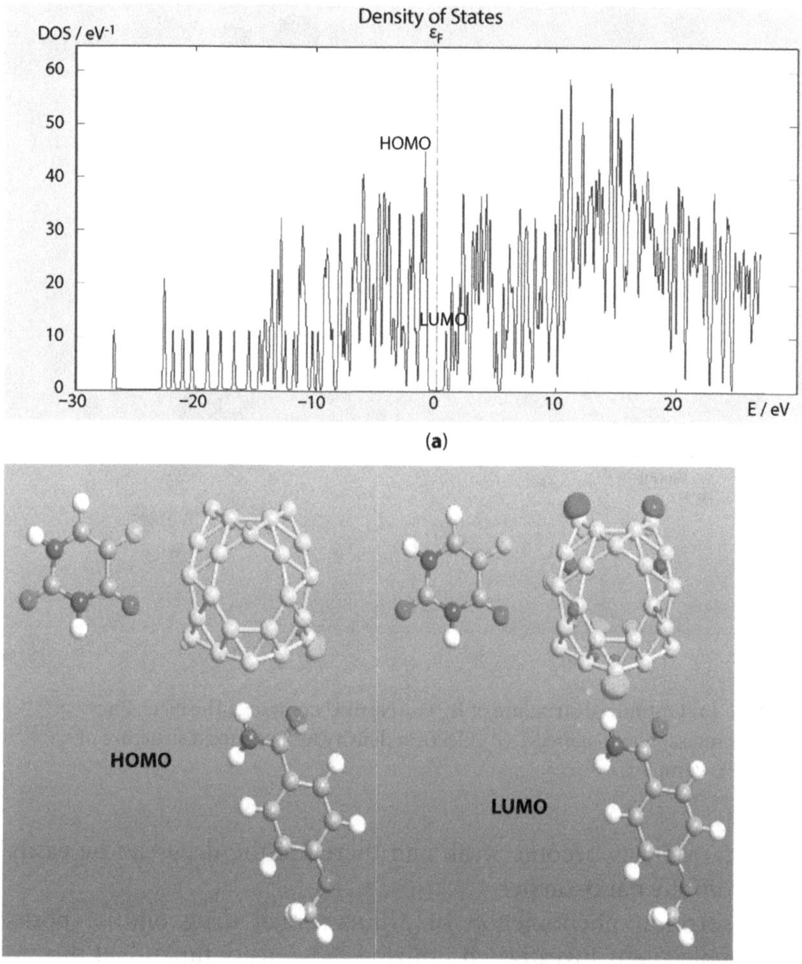

(a)

(b) (c)

Figure 12.16 (a) DOS, (b) HOMO and (c) LUMO for the B_{40}+6AN+7FU complex.

F atoms of FU in the acidic medium as they are highly electronegative. Thus both these atoms were protonated and the structure was optimized till forces converge to less than 0.05eV/Å (Figure 12.20). Post optimization the hydrogen ions get transferred to the boron cage and the adsorption energy decreases from -0.76eV to -0.63eV. Also the interaction distance increases from 2.73Å to 2.90Å and 2.77Å to 2.88Å for O-B and F-B respectively. As the adsorption energy has decreased and interaction distance has increased in the acidic medium, thus the interaction of the drug with

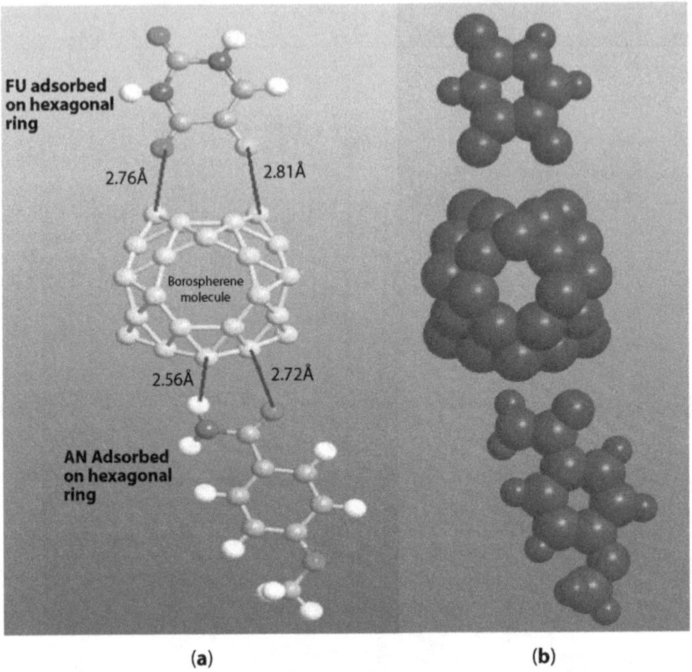

Figure 12.17 (a) Optimized structure of B_{40}+6AN+6FU complex (The black lines represent the interaction distances). (b) Electron density for optimized structure of B_{40}+6AN+6FU complex.

the nano-carrier has become weak and therefore the drug can be easily released from the nano-carrier.

The adsorption phenomenon of 5-fluorouracil drug on the borospherene molecule is investigated utilizing the density functional theory. Anisamide molecule is attached as ligand to the fullerene to achieve targeted drug delivery to prostate cancer cells. Various configurations are exploited by studying the interaction of the drug and ligand with both the 6- and 7- membered rings of the borospherene. It is deduced that the interaction of the ligand with the heptagonal ring and the drug with the hexagonal ring gives the most stable complex with adsorption energy of -0.76eV. This complex is found to be stable even in aqueous solution. The HLG is altered when the ligand and drug interact with the B_{40}. Drug release based on the pH-stimulated mechanism is also suggested. It is inferred that the adsorption energy significantly decreases while the interaction distances between the drug and borospherene increase in the acidic medium thus implying the feasibility of drug release. Thus, the results obtained in the

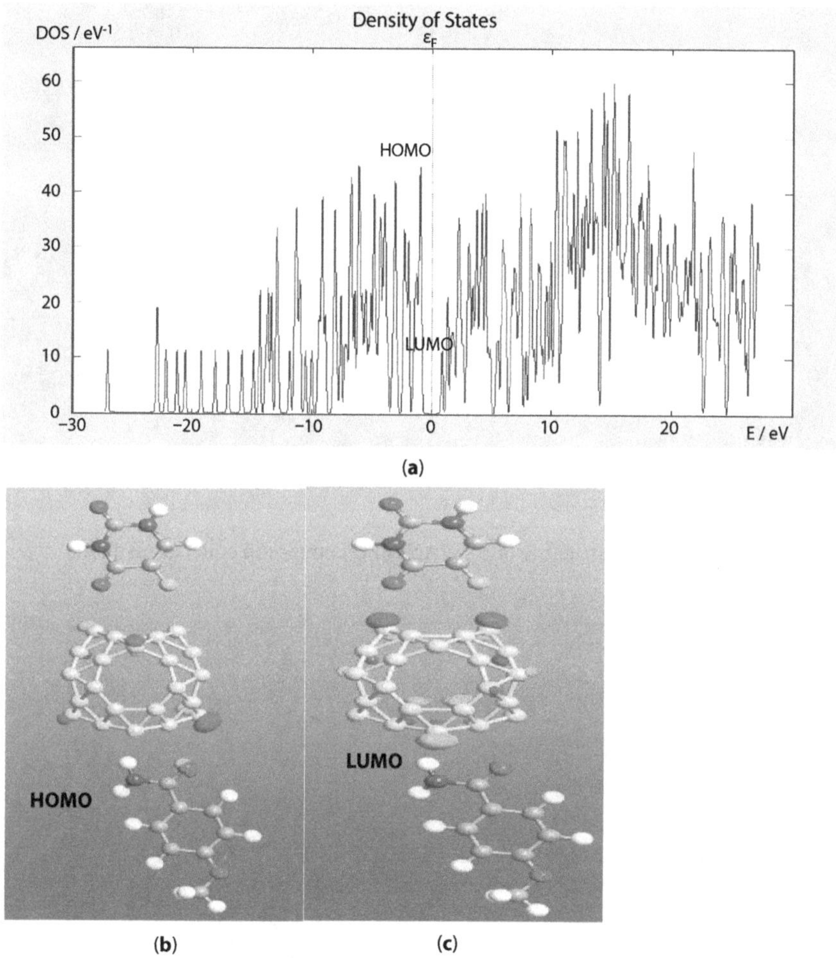

Figure 12.18 (a) DOS, (b) HOMO and (c) LUMO for the B_{40}+6AN+6FU complex.

present work confirm that B_{40} can be exploited as a nano-carrier in drug delivery of 5-Fluorouracil to the prostate cancer cells.

This work is of significance as the research in the biomedical field for potential applications of nanomaterials is garnering attention as nanomaterials have extraordinary structural characteristics, non-toxic behavior, greater surface area and good affinity towards various drugs. Various carbon-based nanomaterials have already been explored for such biomedical applications in the past but much has not been done in the field of boron-based nanostructures. These structures need to be exploited more as previous studies have proved that nanomaterials comprising of boron

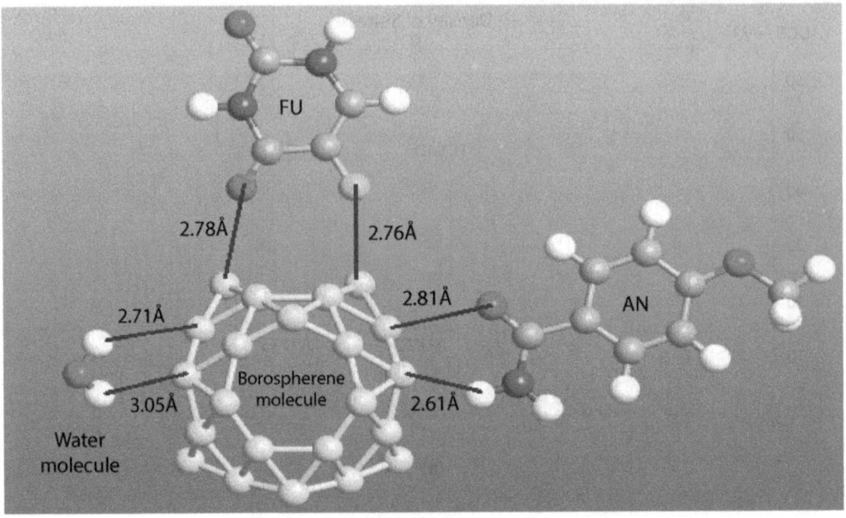

Figure 12.19 Optimized structure of B_{40}+7AN+6FU complex in aqueous medium.

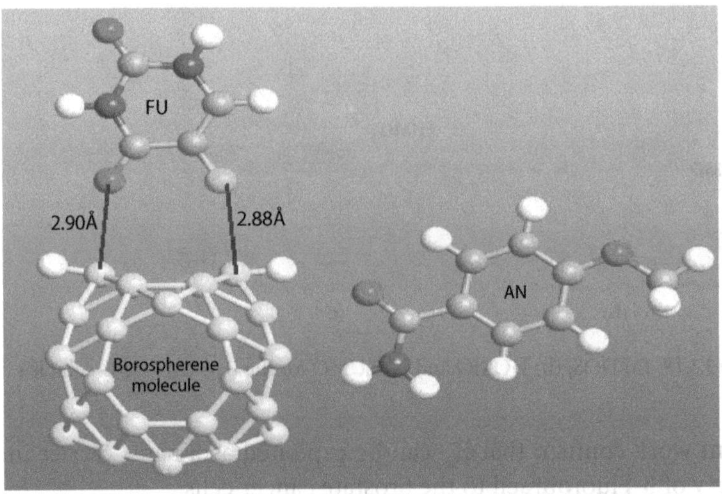

Figure 12.20 The optimized structure of B_{40}+7AN+6FU in the acidic medium.

have greater potential in drug-delivery applications due to their greater surface area and better binding affinity with various drugs that are widely used to cure various categories of cancer [35, 36, 42, 43]. In addition, these structures can be exploited to aim for targeted delivery of drugs to the affected cells without causing any damage to the healthy tissues and cells.

The future research also needs to focus on the ADMET profiles of these nanostructure-drug combinations to study their toxicity properties and molecular docking studies to garner information about the interaction of these drugs with the various proteins in the human body.

Acknowledgement

We acknowledge the Virtual Nano Lab at Guru Nanak Dev University, Amritsar, for providing the necessary computational facilities.

Conflict of Interest

The authors have no conflict of interest.

References

[1]. Xu, Y., Bai, H., Lu, G., Li, C. and Shi, G., 2008. Flexible graphene films via the filtration of water-soluble noncovalent functionalized graphene sheets. *J. Am. Chem. Soc.*, 130(18), pp. 5856-5857.

[2]. Omidvar, A.,2018. Design of a novel series of donor-acceptor frameworks via superalkali-superhalogen assemblage to improve the nonlinear optical responses. *Inorg. Chem.*, 57, pp. 9335–9347.

[3]. Omidvar, A., **2018**. Indium-doped and positively charged ZnO nanoclusters: versatile materials for CO detection. *Vacuum,* 147, pp. 126–133.

[4]. Omidvar, A., Anafcheh M., Hadipour N., **(2013).** Computational studies on carbon nanotube-graphene nanoribbon hybrids by density functional theory calculations. *Sci. Iran.,* 20, pp. 1014–1017.

[5]. Omidvar, A., **2017**. Electronic structure tuning and band gap opening of nitrogen and boron doped holey graphene flake: the role of single/dual doping. *Mater. Chem. Phys.*, 202, pp. 258–265.

[6]. Omidvar, A., **2017**. Catalytic activation of O2 molecule by transition metal atoms deposited on the outer surface of BN nanocluster. *J. Mol. Graph Mode,* 77, pp. 218–224.

[7]. Omidvar, A., **2017**. Borophene: a novel boron sheet with a hexagonal vacancy offering high sensitivity for hydrogen cyanide detection. *Comput. Theor. Chem.*, 1115, pp. 179–184.

[8]. Liu Z, Robinson JT, Sun XM, Dai HJ **2008**. PEGylated nanographene oxide for delivery of water-insoluble cancer drugs. *J. Am. Chem. Soc.,* 13, pp. 10876–10877.

[9]. Zhang, LM., Xia, JG., Zhao, QH., Liu, LW., Zhang, Z., **2010**. Functional graphene oxide as a nanocarrier for controlled loading and targeted delivery of mixed anticancer drugs. *Small,* 6, pp. 537–544.

[10]. Zhou, T., Zhou, X., Xing, D., **2014**.Controlled release of doxorubicin from graphene oxide based charge-reversal nanocarrier. *Biomaterials,* 35, pp. 4185–4194.

[11]. Webster, D., Sundaram, P., Byrne, M., **2013**. Injectable nanomaterials for drug delivery: carriers, targeting moieties, and therapeutics. *Eur. J. Pharm. Biopharm.,* 84, pp. 1–20.

[12]. Farokhzad, O., Langer, R., **2009**. Impact of nanotechnology on drug delivery. *ACS Nano.,* 3, pp. 16–20.

[13]. Zhuqing, W., Lucio, C., Gang, W.,**2017**. Recent advances in the synthesis of graphene-based nanomaterials for controlled drug delivery. *Appl. Sci.,* 7, pp. 1175–111193.

[14]. Nicolardi, S., Yur,i EM., Burgt, V., Jeroen, D., Wuhrer M., Hokke, C., Chiodo F., **2017**. Structural characterization of biofunctionalized gold nanoparticles by ultrahigh-resolution mass spectrometry. *ACS Nano., 11, pp.8257–8264.*

[15]. Howorka, S., **2017**. Building membrane nanopores. *Nat. Nanotechnol.,* 12, pp. 619–630.

[16]. Friedman, S., DeCamp, D., Sijbesma, R., Srdanov, G., Wudl, F., Kenyon, G., **1993**. Inhibition of the HIV-1 protease by fullerene derivatives: model building studies and experimental verifcation. *J. Am. Chem. Soc.,* 115, pp. 6506–6509.

[17]. Dugan, L., Turetsky, D., Du, C., Lobner, D., Wheeler, M., Almli, C., Shen, C., Luh, T., Choi, D., Lin, T., **1997**. Carboxyfullerenes as neuroprotective agents. *Proc. Natl. Acad. Sci.,* 9, pp. 9434–9439.

[18]. Cagle DW, Kennel SJ, Mirzadeh S, Alford JM, Wilson LJ (1999) *In vivo* studies of fullerene-based materials using endohedral metallofullerene radiotracers. *Proc. Natl. Acad. Sci.,* 96, pp. 5182–5187.

[19]. Montellano, A., Da, RT., Bianco, A., Prato, M., **2011**. Fullerene C60 as a multifunctional system for drug and gene delivery. *Nanoscale,*3, pp. 4035–4041.

[20]. Andujar, SA., Lugli, F., Höfnger, S., Enriz, R., Zerbetto, F., **2012**. Amyloid-β fibril disruption by C60- molecular guidance for rational drug design. *Phys. Chem. Chem. Phys.,* 14, pp. 8599–8607.

[21]. Ahmed, L., Rasulev, B., Turabekova, M., Leszczynska, D., Leszczynski, J., **2013**. Receptor- and ligand-based study of fullerene analogues: comprehensive computational approach including quantum-chemical, QSAR and molecular docking simulations. *Org. Biomol. Chem.,* 11, pp. 5798–5808.

[22]. Gökpek, Y., Bilge, M., Bilge, D., Alver, Ö., Parlak, C., **2017**. Adsorption mechanism, structural and electronic properties: 4-phenylpyridine & undoped or doped (B or Si) C60. *J. Mol. Liq.,* 238, pp. 225–228.

[23]. Zacchigna, M., Klumpp, C., Prato, M., Bianco, A., **2009**. *In vitro* behavior of multi functionalized fullerene-warfarin conjugates. *J. Nanosci. Nanotechnol.,* 9, pp. 6210-6221.

[24]. Alipour, E., Alimohammady, F., Yumashev, A., Maseleno, A., **2020.** Fullerene C_{60} containing porphyrin-like metal center as drug delivery system for ibuprofen drug. *J. Mol. Model.*, 26, pp. 7.

[25]. Parlak, C., Alver, O., Senyel, M., **2017**. Computational study on favipiravir adsorption onto undoped and silicon-decorated C_{60} fullerenes. *J. Theory. Comput. Chem.*, 16(2), pp. 1750011.

[26]. Ali-Anber, M., Al-Mowali, AH., Ali, AM., **2014**. Theoretical semiempirical study of the nitrone (anticancer drug) interaction with fullerene C_{60} (as delivery). *Acta Phys. Pol. A,* 126(3), pp.845-848.

[27]. Rik, BF., Khojasteh, R., Ahmadi R., Razi, MK., **2019**. Evaluation of C_{60} nano-structure performance as nano-carriers of procarbazine anti-cancer drug using density functional theory methods. *Iran. Chem. Commun.*, 7, pp. 405-414.

[28]. Zhai, H.J., Zhao, Y.F., Li, W.L., Chen, Q., Bai, H., Hu, H.S., Piazza, Z.A., Tian, W.J., Lu, H.G., Wu, Y.B., Mu, Y.W., Wei, G.F., Liu, Z.P., Li, J., Li, S.D. and Wang, L.S., **2014**. Observation of an all-boron fullerene. *Nat. Chem., 6,* pp. 727-731.

[29]. Yang Z, Ji YL, Lan G, Xu LC, Liu X, Xu B., **2015**. Design molecular rectifier and photodetector with all boron fullerene. *Solid State Commun., 217, pp.38.*

[30]. Dong, H., Hou, T., Lee, S., Li, Y., **2015**. New Ti-decorated B_{40} fullerene as a promising hydrogen storage material. *Sci. Rep.*, 5, pp. 9952.

[31]. Dong, H., Lin, B., Gilmore, K., Hou, T., Lee, ST., Li, Y., **2015**. B_{40} fullerene: An efficient material for CO_2 captures, storage and separation. *Curr. Appl Phys.*, 15, pp. 1084.

[32]. Gao, G., Ma, F., Jiao, Y., Sun, Q., Jiao, Y., Waclawik, E., Du, A., **2015**. Modelling CO_2 adsorption and separation on experimentally-realized B_{40} fullerene. *Comput. Mater. Sci.*, 108, pp. 38–41.

[33]. Moradi, M., Vahabi, V., Bodaghi, A., **2016**. Computational study on the fullerene-like B_{40} Borospherene properties and its interaction with ammonia. *J. Mol. Liq., 223,* pp. 315–320.

[34]. Tang, C., Zhang, X., **2016**. The hydrogen storage capacity of Sc atoms decorated porous boron fullerene B_{40}: A DFT study. *Int. J. Hydrogen Energy,* 41, ppl. 16992-16999.

[35]. Maniei, Z., Shakerzadeh, E., Mahdavifar, Z., **2018**. Theoretical approach into potential possibility of efficient NO_2 detection via B_{40} and Li-B_{40} fullerenes. *Chem. Phys. Lett.*, 691, pp. 360–365.

[36]. Lin, B., Dong, H., Du, C., Hou, T., Lin, H., Li, Y., **2016**. B_{40} fullerene as a highly sensitive molecular device for NH_3 detection at low bias: a first-principles study. *Nanotechnol.*, 27, pp. 075501.

[37]. Kaur, J., Kumar, R., Vohra, R., **2020**. A pursuit to design highly sensitive fullerene-based sensors: adsorption and dissociation phenomenon of toxic sulfur gases on B_{40} fullerene. *J. Mol. Model,* 26, pp. 17.

[38]. Shakerzade, E., Biglari, Z., Tahmasebi, E., **2016**. M-B_{40} (M = Li, Na, K) serving as a potential promising novel NLO nanomaterial. *Chem. Phys. Lett., 65,* pp. 76-80.

[39]. Li Z, Yu., G, Zhang, X., Huang, X., Chen, W., **2017**. Bonding the Superalkali M_3O (M=Li and K): An Effective Strategy to Improve the Electronic and Nonlinear Optical Properties of the Inorganic B_{40} Nanocage. *Physica E,* 94, pp. 204-210.

[40]. Wang, W., Guo, Y., Yan, X., **2016**. The spin-dependent transport of transition-metal-encapsulated B_{40} fullerene. *RSC Adv.*

[41]. Fa, W., Chen, S., Pande, S., Zeng, X., **2015**. Stability of metal encapsulating boron fullerene B_{40}. *J. Phys. Chem. A,* 119(45), pp. 11208–11214.

[42]. Shah, EV., Roy, D., **2016**. Sc_3N and Sc_2C_2 encapsulated B_{40}: Smarter than its carbon analogue. *Physica E,* 8, pp.: 354–360.

[43]. Kaur, R., Kaur, J., **2017**. The electronic transport properties of B_{40} fullerenes with chalcogens as anchor atoms. *J. Mol. Model.,* 23, pp. 351.

[44]. Kaur, J., Kaur, R., **2017**. To delve about the charge transport properties of p-block elements doped M-B_{40} (M= Al, Si, P, S) molecular device, *8th ICCCNT* 1-4.

[45]. Pardini, B., Kumar, R., Naccarati, A., Novotny, J., Prasad, R., Forsti, A., Hemminki, K., Vodicka, P., Lorenzo, B., **2011**. %-Fluorouracil-based chemotherapy for colorectal cancer and MTHFR/MTRR genotypes. *Br. J. Clin. Pharmacol.,* 72(1), pp. 162-163.

[46]. Afzal, S., Jensen, S., Vainer, B., Vogel, U., Matsen, J., Sorensen, J., Andersen, P., Poulsen, H. **2009**. MTHFR polymorphisms and 5-FU-based adjuvant chemotherapy in colorectal cancer. *Ann. Oncol.,* 20, pp. 1660-6.

[47]. https://www.drugs.com/monograph/fluorouracil.html.

[48]. https://www.reference.medscape.com/drug/adrucil-fluorouracil-342092#showall.

[49]. Ha, J., Hwang, D., YU, J., Park, D., Ryu, S., **2011**. Onset of manic episode during chemotherapy with 5-fluorouracil. *Psychiatry Investig.,* 8(1), pp.71-3.

[50]. Longley, D., Harkin, D., Johnston, P., ***2003***. "5-Fluorouracil: Mechanisms of action and clinical strategies", *Nature,* 3, pp. 330-338.

[51]. Dou, Z., Xu, Y., Sun, H., Liu, Y., ***2012***. "Synthesis of PEGylated fullerene–5-fluorouracil conjugates to enhance the antitumor effect of 5-fluorouracil", *Nanoscale* 4, 4624–4630.

[52]. Zare, K., Sadegh H., Ghoshekandi, R., ***2013***. "Studies of *ab initio* and Monte Carlo simulation on interaction of fluorouracil anticancer drug with carbon nanotube", *J. Nanostruc. Chem.,* 3, 71–78.

[53]. Arsawang, U., Saengsawang, O., Rungrotmongkol, T., Sornmee, P., Wittayanarakul, K., Remsungnen, T., Hannongbua, S., ***2011***. How do carbon nanotubes serve as carriers for gemcitabine transport in a drug delivery system? *J. Mol. Graphics Modell.,* 29, 591-596.

[54]. Hazrati M., Hadipour, N., Adsorption behaviorof 5-fluorouracil on pristine, B-, Si- and Al-doped C60 fullerenes: A first-principles study, *Phys. Lett. A*, 380 (2016) 937-941.

[55]. Hazrati, M., Javanshir, Z., Bagheri, Z., **2017**. $B_{24}N_{24}$ fullerene as carrier for 5-fluorouracil anti-cancer drug delivery: DFT studies. *J. Mol. Graphics Modell.*

[56]. Lang, N., ***1995***. Resistance of atomic wires, Phys. Rev. B: Condens. *Matter Mater. Phys.*, 52, 5335.

[57]. Xue, Y., Datta, S., Ratner, M., ***2002***. First-principles based matrix Green's function approach to molecular electronic devices, General Formulism. *Chem. Phys.*, 281, 151.

[58]. Brandbyge, M., Mozos, J., Ordejon, P., Taylor, J., Stockbro, K., ***2002*** Densityfunctional method for non-equilibrium electron transport, *Phys. Rev. B: Condens. Matter Mater. Phys.*, 65, 165401.

[59]. Taylor, J., Guo, H., Wang, J., ***2001***. *Ab initio* modeling of quantum transport properties of molecular electronic devices, *Phys. Rev. B: Condens. Matter Mater. Phys.* 6, 245407.

[60]. Taylor, J., Brandbyge, M., Stockbro, K., ***2002***. Theory of rectification in tour wires: the role of electrode coupling. *Phys. Rev. Lett.*, 89(13), 138301.

[61]. Swanson, G., Faulkner, J., Smalley, S., Noble, M., Stephens, R., O'Rourke, T., Weiss, G., Quick, D., Thompson, J., Crawford, E., **2006**. Locally advanced prostate cancer treated with concomitant radiation and 5-fluorouracil: Southwest oncology group study 9024. *J. Urol.*, 17692, pp.548-53. 10.1016/j.juro.2006.03.068.

[62]. Berlin, J., Propert, K., Trump, D., Wilding, G., Hudes, G., Glick, J., Burch, P., Keller, A., Loehrer, P., **1998**. 5-Fluorouracil and leucovorin therapy in patients with hormone refractory prostate cancer: An eastern group Phase II study (E1889). *Am. J. Clin. Oncol.*, 21(2), pp.171-6. 10.1097/00000421-199804000-00016.

[63]. Manogue, C., Cotogno, P., Moses, M., Barata, P., Layton, J., Sartor, A., Lewis, B., **2019**. Continuous infusion 5-fluorouracil (5FU) as a novel treatment for heavily pretreated prostate cancer patients: An update. *J. Clin. Oncol.*, 37(7), pp. 319.

[64]. Perdew, J., Burke, K., Ernzerhof, M., ***1996***. Generalized gradient approximation made simple, *Phys. Rev. Lett.*, 77, 3865.

[65]. Atomistic Toolkit Manual, Quantumwise Inc. Atomistix toolkit version 13.8.0, Quantumwise A/S (http://quantumwise.com).

[66]. Toporkiewicz, M., Meissner, J., Matusewicz, L., Czogalla, A., Sikorski, A., **2015**. Toward a magic or imaginary bullet? Ligands for drug targeting to cancer cells: principles, hopes, and challenges. *Int. J. Nanomed.*,10, pp. 1399-1414.

[67]. Sutradhar, K., Amin, M. **2014**. Nanotechnology in cancer drug delivery and selective targeting. *ISRN Nanotechnology*. 939378. 10.1155/2014/939378.

[68]. Hobbs, S., Monsky, W., Yuan, F., et, al., **1998.** "Regulation of transport pathways in tumor vessels: role of tumor type and microenvironment," *Proc. Natl. Acad. Sci. U.S.A.*, vol. 95, no. 8, pp. 4607–4612.

[69]. Rubin, P., Casarett, G., ***1966.*** "Microcirculation of tumors Part I: anatomy, function, and necrosis," *Clin. Radiol.*, vol. 17, no. 3, pp. 220–229.

[70]. Shubik, P., **1982**. "Vascularization of tumors: a review," *Journal of Cancer Research and Clinical Oncology*, vol. 103, no. 3, pp. 211–226.

[71]. Jain, R., Stylianopoulos, T., **2010** "Delivering nanomedicine to solid tumors," *Nat. Rev. Clin. Oncol.*, vol. 7, no. 11, pp. 653–664.

[72]. Jang, S., Wientjes, Lu, D., J. L.-S. Au., ***2003.*** "Drug delivery and transport to solid tumors," *Pharm. Res.*, vol. 20, no. 9, pp. 1337–1350,

[73]. Banerjee, R., Tyagi, P., Li, S., Huang L **2004.** Anisamide-targeted stealth liposomes: A potent carrier for targeting doxorubicin to human prostate cancer cells. *Int. J. Cancer,* 112:693-700. 10.1002/ijc.20452.

[74]. Dasargyri A, Kumin CD, Leroux JC **2016**. Targeting nanocarriers with anisamide: Fact or Artifact? *Adv. Mater.*, 10.1002/adma.201603451.

[75]. Wang, Z., Ciacchi, L., Wei, G., **2017**. Recent advances in the synthesis of grapheme based nanomaterials for controlled drug delivery. *Appl. Sci. 7:1175.* 10.3390/app7111175.

[76]. Swietach, P., Vaughan-Jones, R., Harris, A., Hulikova, A., **2014**. The chemistry, physiology and pathology of pH in cancer. *Phil. Trans. R. Soc. B*, 369:20130099-20130110.

[77]. N.Kandasamy, F.Ahmad, D.Ajitha, B.Raj, and N.Telagam, "Quantum Dot Cellular Automata Based Scan Flip Flop and Boundary Scan Register", *IETE J. Res.*, Vol.69, PP. 535-548, 2023.

[78]. S. K.Sharma, B.Raj, and M.Khosla, "Enhanced Photosensitivity of Highly Spectrum Selective Cylindrical Gate In1-xGaxAs Nanowire MOSFET Photodetector", *Mod. Phys. Lett. B*, Vol. 33, No. 12, p. 1950144, 2019.

[79]. J.Singh, and B.Raj, "Design and Investigation of 7T2M NVSARM with Enhanced Stability and Temperature Impact on Store/Restore Energy", *IEEE Transactions on Very Large Scale Integration Systems*, Vol. 27, No. 6, pp. 1322–1328, 2019.

[80]. A. K.Bhardwaj, S.Gupta, B.Raj, and A.Singh,"Impact of Double Gate Geometry on the Performance of Carbon Nanotube Field Effect Transistor Structures for Low Power Digital Design", *Comput. Theor. Nanosci.*, ASP USA, Vol. 16, pp. 1813–1820, 2019.

[81]. N.Jain, and B.Raj, "Thermal Stability Analysis and Performance Exploration of Asymmetrical Dual-k Underlap Spacer (ADKUS) SOI FinFET for Security and Privacy Applications", *Indian J. Pure Appl. Phys. (IJPAP)*, Vol. 57, pp. 352–360, 2019.

[82]. A.Singh, M.Khosla, and B.Raj, "Design and Analysis of Dynamically Configurable Electrostatic Doped Carbon Nanotube Tunnel FET", *Microelectron. J.*, Elsevier, Vol. 85, pp. 17–24, 2019.

[83]. N.Jain, and B.Raj, "Dual-k Spacer Region Variation at the Drain Side of Asymmetric SOI FinFET Structure: Performance Analysis towards the Analog/RF Design Applications", *J. Nanoelectron. Optoelectron.*, American Scientific Publishers, Vol. 14, pp. 349–359, 2019.

[84]. J.Singh, S.Sharma, B.Raj, and M.Khosla, "Analysis of Barrier Layer Thickness on Performance of In1-xGaxAs Based Gate Stack Cylindrical Gate Nanowire MOSFET", *JNO*, ASP USA, Vol. 13, pp. 1473–1477, 2018.

[85]. N.Jain, and B.Raj, "Analysis and Performance Exploration of High-k SOI FinFETs over the Conventional Low-k SOI FinFET toward Analog/ RF Design", *J. Semicond. (JoS)*, IOP Science,Vol. 39, No. 12, pp. 124002-1-7, 2018.

[86]. C.Goyal, J. S.Ubhi, and B.Raj, "A Reliable Leakage Reduction Technique for Approximate Full Adder with Reduced Ground Bounce Noise", *J. Math. Prob. Eng.*, Hindawi, Vol. 2018, Article ID 3501041, PP.1-16, 2018.

[87]. Anuradha, J. Singh, B.Raj, and M.Khosla, "Design and Performance Analysis of Nano-Scale Memristor-Based Nonvolatile SRAM", *J. Sens. Lett.*, American Scientific Publishers USA, Vol. 16, pp. 798–805, 2018.

[88]. G.Wadhwa, and B.Raj, "Parametric Variation Analysis of Charge-PlasmaBased Dielectric Modulated JLTFET for Biosensor Application", *IEEE Sens. J.*, Vol. 18, No. 15, PP. 6070 - 6077, 2018.

[89]. J.Singh, and B.Raj, "Comparative Analysis of Memristor Models for Memories Design", *J. Semicond. (JoS)*, IoP, Vol. 39, No. 7, PP. 074006-1-12, 2018.

[90]. D.Yadav, S. S.Chouhan, S. K.Vishvakarma, and B.Raj, "Application Specific Microcontroller Design for IoT Based WSN", *Sens. Lett.*, ASP, Vol. 16, pp. 374–385, 2018.

[91]. G.Singh, R. K.Sarin, and B.Raj, "Fault-Tolerant Design and Analysis of Quantum-Dot Cellular Automata Based Circuits", *IEEE/IET Circuits, Devices Syst.*, Vol. 12, pp.638–664, 2018.

[92]. J.Singh, and B.Raj, "Modeling of Mean Barrier Height Levying Various Image Forces of Metal Insulator Metal Structure to Enhance the Performance of Conductive Filament Based Memristor Model", *IEEE Nanotechnol.*, Vol. 17, No. 2, pp. 268–275, 2018.

[93]. A.Jain, S.Sharma, and B.Raj, "Analysis of Triple Metal Surrounding Gate (TM-SG) III-V Nanowire MOSFET for Photosensing Application", *Opto. Electron. J.*, Elsevier, Vol. 26, No. 2, PP. 141–148, 2018.

[94]. A.Jain, S.Sharma, and B.Raj, "Design and Analysis of High Sensitivity Photosensor Using Cylindrical Surrounding Gate MOSFET for Low Power Sensor Applications", *Eng. Sci. Technol. Int. J.*, Elsevier's, Vol. 19, No. 4, pp. 1864–1870, 2016.

[95]. A.Singh, M.Khosla, and B.Raj, "Analysis of Electrostatic Doped Schottky Barrier Carbon Nanotube FET for Low Power Applications," *J. Mater. Sci. - Mater. Electron.*, Springer,Vol. 28, PP. 1762–1768, 2017.

[96]. G.Saiphani Kumar, A.Singh, B.Raj, "Design and Analysis of Gate All around CNTFET Based SRAM Cell Design", *J. Comput. Electron.*, Springer, Vol. 17, No.1, PP. 138–145, 2018.

[97]. G. P.Singh, B. S.Sohi, and B.Raj, "Material Properties Analysis of Graphene Base Transistor (GBT) for VLSI Analog Circuits", *Indian J. Pure Appl. Phys. (IJPAP)*, Vol. 55, pp. 896–902, 2017.

[98]. S.Kumar and B.Raj, "Estimation of Stability and Performance Metric for Inward Access Transistor Based 6T SRAM Cell Design Using n-type/p-type DMDG-GDOV TFET", *IEEE VLSI Circuits and Systems Letter*, Vol. 3, No. 2, PP. 25–39, 2017.

[99]. S.Sharma, A.Kumar, M.Pattanaik, and B.Raj, "Forward Body Biased Multimode Multi-Threshold CMOS Technique for Ground Bounce Noise Reduction in Static CMOS Adders", *Int. J. Inform. Electron. Eng.*, Vol. 3, No. 3, PP. 567–572, 2013.

[100]. H.Singh, P.Kumar, and B.Raj, "Performance Analysis of Majority Gate SET Based 1-bit Full Adder", *IJCCE*, IACSIT Press Singapore, ISSN: 20103743, Vol. 2, No. 4, PP.1-8, 2013.

[101]. A. K.Bhardwaj, S.Gupta, and B.Raj, "Investigation of Parameters for Schottky Barrier (SB) Height for Schottky Barrier Based Carbon Nanotube Field Effect Transistor Device", *J. Nanoelectron. Optoelectron.*, ASP, Vol. 15, PP. 783–791, 2020.

[102]. P.Bansal, and B.Raj, "Memristor: A Versatile Nonlinear Model for Dopant Drift and Boundary Issues", *JCTN*, American Scientific Publishers, Vol. 14, No. 5, PP. 2319–2325, 2017.

[103]. N.Jain, and B.Raj, "An Analog and Digital Design Perspective Comprehensive Approach on Fin-FET (Fin-Field Effect Transistor) Technology - A Review", *Reviews in Advanced Sciences and Engineering (RASE)*, ASP, Vol. 5, PP. 1–14, 2016.

[104]. S.Sharma, B.Raj, and M.Khosla, "Subthreshold Performance of In1-xGaxAs Based Dual Metal with Gate Stack Cylindrical/Surrounding Gate Nanowire MOSFET for Low Power Analog Applications", *J. Nanoelectron. Optoelectron.*, American Scientific Publishers, USA, Vol. 12, PP. 171– 176, 2017.

[105]. B.Raj, A. K.Saxena, and S.Dasgupta, "Analytical Modeling for the Estimation of Leakage Current and Subthreshold Swing Factor of Nanoscale Double Gate FinFET Device" *Microelectron. Int.*, UK, Vol. 26, PP. 53–63, 2009.

[106]. S. S.Soniya, G.Wadhwa, and B.Raj, "An Analytical Modeling for Dual Source Vertical Tunnel Field Effect Transistor", *IJRTE*, Vol. 8, No. 2, 2019.

[107]. S.Singh, and B.Raj, "Design and Analysis of Hetrojunction Vertical T-Shaped Tunnel Field Effect Transistor", *Journal of Electronics Material*, Springer, Vol. 48, No. 10, PP. 6253–6260, 2019.

[108]. C.Goyal, J. S.Ubhi, and B.Raj, "A Low Leakage CNTFET Based Inexact Full Adder for Low Power Image Processing Applications", *Int. J. Circuit Theory Appl.*, Wiley, Vol. 47, No. 9, PP. 1446–1458, 2019.

[109]. B.Raj, A. K.Saxena, and S.Dasgupta, "A Compact Drain Current and Threshold Voltage Quantum Mechanical Analytical Modeling for FinFETs", *J. Nanoelectron. Optoelectron. (JNO)*, USA, Vol. 3, No. 2, PP. 163–170, 2008.

[110]. G.Wadhwa, and B.Raj, "An Analytical Modeling of Charge Plasma Based Tunnel Field Effect Transistor with Impacts of Gate Underlap Region", *Superlattices and Microstruct.*, Elsevier, Vol. 142, PP. 106512-22, 2020.

[111]. S.Singh, and B.Raj, "Modeling and Simulation Analysis of SiGe Heterojunction Double GateVertical t-Shaped Tunnel FET", *Superlattices Microstruct.*, Elsevier,Vol. 142, PP. 106496-104, 2020.

[112]. A.Singh, D. K.Saini, D.Agarwal, S.Aggarwal, M.Khosla, and B.Raj, "Modeling and Simulation of Carbon Nanotube Field Effect Transistor and Its Circuit Application", *J. Semicond. (JoS)*, IOP Science, Vol. 37, PP. 074001-074006, 2016.

[113]. N.Jain, and B.Raj, "Device and Circuit Co-Design Perspective Comprehensive Approach on FinFET Technology - A Review", *J. Electron. Devices*, Vol. 23, No. 1, PP. 1890–1901, 2016.

[114]. S.Kumar and B.Raj, "Analysis of ION and Ambipolar Current for DualMaterial Gate-Drain Overlapped DG-TFET", *J. Nanoelectron. Optoelectron.*, American Scientific Publishers, USA, Vol. 11, PP. 323- 333, 2016.

[115]. N.Anjum, T.Bali, and B.Raj, "Design and Simulation of Handwritten Multiscript Character Recognition", *IJARCCE*, Vol. 2, No. 7, PP. 2544–2549, 2013.

[116]. S.Sharma, B.Raj, and M.Khosla, "A Gaussian Approach for Analytical Subthreshold Current Model of Cylindrical Nanowire FET with Quantum Mechanical Effects", *Microelectron. J.*, Elsevier, Vol. 53, PP. 65–72, 2016.

[117]. K.Singh, and B.Raj, "Performance and Analysis of Temperature Dependent Multi-Walled Carbon Nanotubes as Global Interconnects at Different Technology Nodes", *J. Comput. Electron.*, Springer, Vol. 14, No. 2, PP. 469–476, 2015.

[118]. S.Kumar and B.Raj, "Compact Channel Potential Analytical Modeling of DG-TFET Based on Evanescent–Mode Approach", *J. Comput. Electron.*, Springer, Vol. 14, No. 2, PP. 820–827, 2015.

[119]. K.Singh, and B.Raj, "Temperature Dependent Modeling and Performance Evaluation of Multi-Walled CNT and Single-Walled CNT as Global Interconnects", *J. Electron. Mater*, Springer,Vol. 44, No. 12, PP. 4825–4835, 2015.

[120]. V. K. Sharma, M. Pattanaik, and B. Raj, "INDEP Approach for Leakage Reduction in Nanoscale CMOS Circuits", *Int. J. Electron.*, Taylor & Francis, Vol. 102, No. 2, PP. 200–215, 2014.

[121]. K.Singh, and B.Raj, "Influence of Temperature on MWCNT Bundle, SWCNT Bundle and Copper Interconnects for Nanoscaled Technology Nodes", *J. Mater. Sci.: Mater. Electron.*, Springer, Vol. 26, No. 8, PP. 6134–6142, 2015.

[122]. N.Anjum, T.Bali, and B.Raj, "Design and Simulation of Handwritten Gurumukhi and Devanagri Numerical Recognition", *Int. J. Comput. Appl.*, Published by Foundation of Computer Science, New York, USA,Vol. 73, No. 12, PP. 16–21, 2013.

[123]. S.Khandelwal, V.Gupta, B.Raj, and R. D.Gupta, "Process Variability Aware Low Leakage Reliable Nano Scale DG-FinFET SRAM Cell Design Technique", *J. Nanoelectron. Optoelectron.*, Vol. 10, No. 6, PP. 810–817, 2015.

[124]. V. K. Sharma, M. Pattanaik, and B.Raj, "ONOFIC Approach: Low Power High Speed Nanoscale VLSI Circuits Design", *Int. J. Electron.*, Taylor & Francis, Vol. 101, No. 1, PP. 61–73, 2014.

[125]. S.Khandelwal, B.Raj, and R. D.Gupta, "FinFET Based 6T SRAM Cell Design: Analysis of Performance Metric, Process Variation and Temperature Effect", *J. Comput. Theor. Nanosci.*, ASP, USA, Vol. 12, PP. 2500–2506, 2015.

[126]. S.Singh, Y.Shekhar, R.Jagdeep, S.Anurag, and B.Raj, "Impact of HfO2 in Graded Channel Dual Insulator Double Gate MOSFET", *J. Comput. Theor. Nanosci.*, American Scientific Publishers, Vol. 12, No. 6, PP. 950–953, 2015.

[127]. V. K.Sharma, M.Pattanaik, and B.Raj, "PVT Variations Aware Low Leakage INDEP Approach for Nanoscale CMOS Circuits", *Microelectron. Reliab.*, Elsevier, Vol. 54, PP. 90–99, 2014.

[128]. B.Raj, A. K.Saxena and S.Dasgupta, "Quantum Mechanical Analytical Modeling of Nanoscale DG FinFET: Evaluation of Potential, Threshold Voltage and Source/Drain Resistance", *Elsevier's J. Mater. Sci. Semicond. Process.*, Elsevier, Vol. 16, No. 4, PP. 1131–1137, 2013.

[129]. M.Gopal, S. S. D.Prasad, and B.Raj, "8T SRAM Cell Design for Dynamic and Leakage Power Reduction", *Int. J. Comput. Appl.*, Published by Foundation of Computer Science, New York, USA,Vol. 71, No. 9, PP. 43–48, 2013.

[130]. M.Pattanaik, B.Raj, S.Sharma, and A.Kumar, "Diode Based Trimode MultiThreshold CMOS Technique for Ground Bounce Noise Reduction in Static CMOS Adders", *Adv. Mat. Res.*, Trans Tech Publications, Switzerland, Vol. 548, PP. 885–889, 2012.

[131]. B.Raj, A. K.Saxena, and S.Dasgupta, "Nanoscale FinFET Based SRAM Cell Design: Analysis of Performance Metric, Process Variation, Underlapped FinFET and Temperature Effect", *IEEE Circuits Syst. Mag.*, Vol. 11, No. 2, PP. 38–50, 2011.

[132]. V. K. Sharma, M. Pattanaik, and B.Raj, "Leakage Current ONOFIC Approach for Deep Submicron VLSI Circuit Design", *International Journal of Electrical, Computer, Electronics and Communication Engineering*, World Academy of Sciences, Engineering and Technology, Vol. 7, No. 4, PP. 239–244, 2013.

[133]. T.Chawla, M.Khosla, and B.Raj, "Design and Simulation of Triple Metal Double-Gate Germanium on Insulator Vertical Tunnel Field Effect Transistor", *Microelectron. J.*, Elsevier, Vol. 114, PP. 105125-32, 2021.

[134]. P.Kaur, S. S.Gill, and B.Raj, "Comparative Analysis of OFETs Materials and Devices for Sensor Applications", *J. Silicon*, Springer,Vol. 14, PP. 4463–4471, 2022.

[135]. S. K.Sharma, P.Kumar, B.Raj, and Balwinder Raj, "In1-xGaxAs Double Metal Gate-Stacking Cylindrical Nanowire MOSFET for Highly Sensitive Photo Detector", *J. Silicon*, Springer, Vol. 14, PP. 3535–3541,2022.

[136]. B.Raj, A. K.Saxena, and S.Dasgupta, "Analytical Modeling of Quasi Planar Nanoscale Double Gate FinFET with Source/Drain Resistance and Field Dependent Carrier Mobility: A Quantum Mechanical Study", *J. Comput. (JCP)*, Academy Publisher, FINLAND, Vol. 4, No. 9, PP.1-8, 2009.

[137]. S.Bhushan, S.Khandelwal, and B.Raj, "Analyzing Different Mode FinFET Based Memory Cell at Different Power Supply for Leakage Reduction", *Seventh International Conference on Bio-Inspired Computing: Theories and Application, (BIC-TA 2012)*, organized IIITM Gwalor, PP. 89–100, 14-16 Dec 2012.

[138]. J.Singh, and B.Raj, "Temperature Dependent Analytical Modeling and Simulations of Nanoscale Memristor", *Eng. Sci. Technol.*, an *International Journal*, Elsevier's, Vol. 21, PP. 862–868, 2018.

[139]. S.Singh, S.Bala, B.Raj, and Balwinder Raj, "Improved Sensitivity of Dielectric Modulated Junctionless Transistor for Nanoscale Biosensor Design", *Sens. Lett.*, ASP, Vol. 18, PP. 328–333, 2020.

13

Design and Analysis of Rectangular Wave Guide Using an HFSS Simulator

Srilali Siragam

Department of Electronics & Communication Engineering, Swarnandhra College of Engineering and Technology, Seetharamapuram, Narsapur, A.P., India

Abstract
When a rectangular waveguide is in free space, the waveguide's characteristics and field distribution for different modes are computed using HFSS software. The analysis is performed for several parameters. The electric and magnetic field intensities, as well as the fundamental modal distributions, are investigated within a rectangular waveguide. Simulated analysis for the X-band is performed on a single size that has dimensions of 22.76, 10.15, 0.2, and 60 mm for breadth, width, wall thickness, and waveguide lengths. The performance of the rectangular waveguide is examined together with characteristic profiles. The waveguide is simulated, and the outcomes are confirmed. An HFSS (High Frequency Structural Simulator) simulation platform is shown in the figure for the analysis of aperture radiation from a rectangular waveguide made of dielectric material or air.

Keywords: Wave guide, HFSS simulator, microwaves, cut-off frequencies, impedance matching, numerical analysis

13.1 Background

For distinct kinds of waves, there are many waveguide types. The most common and original denotation is a hollow metallic conductor tube used to convey radio waves, notably microwaves (μ-waves). At higher radio frequencies, dielectric waveguides are employed, whilst optical fibers and

Email: srilalisep7@gmail.com

Balwinder Raj, Suman Lata Tripathi, Tarun Chaudhary, K. Srinivasa Rao and Mandeep Singh (eds.) Integrated Devices for Artificial Intelligence and VLSI, (329–354) © 2024 Scrivener Publishing LLC

transparent dielectric waveguides are used as light waveguides. In the absence of a physical constraint such as a waveguide, wave intensities decline as they propagate into 3-D space utilizing the square law of inversion [1].

Each waveguide has a cut-off wavelength set by its dimensions and will not carry waves of larger wavelength. Similarly, an optical fiber that directs light will not carry waves of much longer wavelength. Natural structures can act as waveguides as well. Any shape of waveguide cross section can support electromagnetic waves. It is difficult to analyze irregular shapes. Waveguides are frequently employed in rectangular and circular configurations.

Long before the term was coined, waveguides were utilized for signal transmission. Sound waves steered through a tight wire, as well as sound waves guided through a hollow tube-like cave or stethoscope, have long been recognized. Waveguides also carry electricity among radio, optical and radar equipment. Waveguides are the primary idea behind wave guide testing, one of many non-destructive testing techniques.

The following examples are:

Light and signals can be transported over long distances using optical fibers, which have minimal reduction and a varied usable wavelength range.
The impedance of a waveguide in a radar must be coordinated for effective power transmission.
Parabolic dish feeds are frequently connected to their electronics, which could be receivers or emitters, using both circular and rectangular waveguides.
Waveguides are employed in scientific instruments to assess a material's properties. The specimen can be tested with the waveguide in touch with it to save the power, or it can be tested waveguide inside to test tiny objects more precisely. One common kind of waveguide is transmission lines [1].

(*i*) Properties of Modes and Cut-Off Frequencies

One solution of the shape of the wave or wave equations is a propagation mode in a waveguide. Because of the boundary condition constraints, the wave function in the waveguide can only propagate in a limited number of frequencies and forms. A mode's cut-off frequency (f_c) is the minimum frequency at which it can broadcast. The waveguide's fundamental mode and its lesser f_c is the waveguide cut-off frequency [2].

To find propagation modes, the Helmholtz equation is solved along with boundary conditions based on the form and bordering the region. The conventional assumption for infinitely elongated uniform waveguides permits us to assume that the wave has a propagating form. Commonly, all unknown time-varying fields are first replaced with their complex phasor representation, which is adequate to properly define any infinitely long tone signal at frequency, and then boundary conditions are rewritten in accordance with this.

The guided wave's propagation constant is difficult in general. Depending on the eigenvalue equation solution and angular frequency, the propagation constant in the lossless case might have either real or imaginary values [3].

(*ii*) Impedance Matching

According to the network concept, impedance, which is measured in Ω, is a generalization of resistance in AC version [2]. According to circuit theory, a waveguide is a transmission line with a particular length and impedance. In different way, during wave propagation, the impedance represents the circuit component's voltage-to-current ratio. Once the wave and material parameters are appropriately translated into electrical terms, this waveguide description, which was initially meant for alternating current, also applies to electromagnetic and sound waves.

Impedance matching is essential when connecting the parts of an electric circuit because it controls the wave transmitted in forward and reverse directions. A full transmission is often required when mounting a waveguide to an antenna; therefore an attempt is finished impedance matching.

The coefficient of reflection (ρ) can be found using: (Gamma) = $(Z_2-Z_1)/(Z_2+Z_1)$, where Z_1 and Z_2 are the first and second components' respective impedances, and 0 and 1 represent full transmission and reflection, respectively, and ½ represents a input voltage is half reflected.

A standing wave is caused by a reflected wave and incoming waves mixed with an impedance (Z) mismatch. The standing wave ratio also indicates impedance mismatch, which is correlated with the ratio of Z and ρ by the formula: $VSWR=|Vmax|/|Vmin|=(1+|\rho|)/(1-|\rho|)$, where value 1 indicates transmission full.

13.2 Introduction

A hollow tube called a waveguide is used to transmit electromagnetic waves in the microwave spectrum. An enclosed conducting variant of the WR-90 rectangular waveguide is used [4, 5]. Depending on how they are shaped, waveguides are classified into many categories. Rectangular waveguides are defined as waveguides with a rectangular shape. The first kind of transmission line utilized for compact devices like radars and within equipment shelters is rectangular waveguide. Electromagnetic energy or power is sent from one point in space to another using rectangular waveguides. These are employed in numerous applications, like radars, slotted lines, isolators, and attenuators. A rectangular waveguide can only support transverse magnetic (TM) and transverse electric (TE) modes and not transverse electromagnetic (TEM) waves because there is only one conductor in the waveguide [6–8]. HFSS simulation software is used to simulate an air-filled rectangular waveguide in order to obtain various parameters. Analysis of the magnetic and electric field patterns has been done. The performance of the waveguide is investigated together with the analysis of characteristic profiles. The outcomes of theoretical analyses have been examined, compared, and validated. The cutoff wavelength has also been identified, and after being contrasted with the theoretical value, its relevance has been investigated. These factors can be useful for producing components that will be used in real-world applications. Below a particular frequency, known as f_c, a rectangular waveguide cannot propagate. The waveguide's rectangular dominant mode, TE_{10}, has the lowest cutoff frequency [9–11].

The rectangular waveguide characteristics were simulated using Ansoft HFSS 16.2, a simulation program. This program is used to compute the electrical behavior of high speed and high frequency components by simulating complicated 3D geometries using the finite element approach. The electrical performance of the components is best characterized and numerous characteristics are efficiently evaluated using the HFSS. It aids in the analysis of a variety of electromagnetic structural characteristics, including S-parameters, propagation constants, input impedances, radiation patterns, etc. It is a potent post-processor that provides unmatched insight into electrical performance, and the results are more precise and useful prior to the design of any component [21–26].

In this article, the rectangular waveguide's characteristics are predicted using HFSS simulation. This software is used to investigate the cutoff frequency, radiation pattern, and other near field phenomena [12–14].

The following process is used to design the waveguide using HFSS, as shown in Figure 13.1. This section describes how to design a waveguide and the procedures involved in the design of a specific waveguide. The designed waveguide is then simulated using HFSS, and the boundaries, excitations, and radiation of the waveguide are observed. The following design consists primarily of project menu, insert HFSS design, solution type, driven model in the first stage, and boundary assignments, material assignment, model drawing, and unit of measurement in the second stage [27–33]. The third stage includes excitation assignment, solution setting, and frequency sweep preparation. Finally, the simulation runs are contained in the final step.

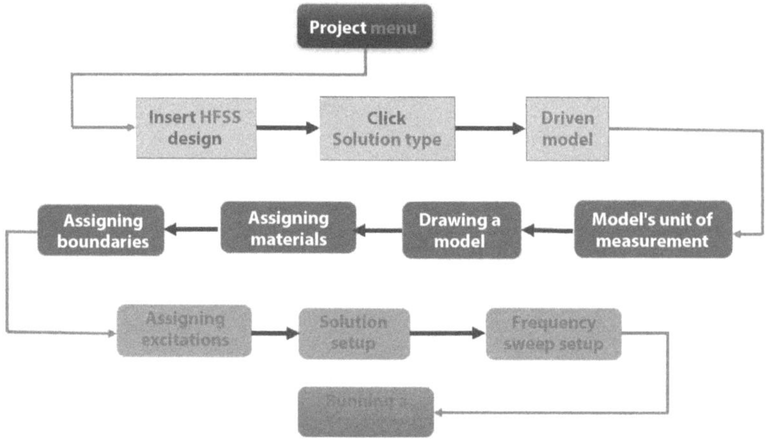

Figure 13.1 HFSS flow graph.

13.3 Mathematical Computations

We investigate a hollow rectangular waveguide with the internal dimensions ab (a>b), which has its length and width in the y and x-axes and is located in the rectangular coordinate system, as depicted in Figure 13.2 (a). The dielectric medium is used to be air as shown in Figure 13.2 (b). The fact that electromagnetic waves propagate in the z direction suggests that the magnetic field H_z's H component is necessary for energy transmission in the waveguide [34–40]. Different field patterns that can propagate independently are present in the electromagnetic waves inside the waveguide and are referred to as modes. It makes sense to categorize two-dimensional fields as TM or TE if E or H were parallel to the direction of propagation. $E_z = 0$ is a property of the TE_{mn} mode, meaning that the H_z component is necessary for waveguide transmission. In TM_{mn} modes, both the E_z component and $H_z = 0$ are required for waveguide transmission [15–17]. The following are the wave equations for waves moving down the z-axis:

$$\gamma = \sqrt{(R + j\omega L)(R + j\omega C)}$$

$$\gamma = \alpha + j\beta$$

$$\Delta^2 = H_z = \mu\varepsilon \frac{\partial^2 Hz}{\partial t^2} \; and \; E_z = 0 \; for \; TE \; waves$$

$$\Delta^2 = E_z = \mu\varepsilon \frac{\partial^2 Ez}{\partial t^2} \; and \; H_z = 0 \; for \; TM \; waves$$

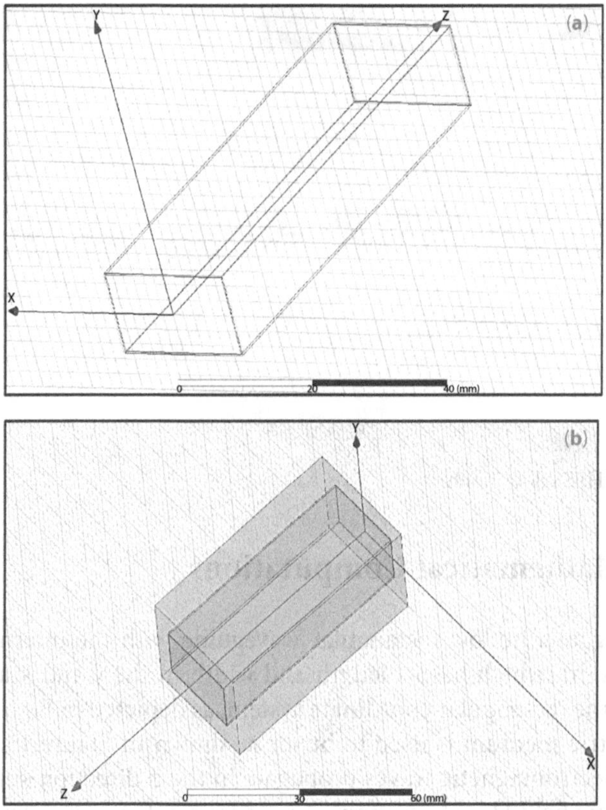

Figure 13.2 (a) Rectangular wave guide with coordinate system. (b) Air filled wave guide.

Here, the electromagnetic wave's magnetic and electric z components inside the waveguide are denoted by H_z and E_z, respectively. The propagation constant is taken to be 1 and the dimensions are 22.86 and 10.16 mm for breadth (a) and width (b), are used to simulate the rectangular waveguide. The cutoff frequency for a TM_{mn} mode is the same as the TE_{mn} mode [41–46]. It is defined as the lowest frequency necessary for wave transmission inside the waveguide. The only distinction is that neither m nor n can be zero. The only distinction is that neither m nor n can be zero. The dominant mode in a given mode is the one with the lowermost frequency. The TE_{10} mode, which is the dominant mode in the rectangular waveguide, is often employed for all transmissions [18, 19].

The various modes of *fc* in rectangular waveguide are given as:

$$f_c = \frac{1}{2\sqrt{\mu\varepsilon}}\sqrt{\frac{m^2}{a^2}+\frac{n^2}{b^2}}$$

Where a and b are the waveguide's predetermined dimensions, and since air is the medium, the permittivity is 1. The guided wavelength is defined as the distance between the two equal phase planes along the waveguide as follows.

$$\lambda_g = \frac{\lambda}{\sqrt{1-\left(\frac{f_c}{f}\right)^2}}$$

$$\gamma_{mn} = j\beta_{mn},$$

$$\beta = \omega\sqrt{\mu\varepsilon\left(1-\left(\frac{f_c}{f}\right)^2\right)} \text{ for } f > f_c$$

If the operating frequency is higher than the cutoff frequency, or when $f > f_c$, the wave propagates in the waveguide. The cutoff frequency is the frequency at which the propagation constant's value shifts from real to imaginary. The characteristic impedance is provided by

$$Z = \frac{\eta}{\sqrt{1-\left(\frac{f_c}{f}\right)^2}} \text{ for TE wave}$$

$$Z = \eta\sqrt{\left(1-\left(\frac{f_c}{f}\right)^2\right)} \text{ for TM wave}$$

Here $\eta = \frac{1}{\sqrt{\mu\varepsilon}}$ is the intrinsic impedance.

13.4 Numerical Analysis

i) Electric field intensity

Source charge Q produces an electric field that is inversely proportional to the square of the source's distance. The inverse square law is applied here [47–52]. The strength of the electric field weakens with increasing separation distance and vice versa. The electric field intensity variations are as shown in Figures 13.6 and 13.7.

The formula is provided by

$$E = \frac{K.Q}{d^2}, \text{Here Q= source charge and d=distance.}$$

ii) Magnetic field intensity

The ability to exert a magnetic force and induce magnetic induction in the matter it is put in can be used to define a magnetic field. The formula is presented as

$$\mathrm{B} = \frac{\mu_0 m}{4\prod d^2},$$

where B= magnetic field, and μ_0=4∏×10^{-7} H/m. The magnetic field intensity variations are also shown in the Figures 13.8 and 13.9.

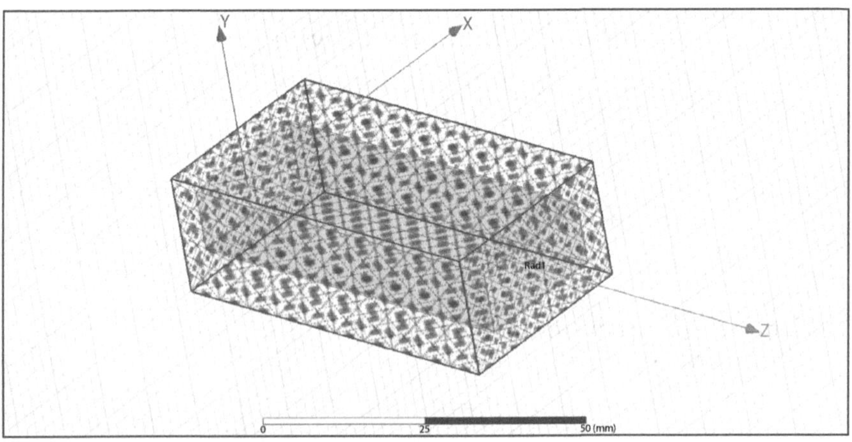

Figure 13.3 Radiation boundary of the waveguide.

Figure 13.3 depicts the radiation boundary of the planned waveguide and its coordinates. It refers to the wave motion from the domain's inner side via the boundary with modest reflections.

The excitation of waveguide ports 1 and 2 is shown in Figures 13.4 and 13.5. The excitation of the rectangular waveguide suggests that the probe should be placed to stimulate the electric field intensity of the mode and that a coupling loop should be placed to produce the desired magnetic field intensity of the mode.

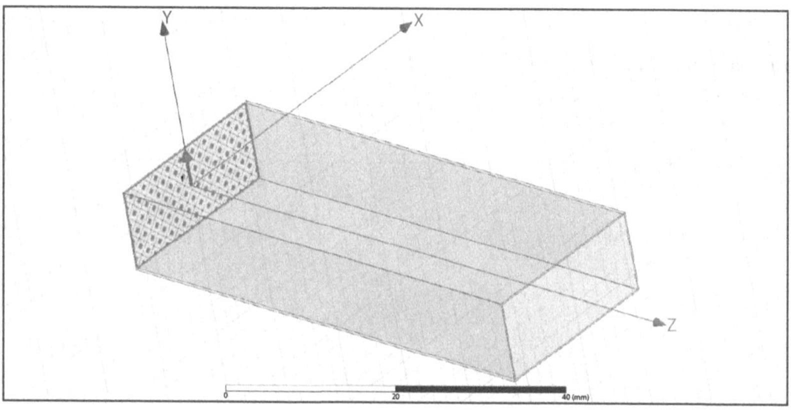

Figure 13.4 Excitation of the waveguide waveport-1.

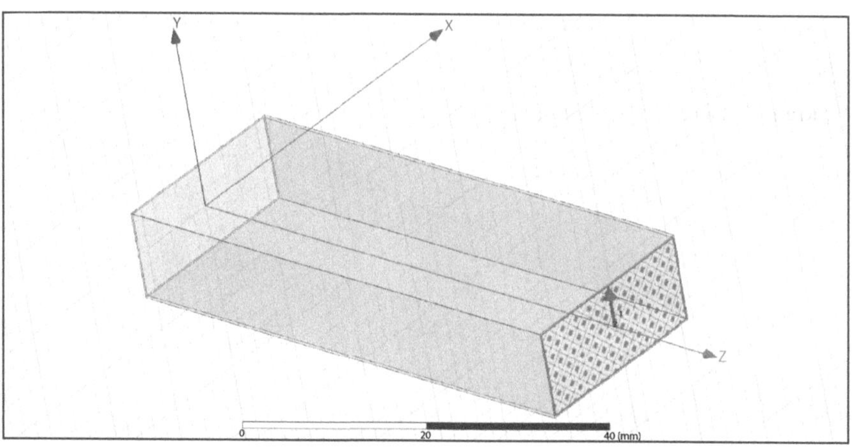

Figure 13.5 Excitation of the waveguide waveport-2.

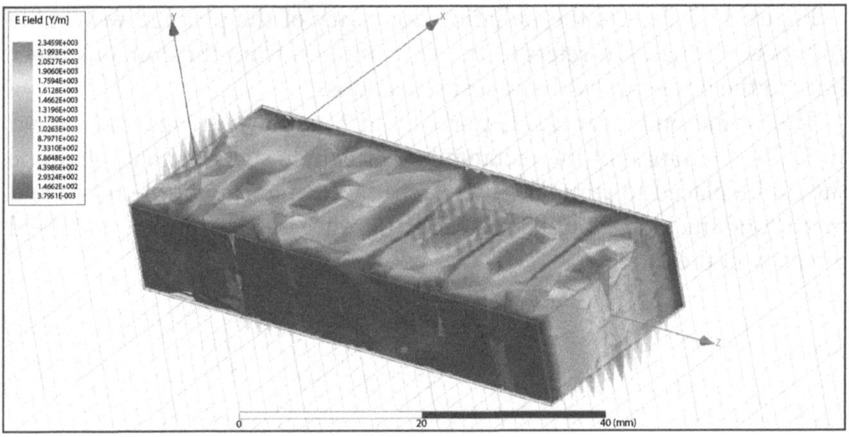

Figure 13.6 Electric field intensity of Mag_E1.

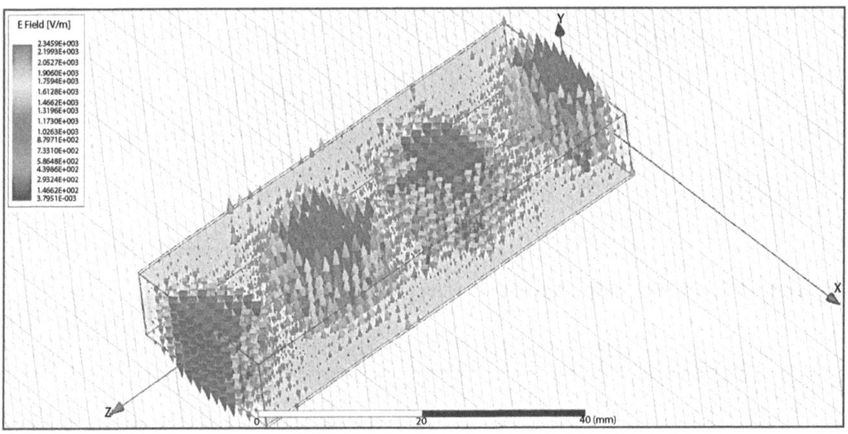

Figure 13.7 Electric field intensity of Vect_E1.

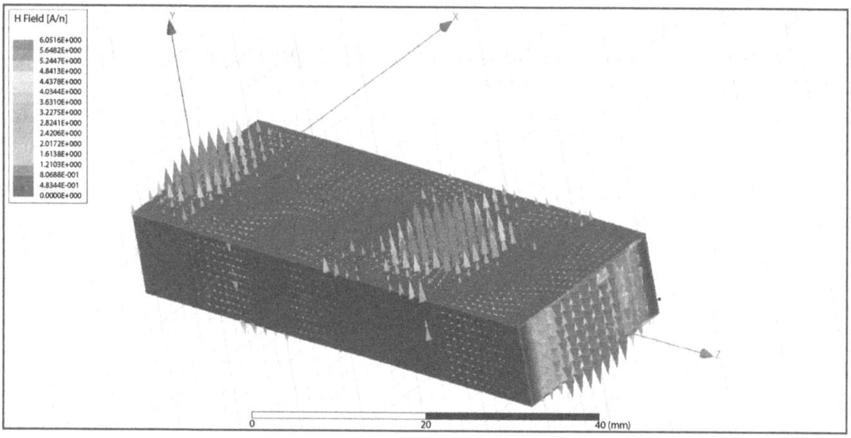

Figure 13.8 Magnetic field intensity of Mag_H1.

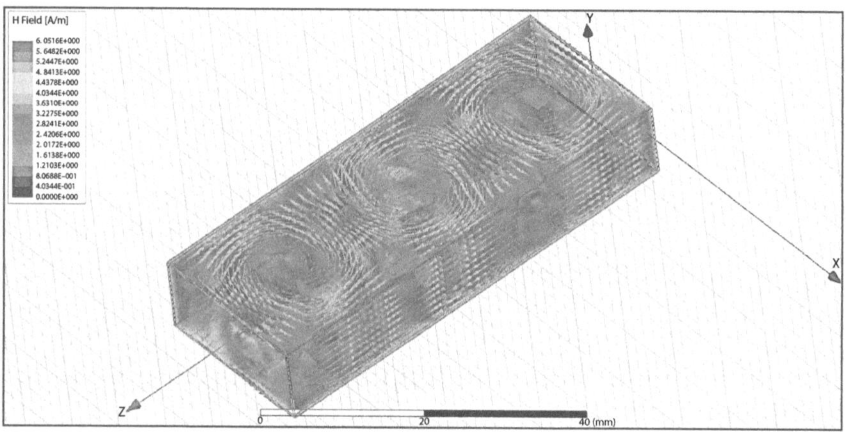

Figure 13.9 Magnetic field intensity of Vect_H1.

A symmetric slab waveguide contains the guided modes TE mode and TM mode. There is no electric field component in the direction of propagation in the TE mode of propagation [53–58]. The magnetic field component is not present in the direction of propagation in the TM mode. TE mode and TM mode operate similarly as a result. The field patterns of the TE and TM modes are depicted in Figure 13.10 as shown below.

E-field

In E-field there are various kinds of transmitting modes are available like TE10, TE20, TE11 and TE, TM11.

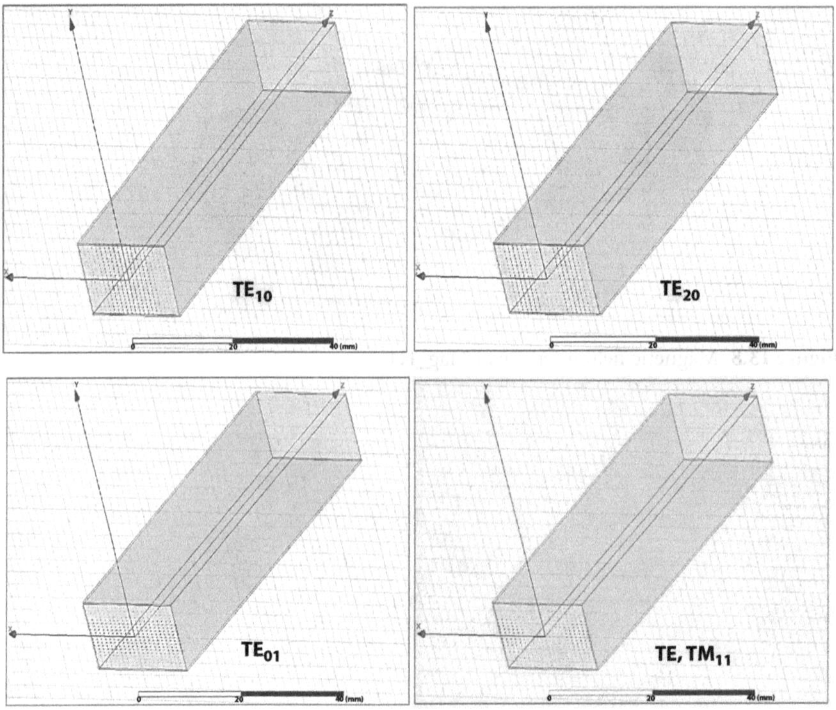

sH-field

Similarly, in H-field there are various kinds of transmitting modes are available like TM10, TM20, TM11 and TM, TE11.

iii) Propagation constant

As seen in Figure 13.11, with an increase in frequency, the propagation constant is seen to increase as well [59–65]. Because the propagation constant varies with frequency.

$$\gamma_{mn} = \alpha_{mn} + j\beta_{mn}$$
$$\gamma_{mn} = j\beta_{mn}$$
$$\beta = \omega = \sqrt{\mu\varepsilon\left(1-\left(\frac{f_c}{f}\right)^2\right)} \quad for\ f > f_c$$

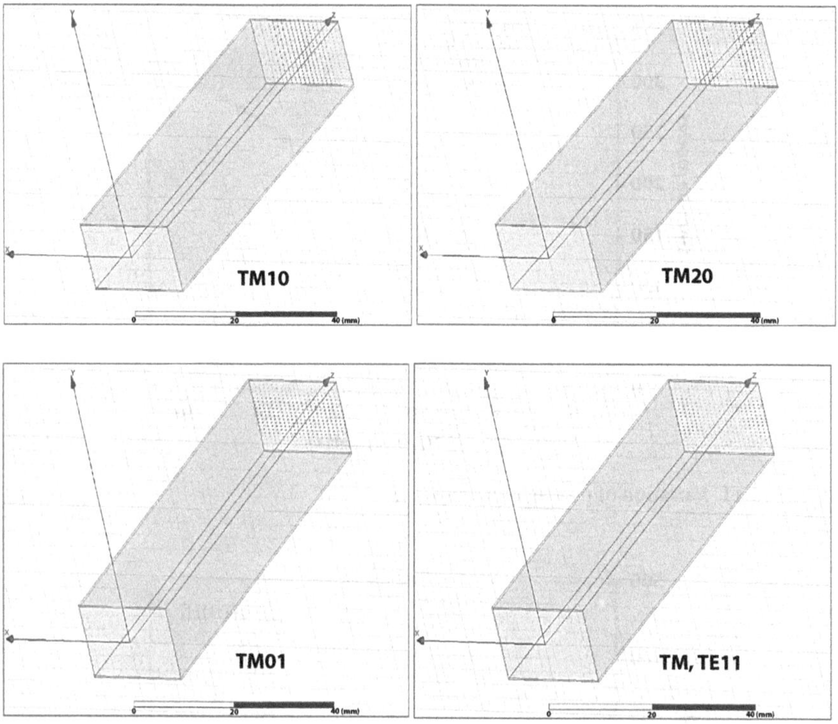

Figure 13.10 TE and TM modes field patterns.

iv) Impedance

The waveguide's typical impedance is depicted in Figure 13.12. As frequency rises, waveguide impedance falls, but at lower frequencies it doesn't change [66–74].

$$Z = \eta\sqrt{\left(1-\left(\frac{f_c}{f}\right)^2\right)} \text{ for transverse magnetic wave}$$

$$Z_{TE} = \frac{\eta}{\sqrt{1-\left(\frac{f_c}{f}\right)^2}} \text{ for transverse electric wave}$$

This Figure 13.13 demonstrates that for frequencies over a mode's cutoff frequency, there is significant power transmission across the waveguide.

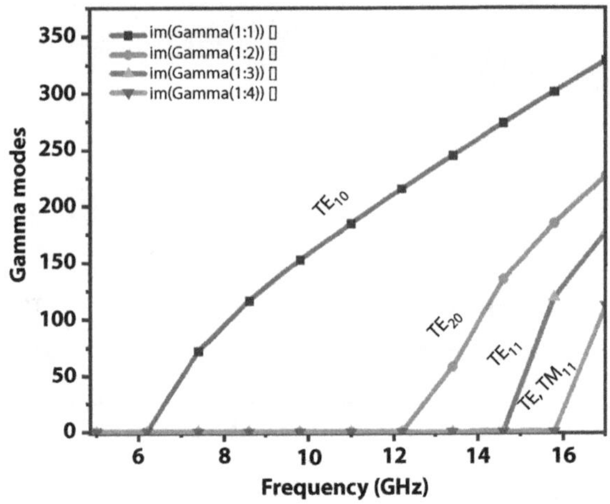

Figure 13.11 Variation of propagation constant.

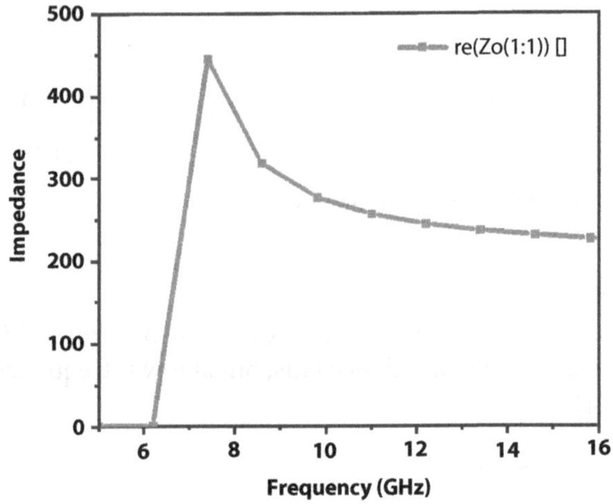

Figure 13.12 Variation of characteristic impedance.

In contrast, relatively little propagation in the waveguide happens below a mode's cut off frequency [75–83]. Without even establishing the integration line, we discovered four modes. HFSS computes all modes and lines to excite the waveguide on the 2-dimensional cross-sectional area. We have practically minimal damping (alpha) for the first three modes in Figure 13.13. Those modes will continue to propagate [84–92]. The damping in the fourth mode is 45.044 Np/meter and propagates slowly as seen in the Figure 13.14 [20].

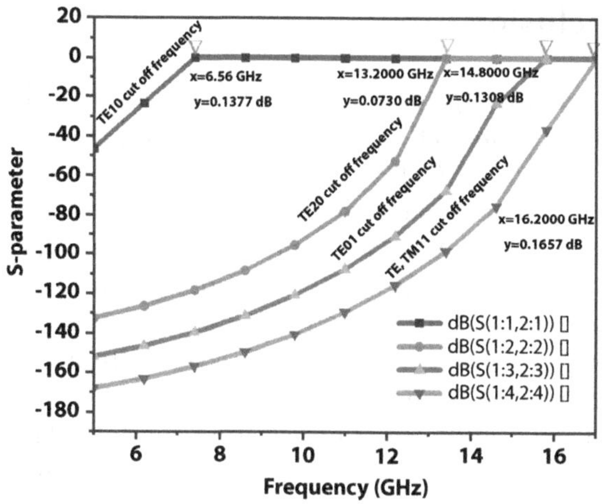

Figure 13.13 Variation of S-parameter.

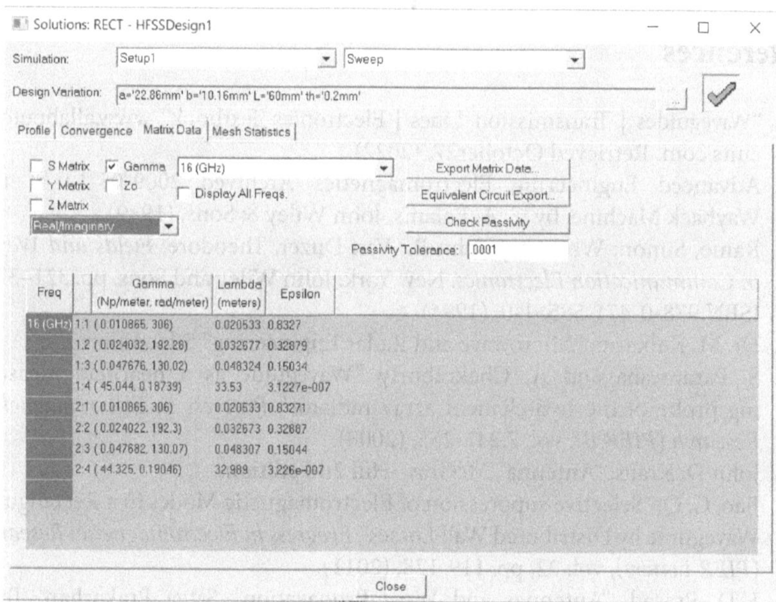

Figure 13.14 Simulated proof of the four modes.

13.5 Conclusion

In this article, we investigated the rectangular waveguide with dimensions of 22.76×10.15 mm and TE_{10}, TE_{20}, TE_{01}, and TE_{11} cutoff frequencies of 6.56, 13.2, 14.8, and 16.2 GHz, respectively [93–101]. While transmitting the information, this strong electromagnetic radiation may cause thermal consequences on people. We have found that the damping (alpha) for the first three modes is essentially nil. These modes will keep spreading [102–110]. We determined that the fourth mode's damping was 45.044 Np/meter and that it propagates slowly. The cutoff wavelength or cutoff frequency grew or reduced as the waveguide's size increased. Lower frequencies can propagate as a result of the good propagation constant [111–116]. As the size increases, characteristic impedance also greatly decreased. It causes information to be transmitted without any loss [117–120]. Thus, several waveguide properties are researched. To spread the waves, we can pick a waveguide with greater performance.

References

1. "Waveguides | Transmission Lines | Electronics Textbook". www.allaboutcircuits.com. Retrieved October27, (2022).
2. Advanced Engineering Electromagnetics Archived 2009-05-14 at the Wayback Machine, by C. A. Balanis, John Wiley & Sons, (1989).
3. Ramo, Simon; Whinnery, John R.; Van Duzer, Theodore. *Fields and Waves in Communication Electronics*. New York: John Wiley and Sons. pp. 321–324. ISBN 978-0-471-58551-0, (1994).
4. Dr. M. Kulkarni, "Microwave and Radar Engineering" 5th edition.
5. S. Paramesha and A, Chakraborty "Waveguide as a nearfield measuring probe of the two-element array radiator" *Progress in Electromagnetics Research (PIER B)*, vol. 7,245-255, (2008).
6. John D. Kraus, "Antenna", McGraw-Hill 2nd edition.
7. Jiao, C. Q, "Selective suppression of Electromagnetic Modes in a Rectangular Waveguide by Distributed Wall Losses", *Progress in Electromagnetics Research (PIER Letters)*, vol. 22, pp. 119-128, (2011).
8. K.D. Prasad, "Antennas and Wave Propagation", Satya Prakashan, Tech Publications, New Delhi, India, (2001).
9. Lee, C. S, Lee, S. W. and Chuang, S. L, "plot of modal field distribution in rectangular and circular waveguides", *IEEE Trans. Microwave Theory Tech.*, vol. 33, pp.271-274, (1985).
10. Matthew N.O. Sadiku, "Elements of Electromagnetics", 4th edition, Oxford University Press.

11. Samuel Y Liao, "Microwave Devices and Circuits" 3rd Edition Pearson Prentice Hall, (2008).
12. Simon, w, wert hen, Mandy wolf F, I, "A novel coplanar transmission line to rectangular waveguide transition, *IEEE MTT-Society International Symposium Digest*. pp .257-260, (1998).
13. HFSS; High frequency structure simulator based on the finite element method, v. 9.2.1, and Soft Corporation, (2004).
14. Harrington, R. F, "Time harmonic Electromagnetic Fields," McGraw Hills, New York, (1961).
15. C.A Balanis "Advanced Engineering Electromagnetic," John Wiley and Sons Inc, (1989).
16. R. E Collin, "Foundation for microwave engineering", 2nd edition, McGraw-Hills New York, (1992).
17. Baudrand, H, Tao, J.W, Atechian, J.," Study of radiating properties of open-ended rectangular wave guides", Antennas and Propagation, *IEEE Transactions* on pages; 1071-1077 Vol;36, Issue;8, Aug. (1988).
18. Pozar, D. M, "Microwave Engineering" 2nd Edition, New York: John Wiley and sons Inc., (1998).
19. Marcueitz, N, "Waveguide handbook", red lab. Series, vol.10, McGraw hill book co, New York, (1951).
20. About HFSS www.ansys.com/products/electronics/ansys-hfss
21. A. Bhardwaj, P. Kumar, B. Raj, and S. Anand, "Design and Performance Optimization of Doping-Less Vertical Nanowire TFET Using Gate Stack Technique," *J. Electron. Mater. (JEMS)*, Springer, Vol. 41, No. 7, pp. 4005–4013, 2022.
22. J. Singh, and B. Raj, "Tunnel Current Model of Asymmetric MIM Structure Levying Various Image Forces to Analyze the Characteristics of Filamentary Memristor", *Appl. Phys. A*, Springer, Vol. 125, No. 3, pp. 203.1 to 203.11, 2019.
23. C. Goyal, J. S. Ubhi, and B. Raj, "Low Leakage Zero Ground Noise Nanoscale Full Adder Using Source Biasing Technique", *J. Nanoelectron. Optoelectron.*, American Scientific Publishers, Vol. 14, pp. 360–370, 2019.
24. G. Singh, R. K. Sarin, and B. Raj, "A Novel Robust Exclusive-OR Function Implementation in QCA Nanotechnology with Energy Dissipation Analysis", *J. Comput. Electron.*, Springer, Vol. 15, No. 2, pp. 455–465, 2016.
25. T. Wadhera, D. Kakkar, G. Wadhwa, and B. Raj, " Recent Advances and Progress in Development of the Field Effect Transistor Biosensor: A Review", *J. Electron. Mater.*, Springer, Vol. 48, No. 12, pp. 7635–7646, 2019.
26. G. Wadhwa, P. Kamboj, and B. Raj, "Design Optimisation of Junctionless TFET Biosensor for High Sensitivity", *Adv. Nat. Sci.: Nanosci. Nanotechnol.*, Vol. 10, No. 7, p. 045001, 2019.
27. P. Bansal, and B. Raj, "Memristor Modeling and Analysis for Linear Dopant Drift Kinetics", *J. Nanoeng. Nanomanuf.*, American Scientific Publishers, Vol. 6, pp. 1–7, 2016.

28. A. Singh, M. Khosla, and B. Raj, "Circuit Compatible Model for Electrostatic Doped Schottky Barrier CNTFET," *J. Electron. Mater.*, Springer, Vol. 45, No. 12, pp. 4825–4835, 2016.
29. Ashima, D. Vaithiyanathan, and B. Raj, "Performance Analysis of Charge Plasma induced Graded Channel Si Nanotube", *J. Eng. Res. (JER)*, EMSME Special Issue, Vol. 9, pp. 146–154, 2021.
30. A. S. Tomar, V. K. Magraiya, and B. Raj, "Scaling of Access and Data Transistor for High Performance DRAM Cell Design", *Quantum Matter*, Vol. 2, pp. 412–416, 2013.
31. N. Jain, and B. Raj, "Parasitic Capacitance and Resistance Model Development and Optimization of Raised Source/Drain SOI FinFET Structure for Analog Circuit Applications", *J. Nanoelectron. Optoelectron.*, ASP, USA, Vol. 13, pp. 531–539, 2018.
32. S. Singh, S. K. Vishvakarma, and B. Raj, "Analytical Modeling of Split-Gate Junction-Less Transistor for a Biosensor Application", *Sens. Bio-Sens.*, Elsevier, Vol. 18, pp. 31–36, 2018.
33. M. Gopal, and B. Raj, "Low Power 8T SRAM Cell Design for High Stability Video Applications", *ITSI Trans. Electr. Electron. Eng.*, Vol. 1, No. 5, pp. 91–97, 2013.
34. B. Raj, J. Mitra, D. K. Bihani, V. Rangharajan, A. K. Saxena, and S. Dasgupta, "Analysis of Noise Margin, Power and Process Variation for 32 nm FinFET Based 6T SRAM Cell", *J. Comput. (JCP)*, Academy Publisher, FINLAND, Vol. 5, No. 6, PP.1-6, 2010.
35. D. Sharma, R. Mehra, and B. Raj, "Comparative Analysis of Photovoltaic Technologies for High Efficiency Solar Cell Design", *Superlattices Microstruc.*, Elsevier, Vol. 153, pp. 106861, 2021.
36. P. Kaur, A. S. Buttar, and B. Raj, "A Comprehensive Analysis of Nanoscale Transistor Based Biosensor: A Review", *Indian J. Pure Appl. Phys.*, Vol. 59, pp. 304–318, 2021.
37. D. Yadav, B. Raj, and B. Raj, "Design and Simulation of Low Power Microcontroller for IoT Applications", *J. Sens. Lett.*, ASP USA, Vol. 18, pp. 401–409, 2020.
38. S. Singh, and B. Raj, "A 2-D Analytical Surface Potential and Drain Current Modeling of Double-Gate Vertical t-Shaped Tunnel FET", *J. Comput. Electron.*, Springer, Vol. 19, pp. 1154–1163, 2020.
39. J. Singh, and B. Raj, "An Accurate and Generic Window function for Non-linear Memristor Model" *J. Comput. Electron.*, Springer, Vol. 18, No. 2, pp. 640–647, 2019.
40. M. Kaur, N. Gupta, S. Kumar, B. Raj, and A. K. Singh, "Comparative RF and Crosstalk Analysis of Carbon Based Nano Interconnects", *IET Circuits, Devices & Systems*, Vol. 15, No. 6, pp. 493–503, 2021.
41. N. Kandasamy, F. Ahmad, D. Ajitha, B. Raj, and N. Telagam, "Quantum Dot Cellular Automata Based Scan Flip Flop and Boundary Scan Register", *IETE J. Res.*, Vol. 69, PP. 535-548, 2023.

42. S. K. Sharma, B. Raj, and M. Khosla, "Enhanced Photosensitivity of Highly Spectrum Selective Cylindrical Gate In1-xGaxAs Nanowire MOSFET Photodetector", *Mod. Phys. Lett. B*, Vol. 33, No. 12, p. 1950144, 2019.
43. J. Singh, and B. Raj, "Design and Investigation of 7T2M NVSARM with Enhanced Stability and Temperature Impact on Store/Restore Energy", *IEEE Transactions on Very Large Scale Integration Systems*, Vol. 27, No. 6, pp. 1322–1328, 2019.
44. A. K. Bhardwaj, S. Gupta, B. Raj, and A. Singh, "Impact of Double Gate Geometry on the Performance of Carbon Nanotube Field Effect Transistor Structures for Low Power Digital Design", *J. Comput. Theor. Nanosci.*, ASP USA, Vol. 16, pp. 1813–1820, 2019.
45. N. Jain, and B. Raj, "Thermal Stability Analysis and Performance Exploration of Asymmetrical Dual-k Underlap Spacer (ADKUS) SOI FinFET for Security and Privacy Applications", *Indian J. Pure Appl. Phys. (IJPAP)*, Vol. 57, pp. 352–360, 2019.
46. A. Singh, M. Khosla, and B. Raj, "Design and Analysis of Dynamically Configurable Electrostatic Doped Carbon Nanotube Tunnel FET", *Microelectron. J.*, Elsevier, Vol. 85, pp. 17–24, 2019.
47. N. Jain, and B. Raj, "Dual-k Spacer Region Variation at the Drain Side of Asymmetric SOI FinFET Structure: Performance Analysis towards the Analog/RF Design Applications", *J. Nanoelectron. Optoelectron.*, American Scientific Publishers, Vol. 14, pp. 349–359, 2019.
48. J. Singh, S. Sharma, B. Raj, and M. Khosla, "Analysis of Barrier Layer Thickness on Performance of In1-xGaxAs Based Gate Stack Cylindrical Gate Nanowire MOSFET", *JNO*, ASP USA, Vol. 13, pp. 1473–1477, 2018.
49. N. Jain, and B. Raj, "Analysis and Performance Exploration of High-k SOI FinFETs over the Conventional Low-k SOI FinFET toward Analog/ RF Design", *J. Semicond.*, IOP Science, Vol. 39, No. 12, pp. 124002-1-7, 2018.
50. C. Goyal, J. S. Ubhi, and B. Raj, "A Reliable Leakage Reduction Technique for Approximate Full Adder with Reduced Ground Bounce Noise", *Math. Probl. Eng.*, Hindawi, Vol. 2018, Article ID 3501041, PP.1-16, 2018.
51. Anuradha, J. Singh, B. Raj, and M. Khosla, "Design and Performance Analysis of Nano-Scale Memristor-Based Nonvolatile SRAM", *J. Sens. Lett.*, American Scientific Publishers USA, Vol. 16, pp. 798–805, 2018.
52. G. Wadhwa, and B. Raj, "Parametric Variation Analysis of Charge-Plasma-Based Dielectric Modulated JLTFET for Biosensor Application", *IEEE Sens. J.*, Vol. 18, No. 15, PP. 6070 – 6077, 2018.
53. J. Singh, and B. Raj, "Comparative Analysis of Memristor Models for Memories Design", *JoS*, IoP, Vol. 39, No. 7, PP. 074006-1-12, 2018.
54. D. Yadav, S. S. Chouhan, S. K. Vishvakarma, and B. Raj, "Application Specific Microcontroller Design for IoT Based WSN", *Sens. Lett.*, ASP, Vol. 16, pp. 374–385, 2018.

55. G. Singh, R. K. Sarin, and B. Raj, "Fault-Tolerant Design and Analysis of Quantum-Dot Cellular Automata Based Circuits", *IEEE/IET Circuits Devices Syst.*, Vol. 12, pp. 638–664, 2018.
56. J. Singh, and B. Raj, "Modeling of Mean Barrier Height Levying Various Image Forces of Metal Insulator Metal Structure to Enhance the Performance of Conductive Filament Based Memristor Model", *IEEE Nanotechnol.*, Vol. 17, No. 2, pp. 268–275, 2018.
57. A. Jain, S. Sharma, and B. Raj, "Analysis of Triple Metal Surrounding Gate (TM-SG) III-V Nanowire MOSFET for Photosensing Application", *Opto Electron. J.*, Elsevier, Vol. 26, No. 2, PP. 141–148, 2018.
58. A. Jain, S. Sharma, and B. Raj, "Design and Analysis of High Sensitivity Photosensor Using Cylindrical Surrounding Gate MOSFET for Low Power Sensor Applications", *Eng. Sci. Technol. Int. J.*, Elsevier's, Vol. 19, No. 4, pp. 1864–1870, 2016.
59. A. Singh, M. Khosla, and B. Raj, "Analysis of Electrostatic Doped Schottky Barrier Carbon Nanotube FET for Low Power Applications," *J. Mater. Sci.: Mater. Electron.*, Springer, Vol. 28, PP. 1762–1768, 2017.
60. G. Saiphani Kumar, A. Singh, B. Raj, "Design and Analysis of Gate All around CNTFET Based SRAM Cell Design", *J. Comput. Electron.*, Springer, Vol. 17, No. 1, PP. 138–145, 2018.
61. G. P. Singh, B. S. Sohi, and B. Raj, "Material Properties Analysis of Graphene Base Transistor (GBT) for VLSI Analog Circuits", *Indian J. Pure Appl. Phys. (IJPAP)*, Vol. 55, pp. 896–902, 2017.
62. S. Kumar and B. Raj, "Estimation of Stability and Performance Metric for Inward Access Transistor Based 6T SRAM Cell Design Using n-type/p-type DMDG-GDOV TFET", *IEEE VLSI Circuits Syst. Lett.*, Vol. 3, No. 2, PP. 25–39, 2017.
63. S. Sharma, A. Kumar, M. Pattanaik, and B. Raj, "Forward Body Biased Multimode Multi-Threshold CMOS Technique for Ground Bounce Noise Reduction in Static CMOS Adders", *Int. J. Inf. Electron. Eng.*, Vol. 3, No. 3, PP. 567–572, 2013.
64. H. Singh, P. Kumar, and B. Raj, "Performance Analysis of Majority Gate SET Based 1-bit Full Adder", *International Journal of Computer and Communication Engineering (IJCCE)*, IACSIT Press Singapore, ISSN: 2010-3743, Vol. 2, No. 4, PP.1-8, 2013.
65. A. K. Bhardwaj, S. Gupta, and B. Raj, "Investigation of Parameters for Schottky Barrier (SB) Height for Schottky Barrier Based Carbon Nanotube Field Effect Transistor Device", *J. Nanoelectron. Optoelectron.*, ASP, Vol. 15, PP. 783–791, 2020.
66. P. Bansal, and B. Raj, "Memristor: A Versatile Nonlinear Model for Dopant Drift and Boundary Issues", *JCTN*, American Scientific Publishers, Vol. 14, No. 5, PP. 2319–2325, 2017.
67. N. Jain, and B. Raj, "An Analog and Digital Design Perspective Comprehensive Approach on Fin-FET (Fin-Field Effect Transistor) Technology - A Review",

Reviews in Advanced Sciences and Engineering (RASE), ASP, Vol. 5, PP. 1–14, 2016.

68. S. Sharma, B. Raj, and M. Khosla, "Subthreshold Performance of In1-xGaxAs Based Dual Metal with Gate Stack Cylindrical/Surrounding Gate Nanowire MOSFET for Low Power Analog Applications", *J. Nanoelectron. Optoelectron.*, American Scientific Publishers, USA, Vol. 12, PP. 171–176, 2017.
69. B. Raj, A. K. Saxena, and S. Dasgupta, "Analytical Modeling for the Estimation of Leakage Current and Subthreshold Swing Factor of Nanoscale Double Gate FinFET Device" *Microelectron. Int.*, UK, Vol. 26, PP. 53–63, 2009.
70. S. S. Soniya, G. Wadhwa, and B. Raj, "An Analytical Modeling for Dual Source Vertical Tunnel Field Effect Transistor", *International Journal of Recent Technology and Engineering (IJRTE)*, Vol. 8, No. 2, 2019.
71. S. Singh, and B. Raj, "Design and Analysis of Hetrojunction Vertical T-Shaped Tunnel Field Effect Transistor", *J. Electron. Mater.*, Springer, Vol. 48, No. 10, PP. 6253–6260, 2019.
72. C. Goyal, J. S. Ubhi, and B. Raj, "A Low Leakage CNTFET Based Inexact Full Adder for Low Power Image Processing Applications", *Int. J. Circuit Theory Appl.*, Wiley, Vol. 47, No. 9, PP. 1446–1458, 2019.
73. B. Raj, A. K. Saxena, and S. Dasgupta, "A Compact Drain Current and Threshold Voltage Quantum Mechanical Analytical Modeling for FinFETs", *J. Nanoelectron. Optoelectron.*, USA, Vol. 3, No. 2, PP. 163–170, 2008.
74. G. Wadhwa, and B. Raj, "An Analytical Modeling of Charge Plasma Based Tunnel Field Effect Transistor with Impacts of Gate Underlap Region", *Superlattices Microstruc.*, Elsevier, Vol. 142, PP. 106512-22, 2020.
75. S. Singh, and B. Raj, "Modeling and Simulation Analysis of SiGe Hetero-junction Double GateVertical t-Shaped Tunnel FET", *Superlattices Microstruc.*, Elsevier, Vol. 142, PP. 106496-104, 2020.
76. A. Singh, D. K.Saini, D.Agarwal, S.Aggarwal, M.Khosla, and B. Raj, "Modeling and Simulation of Carbon Nanotube Field Effect Transistor and Its Circuit Application", *J. Semicond.*, IOP Science, Vol. 37, PP. 074001–074006, 2016.
77. N. Jain, and B. Raj, "Device and Circuit Co-Design Perspective Comprehensive Approach on FinFET Technology - A Review", *J. Electron Devices*, Vol. 23, No. 1, PP. 1890–1901, 2016.
78. S. Kumar and B. Raj, "Analysis of ION and Ambipolar Current for Dual-Material Gate-Drain Overlapped DG-TFET", *J. Nanoelectron. Optoelectron.*, American Scientific Publishers, USA, Vol. 11, PP. 323–333, 2016.
79. N. Anjum, T. Bali, and B. Raj, "Design and Simulation of Handwritten Multiscript Character Recognition", *International Journal of Advanced Research in Computer and Communication Engineering (IJARCCE)*, Vol. 2, No. 7, pp. 2544–2549, 2013.
80. S. Sharma, B. Raj, and M. Khosla, "A Gaussian Approach for Analytical Subthreshold Current Model of Cylindrical Nanowire FET with Quantum Mechanical Effects", *Microelectron. J.*, Elsevier, Vol. 53, PP. 65–72, 2016.

81. K. Singh, and B. Raj, "Performance and Analysis of Temperature Dependent Multi-Walled Carbon Nanotubes as Global Interconnects at Different Technology Nodes", *J. Comput. Electron.*, Springer, Vol. 14, No. 2, PP. 469-476, 2015.
82. S. Kumar and B. Raj, "Compact Channel Potential Analytical Modeling of DG-TFET Based on Evanescent–Mode Approach", *J. Comput. Electron.*, Springer, Vol. 14, No. 2, PP. 820–827, 2015.
83. K. Singh, and B. Raj, "Temperature Dependent Modeling and Performance Evaluation of Multi-Walled CNT and Single-Walled CNT as Global Interconnects", *J. Electron. Mater.*, Springer, Vol. 44, No. 12, PP. 4825–4835, 2015.
84. V. K. Sharma, M. Pattanaik, and B. Raj, "INDEP Approach for Leakage Reduction in Nanoscale CMOS Circuits", *International Journal of Electronics*, Taylor & Francis, Vol. 102, No. 2, PP. 200–215, 2014.
85. K. Singh, and B. Raj, "Influence of Temperature on MWCNT Bundle, SWCNT Bundle and Copper Interconnects for Nanoscaled Technology Nodes", *J. Mater. Sci.: Mater. Electron.*, Springer, Vol. 26, No. 8, PP. 6134–6142, 2015.
86. N. Anjum, T. Bali, and B. Raj, "Design and Simulation of Handwritten Gurumukhi and Devanagri Numerical Recognition", *International Journal of Computer Applications, Published by Foundation of Computer Science*, New York, USA, Vol. 73, No. 12, PP. 16–21, 2013.
87. S. Khandelwal, V. Gupta, B. Raj, and R. D. Gupta, "Process Variability Aware Low Leakage Reliable Nano Scale DG-FinFET SRAM Cell Design Technique", *J. Nanoelectron. Optoelectron.*, Vol. 10, No. 6, PP. 810–817, 2015.
88. V. K. Sharma, M. Pattanaik, and B. Raj, "ONOFIC Approach: Low Power High Speed Nanoscale VLSI Circuits Design", *Int. J. Electron.*, Taylor & Francis, Vol. 101, No. 1, PP. 61–73, 2014.
89. S. Khandelwal, B. Raj, and R. D. Gupta, "FinFET Based 6T SRAM Cell Design: Analysis of Performance Metric, Process Variation and Temperature Effect", *J. Comput. Theor. Nanosci.*, ASP, USA, Vol. 12, PP. 2500–2506, 2015.
90. S. Singh, Y. Shekhar, R. Jagdeep, S. Anurag, and B. Raj, "Impact of HfO2 in Graded Channel Dual Insulator Double Gate MOSFET", *J. Comput. Theor. Nanosci.*, American Scientific Publishers, Vol. 12, No. 6, PP. 950–953, 2015.
91. V. K. Sharma, M. Pattanaik, and B. Raj, "PVT Variations Aware Low Leakage INDEP Approach for Nanoscale CMOS Circuits", *Microelectron. Reliab.*, Elsevier, Vol. 54, PP. 90–99, 2014.
92. B. Raj, A. K. Saxena and S. Dasgupta, "Quantum Mechanical Analytical Modeling of Nanoscale DG FinFET: Evaluation of Potential, Threshold Voltage and Source/Drain Resistance", *Mater. Sci. Semicond. Process.*, Elsevier, Vol. 16, No. 4, PP. 1131–1137, 2013.
93. M. Gopal, S. S. D. Prasad, and B. Raj, "8T SRAM Cell Design for Dynamic and Leakage Power Reduction", *Int. J. Comput. Appl.*, Published by Foundation of Computer Science, New York, USA, Vol. 71, No. 9, PP. 43–48, 2013.

94. M. Pattanaik, B. Raj, S. Sharmaa, and A. Kumar, "Diode Based Trimode Multi- Threshold CMOS Technique for Ground Bounce Noise Reduction in Static CMOS Adders", *Adv. Mater. Res.*, Trans Tech Publications, Switzerland, Vol. 548, PP. 885–889, 2012.
95. B. Raj, A. K. Saxena, and S. Dasgupta, "Nanoscale FinFET Based SRAM Cell Design: Analysis of Performance Metric, Process Variation, Underlapped FinFET and Temperature Effect", *IEEE Circuits and System Magazine,* Vol. 11, No. 2, PP. 38–50, 2011.
96. V. K. Sharma, M. Pattanaik, and B. Raj, "Leakage Current ONOFIC Approach for Deep Submicron VLSI Circuit Design", *International Journal of Electrical, Computer, Electronics and Communication Engineering*, World Academy of Sciences, Engineering and Technology, Vol. 7, No. 4, PP. 239–244, 2013.
97. T. Chawla, M. Khosla, and B. Raj, "Design and Simulation of Triple Metal Double-Gate Germanium on Insulator Vertical Tunnel Field Effect Transistor", *Microelectron. J.*, Elsevier, Vol. 114, PP. 105125-32, 2021.
98. P. Kaur, S. S. Gill, and B. Raj, "Comparative Analysis of OFETs Materials and Devices for Sensor Applications", *Journal of Silicon*, Springer, Vol. 14, PP. 4463–4471, 2022.
99. S. K. Sharma, P. Kumar, B. Raj, and Balwinder Raj, "In1-xGaxAs Double Metal Gate-Stacking Cylindrical Nanowire MOSFET for Highly Sensitive Photo Detector", *Journal of Silicon*, Springer, Vol. 14, PP. 3535–3541,2022.
100. B. Raj, A. K.Saxena, and S.Dasgupta, "Analytical Modeling of Quasi Planar Nanoscale Double Gate FinFET with Source/Drain Resistance and Field Dependent Carrier Mobility: A Quantum Mechanical Study", *J. Comput.*, Academy Publisher, FINLAND, Vol. 4, No. 9, PP.1-8, 2009.
101. S. Bhushan, S. Khandelwal, and B. Raj, "Analyzing Different Mode FinFET Based Memory Cell at Different Power Supply for Leakage Reduction", *Seventh International Conference on Bio-Inspired Computing: Theories and Application*, (BIC-TA 2012), organized IIITM Gwalor, PP. 89–100, 14-16 Dec 2012.
102. J. Singh, and B. Raj, "Temperature Dependent Analytical Modeling and Simulations of Nanoscale Memristor", *Eng. Sci. Technol. Int. J.*, Elsevier's, Vol. 21, PP. 862–868, 2018.
103. S. Singh, S. Bala, B. Raj, and Balwinder Raj, "Improved Sensitivity of Dielectric Modulated Junctionless Transistor for Nanoscale Biosensor Design", *Sens. Lett.*, ASP, Vol. 18, PP. 328–333, 2020.
104. V. Kumar, S. K. Vishvakarma, and B. Raj, "Design and Performance Analysis of ASIC for IoT Applications", *Sens. Lett.*, ASP, Vol. 18, PP. 31–38, 2020.
105. A. Jaiswal, R. K. Sarin, B. Raj, and S. Sukhija, "A Novel Circular Slotted Microstrip-fed Patch Antenna with Three Triangle Shape Defected Ground Structure for Multiband Applications", *Advanced Electromagnetic (AEM)*, Vol. 7, No. 3, PP. 56–63, 2018.

106. G. Wadhwa, and B Raj, "Label Free Detection of Biomolecules Using Charge-Plasma-Based Gate Underlap Dielectric Modulated Junctionless TFET", *J. Electron. Mater.*, Springer, Vol. 47, No. 8, PP. 4683–4693, 2018.
107. G. Singh, R. K. Sarin, and B. Raj, "Design and Performance Analysis of a New Efficient Coplanar Quantum-Dot Cellular Automata Adder", *Indian J. Pure Appl. Phys.*, Vol. 55, PP. 97–103, 2017.
108. A. Singh, M. Khosla, and B. Raj, "Design and Analysis of Electrostatic Doped Schottky Barrier CNTFET Based Low Power SRAM", *AEU Int. J. Electron. Commun.*, Elsevier,Vol. 80, PP. 67–72, 2017.
109. P. Kaur, V. Pandey, and B. Raj, "Comparative Study of Efficient Design, Control and Monitoring of Solar Power Using IoT", *Sens. Lett.*, ASP, Vol. 18, PP. 419–426, 2020.
110. G. Wadhwa, P. Kamboj, J. Singh, B. Raj, "Design and Investigation of Junctionless DGTFET for Biological Molecule Recognition", *Trans. Electr. Electron. Mater.*, Springer, Vol. 22, PP. 282–289, 2020.
111. T. Chawla, M. Khosla, and B. Raj, "Optimization of Double-Gate Dual Material GeOI-Vertical TFET for VLSI Circuit Design", *IEEE VLSI Circuits Syst. Lett*, Vol. 6, No. 2, PP. 13–25, 2020.
112. S. K. Verma, S. Singh, G. Wadhwa and B. Raj, "Detection of Biomolecules Using Charge-Plasma Based Gate Underlap Dielectric Modulated Dopingless TFET", *Trans. Electr. Electron. Mater.*, Springer, Vol. 21, PP. 528–535, 2020.
113. N. Jain, and B. Raj, "Impact of Underlap Spacer Region Variation on Electrostatic and Analog/RF Performance of Symmetrical High-k SOI FinFET at 20 nm Channel Length", *J. Semiconduct.*, IOP Science, Vol. 38, No. 12, PP. 122002-10, 2017.
114. S. Singh, and B. Raj, "Analytical Modeling and Simulation Analysis of T-Shaped III-V Heterojunction Vertical T-FET", *Superlattices Microstruct.*, Elsevier, Vol. 147, PP. 106717-25, 2020.
115. G. Singh, R. K. Sarin, and B. Raj, "Design and Analysis of Area Efficient QCA Based Reversible Logic Gates", *J. Microprocessors Microsyst.*, Elsevier, Vol. 52, PP. 59–68, 2017.
116. A. Singh, M. Khosla, and B. Raj, "Compact Model for Ballistic Single Wall CNTFET under Quantum Capacitance Limit", *J. Semiconduct*, IOP Science, Vol. 37, PP. 104001–104008, 2016.
117. S. Singh, M. Khosla, G. Wadhwa, and B. Raj, "Design and Analysis of Double-Gate Junctionless Vertical TFET for Gas Sensing Applications", *Appl. Phys. A*, Springer, Vol. 127, No. 16, PP.883-890, 2021.
118. I. Singh, B. Raj, M. Khosla, and B. K. Kaushik, "Potential MRAM Technologies for Low Power SoCs", SPIN World Scientific Publisher, SCIE; Vol. 10, No. 04, PP. 2050027-35, 2020.

119. S. Singh, and B. Raj, "Parametric Variation Analysis on Hetero-Junction Vertical t-Shape TFET for Suppressing Ambipolar Conduction", *Indian J. Pure Appl. Phys.*, Vol. 58, PP. 478–485, 2020.
120. S. S. Soniya, G. Wadhwa, and B. Raj, "Design and Analysis of Dual Source Vertical Tunnel Field Effect Transistor for High Performance", *Trans. Electr. Electron. Mater.*, Springer, Vol. 21, PP. 74–82, 2019.

Index

Also of Interest

By the same editors

DECENTRALIZED SYSTEMS AND DISTRIBUTED COMPUTING, Edited by Sandhya Avasthi, Suman Lata Tripathi, Namrata Dhanda, and Satya Bhushan Verma, ISBN: 978139420436. This book provides a comprehensive exploration of next-generation internet, distributed systems, and distributed computing, offering valuable insights into their impact on society and the future of technology.

ELECTRIC VEHICLE DESIGN: Design, Simulation and Applications, Edited by Krishan Arora, Suman Lata Tripathi, and Himashu Sharma, ISBN: 9781394204373. This book will serve as a definitive guide to conceptual and practical knowledge about the design of hybrid electrical vehicles (HEV), battery electrical vehicles (BEV), fuel cell electrical vehicles (FCEV), plug-in hybrid electrical vehicles (PHEV), and efficient EV charging techniques with advanced tools and methodologies for students, engineers, and academics alike.

INDUSTRIAL CONTROL SYSTEMS, Edited by Vipin Chandra Pal, Suman Lata Tripathi, and Souvik Ganguli, ISBN: 9781119829256. This volume serves as a comprehensive guide in the journey of industrial control systems with a multidisciplinary approach to the key engineering problems in the 21st century.

MODERN AUTOMOTIVE ELECTRICAL SYSTEMS: Theory and Applications, Edited by Pedram Asef, Sanjeevikumar Padmanaban, and Andrew Lapthorn, ISBN: 1119801047. Presenting the concepts and advances of modern automotive electrical systems, this volume, written and edited by a global team of experts, also goes into the practical applications for the engineer, student, and other industry professionals.

NANODEVICES FOR INTEGRATED CIRCUIT DESIGN, Edited by Suman Lata Tripathi, Abhishek Kumar, K. Srinivasa Rao, and Prasantha R. Mudimela, ISBN: 9781394185788. Written and edited by a team of experts in the field, this important new volume broadly covers the design of nano-devices and their integrated applications in digital and analog integrated circuits (IC) design.

EXPLAINABLE MACHINE LEARNING MODELS AND ARCHITECTURES: Real-Time System Implementation, Edited by Suman Lata Tripathi and Mufti Mahmud, ISBN: 9781394185849. This cutting-edge new volume covers the hardware architecture implementation, the software implementation approach, and the efficient hardware of machine learning applications.

MACHINE LEARNING TECHNIQUES FOR VLSI CHIP DESIGN, Edited by Abhishek Kumar, Suman Lata Tripathi, and K. Srinivasa Rao, ISBN: 9781119910398. This cutting-edge new volume covers the hardware architecture implementation, the software implementation approach, and the efficient hardware of machine learning applications with FPGA or CMOS circuits, and many other aspects and applications of machine learning techniques for VLSI chip design.

INTELLIGENT GREEN TECHNOLOGIES FOR SMART CITIES, Edited by Suman Lata Tripathi, Souvik Ganguli, Abhishek Kumar, and Tengiz Magradze, ISBN: 9781119816065. Presenting the concepts and fundamentals of smart cities and developing "green" technologies, this volume, written and edited by a global team of experts, also goes into the practical applications that can be utilized across multiple disciplines and industries, for both the engineer and the student.

DESIGN AND DEVELOPMENT OF EFFICIENT ENERGY SYSTEMS, edited by Suman Lata Tripathi, Dushyant Kumar Singh, Sanjeevikumar Padmanaban, and P. Raja, ISBN: 9781119761631. Covering the concepts and fundamentals of efficient energy systems, this volume, written and edited by a global team of experts, also goes into the practical applications that can be utilized across multiple industries, for both the engineer and the student.

Electrical and Electronic Devices, Circuits, and Materials: Technical Challenges and Solutions, edited by Suman Lata Tripathi, Parvej Ahmad Alvi, and Umashankar Subramaniam, ISBN: 9781119750369. Covering every aspect of the design and improvement needed for solid-state electronic devices and circuit and their reliability issues, this new volume also includes overall system design for all kinds of analog and digital applications and developments in power systems.

Green Energy: Solar Energy, Photovoltaics, and Smart Cities, edited by Suman Lata Tripathi and Sanjeevikumar Padmanaban, ISBN: 9781119760764. Covering the concepts and fundamentals of green energy, this volume, written and edited by a global team of experts, also goes into the practical applications that can be utilized across multiple industries, for both the engineer and the student.

Check out these other related titles from Scrivener Publishing

HYDROGEN ELECTRICAL VEHICLES, Edited by Mehmet Sankir and Nurdan Demirci Sankir, ISBN: 9781394166381. Hydrogen electrical vehicles are an essential component of the "Green New Deal" and this book covers cutting-edge technologies designed for fuel-cell-powered cars.

PAVEMENT ASSET MANAGEMENT, By Ralph Haas, W.R. Hudson, and Lynne Cowe Falls, ISBN: 9781119038702. Whether you are new to managing pavement assets or a seasoned practitioner, this book provides the most up-to-date information in the field on the history, technology, practice and look ahead.

The manufacturer's authorised representative in the EU for product safety is Oxford University Press España S.A. of El Parque Empresarial San Fernando de Henares, Avenida de Castilla, 2 – 28830 Madrid (www.oup.es/en or product.safety@oup.com). OUP España S.A. also acts as importer into Spain of products made by the manufacturer.

Printed in the USA/Agawam, MA
March 26, 2025

884981.023